Ceramic Processing
Industrial Practices

Ceramic Processing

Industrial Practices

Edited by
Debasish Sarkar

CRC Press
Taylor & Francis Group
Boca Raton London New York

CRC Press is an imprint of the
Taylor & Francis Group, an **informa** business

CRC Press
Taylor & Francis Group
52 Vanderbilt Avenue,
New York, NY 10017

First issued in paperback 2020

© 2020 by Taylor & Francis Group, LLC
CRC Press is an imprint of Taylor & Francis Group, an Informa business

No claim to original U.S. Government works

ISBN-13: 978-1-138-50408-0 (hbk)
ISBN-13: 978-0-367-72706-2 (pbk)

Library of Congress Cataloging-in-Publication Data

Names: Sarkar, Debasish, 1972- editor.
Title: Ceramic processing : industrial practices / [edited by] Debasish Sarkar.
Description: Boca Raton : Taylor & Francis a CRC title, part of the Taylor & Francis imprint, a member of the Taylor & Francis Group, the academic division of T&F Informa, plc, 2019. | Includes bibliographical references and index.
Identifiers: LCCN 2019015117 | ISBN 9781138504080 (hardback : acid-free paper) | ISBN 9781315145808 (e-book)
Subjects: LCSH: Ceramics. | Ceramic materials.
Classification: LCC TP807 .C4524 2019 | DDC 620.1/4--dc23
LC record available at https://lccn.loc.gov/2019015117

Visit the Taylor & Francis Web site at
http://www.taylorandfrancis.com

and the CRC Press Web site at
http://www.crcpress.com

This book is dedicated with great sense of love to my wife Shubhra and son Achyut

– Debasish Sarkar

Contents

Preface

Ceramic technology has diverse footprints across multiple fields, such as conventional as well as advanced ceramics, which range from refractories to nano-ceramics. Of late the ceramic industries are also getting a lot of attention for the research and developments, automations to modernize the manufacturing system and integrations related to people, processes and technology. This book is an attempt to capture recent developments in ceramic manufacturing sectors, and the respective sections synchronized to help undergraduate and postgraduate students who are extensively involved in ceramic research and development.

Multiple dignitaries from different fields of specialization have contributed to this book depending on their hands-on work experience and hard-earned knowledge from intense research work. And this is the unique value of this book, which will surely give readers a feel of the real world. This book is intended to bridge the gap between theoretical knowledge and the most comprehensive practical knowledge of the working fields.

With this aim, this book starts with introductory information on *manufacturing excellence in ceramic industry,* which is followed by *particle management,* one of the most important steps in fabricating ceramic products. In brief, the processing protocol has been taken into account for the manufacturing of a wide range of ceramics, starting from extremely high density (transparent ceramics) to low density (porous ceramics) ceramics and *ceramic coatings.* In consideration of this knowledge, traditional and advanced ceramics including *refractories and failures, whitewares and glazes, glasses and properties, miniaturization of complex ceramics* and *structural and functional prototypes* are discussed. The topics were chosen through brain-storming discussions with industrial personalities, academics and researchers around the globe and we received their extensive support to complete this book. A summary of each chapter follows.

"Perfection is not attainable but if we chase perfection, we can catch excellence," said Vince Lombardi. To strive toward this excellence, it is paramount to understand manufacturing challenges, the environment in which they operate and the resource network and the market segment they serve. Chapter 1 addresses manufacturing strategies, approaches and methodologies; experimental design; and tools for process improvements in the context of ceramic manufacturing industries like refractories, glass, whitewares and advanced ceramics. In view of smart manufacturing in the ceramic sectors, it touches the emergence of disruptive new technologies like analytics, artificial intelligence, robotics and more. Ceramic particle processing is one of the foremost criteria, and Chapter 2 discusses this issue. Particle fraction size facilitate different degrees of packing and thus particle processing management including *particle generation and characterization, particle mixing* and *particle scaling up* are important steps to achieve targeted product and properties once the manufacturing strategies are confirmed. From traditional to advanced ceramics, a wide range of particle sizes ranging from 6 mm to 10 nm are desirable for diverse objectives. As an example, large crystalline MgO particles are an excellent choice to avoid early stage wear of black refractory brick during steel manufacturing at 1650°C, whereas nano-scale TiO_2 particles result in highly efficient dye-synthesized solar cells (DSSC) at room temperature. In this context, Chapter 2 also deals with the basic philosophy of a particle processing mechanism, for both natural and synthetic materials.

In manufacturing optimization and particle management, fabrication of ceramics is the next step. Bulk ceramics can be broadly categorized based on their relative density (ρ_r); these are transparent ceramics ($\rho_r > 99.99\%$), dense ceramics ($70\% < \rho_r < 99.99\%$) and porous ceramics ($\rho_r < 70\%$). Polycrystalline transparent ceramics require an extensive understanding and specific processing practice in order to achieve inclusion-free transparent media. Several factors involved in achieving transparency are discussed in detail in Chapter 3. Properties and applications of transparent ceramics are also highlighted. Toward the other end of the spectrum on porosity, higher porosity content transforms into porous or cellular ceramics that are discussed in Chapter 4. In contrast to transparent ceramics, the lower solid loading per unit volume of liquid medium is much more effective for the fabrication of porous ceramics. The synchronization of pore content, size, geometry and distribution are important for successful porous components with desirable permeability, mechanical, thermal and electrical properties. Porous ceramics find importance in a wide range of applications such as thermal insulators, both low- and high-temperature gas/liquid filters, heat exchangers in harsh environments, bioimplants and so on. In this context, Chapter 4 aims to give an overview of a different class of porous ceramics, fabrication protocols, characterization methods and properties and some classical applications. Despite bulk (3D) ceramics, coatings are the 2D-layered structures deposited on the substrate through different deposition protocols in order to scale up the performance of versatile structural and functional components operating in distinct service environments. Failure of the deposited ceramic structure often occurs primarily due to poor intrinsic properties resulting from the incompatible substrate surface characteristics, the coating system, processing parameters and early stage wear processes. Chapter 5 discusses the fundamental processing science and basic principles of various ceramic coating processes involved in the development of durable ceramic coatings and their failure to analyze miscellaneous engineering applications.

Refractories are high temperature-resistant structure components (both unfired and fired) encompassing a wide range of density for the zone of interest. Refractories lead a characteristic interlocking microstructure through the binding of oxide or non-oxide grains of different size fractions that eventually depend on processing protocols. Failure of the installed refractories often occur due to inadequate processing and properties along with practical thermo-mechanical attacks, wear behavior, poor installation and discontinuous operation. Chapter 6 explicitly explores the fundamentals, processing, characterizations and control of the different classical wear process mechanisms along with failure analysis in order to investigate and improve the performance of various unshaped and shaped refractories.

In traditional ceramics, the ancient ceramic products known as whitewares that have a triaxial body (clay, quartz, feldspar) is compiled in Chapter 7. Several aspects including raw materials, their selection through phase diagram, colloidal behavior, body preparation protocols, types of drying mechanisms, glazes and coloring and firing are discussed meticulously in order to understand and develop next-generation whitewares. The influence of thermal energy on phase transformation, microstructural change and mechanical response provides a better insight into whitewares. Classic products covering the major market share of whitewares around the globe, such as sanitary ware, tiles, table wares and electrical insulators, are discussed.

Today, glass, another class of materials, not only comprises ceramic oxides, but also polymer and metallic glasses are manufactured through temperature profile synchronization. In Chapter 8 the prime focus is on oxide-based glasses, as there is an extensive academic interest in them and there is a significant consumption of commercial-grade annealed window glass, toughened windshield glass, coated and mirrored glass, bottle glass, bio-glass, etc. Before elaborating upon the fabrication, imperfection and properties of such glasses, a brief note on the enthalpy-temperature diagram, the basic philosophy of structural theory of glass formation and viscosity are discussed. The inline manufacturing protocol is explained from the authors' experience.

In the advanced technological era, it is difficult to draw a borderline between traditional and advanced ceramics, however, production capacity per annum, raw material resources, precision processing, per capita consumption rate and mode of utility made a certain barrier between these two. Despite the several advanced processing technologies used in ceramic processing, Chapter 9 primarily focuses on some modern approaches such as additive manufacturing, atomic layer deposition, micro-injection molding and the precious machining process from the perspective of miniaturization of complex ceramic components that are usually not part of traditional ceramics. Advances in miniaturization are expected to deliver portable and effective devices to perform in any field, from medicine to space, food to defense, environment to energy, and several to fulfill modern life's comforts. When considering such advanced ceramics, it is customary to share knowledge on ceramic prototypes for a particular application. In fact, developing a ceramic prototype involves common protocols such as synthesis or processing, microstructure, properties analysis and product validity. Considering the chain, the final product is governed by the conditions under which the ceramics are being processed, leading to the desired shape and performance. In this context, Chapter 10 describes the processing and mode of actions for several structural and functional ceramic prototypes that cover our daily activities.

In summary, the author firmly believes that students will enjoy reading this book, and it will certainly help them to get a comprehensive idea about the topics selected. This book contains several unique chapters, which may not be available in the conventional literature in greater detail. There is hardly any book available on the manufacturing strategy and Lean practices in the ceramic industry. This book will give a fair understanding of real-world business challenges, improvement opportunities and the automations that drive this sector. Similarly, particle ceramics, transparent ceramics, porous ceramics, ceramic coatings, refractories, whitewares, oxide glasses, additive manufacturing, etc., have been covered in finer detail, which will be beneficial for the students to gain an in-depth idea.

Acknowledgments

I would like to convey my heartfelt thanks to my parents and family members, Achyut, Shubhra, Diya, Sana, Subhash, Susmita, Amit and Shiuli, for their constant endorsement and motivation throughout the journey of writing of this book. I would like to thank my students and friends, Sarath, Sangeeta, Satish, Sushree, Akbar, Surendra, Barsa, Manasi, Tanaya, Sabayasachi, Siva Prasad, Vivek and Rohit, for their uninterrupted co-operation in manuscript preparation.

I would like to give a sincere thanks to my friends who have been instrumental in composing respective chapters and giving their valuable advice to complete this book: Mr. Mithugopal Mandal, Senior Business Consultant, Global Consulting Practice, Tata Consultancy Services, Bangalore, India; Mr. Susanta Basu, Manager, Furnace and Batch House, Emirates Float Glass Llc 353MR2 ICAD-2 Musaffah Abu Dhabi, United Arab Emirates; Dr. Samuel Paul David, Senior Researcher, HiLASE, Institute of Physics, ASCR, Prague, Czech Republic; and my colleague, Dr. Partha Saha, Assistant Professor, Department of Ceramic Engineering, NIT Rourkela, India. I do not think it could have been possible without their great help. Their inquisitiveness for advancement in the field of ceramics has consistently stimulated me toward further research and, finally, to put effort into this book.

Chapters 3 and 4 were co-financed by the European Regional Development Fund and the state budget of the Czech Republic (project HiLASE CoE: Grant No. CZ.02.1.01/0.0/0.0/ 15_006/0000674) and by the European Union's Horizon 2020 research and innovation programme under grant agreement No. 739573. This work was also supported by the Ministry of Education, Youth and Sports of the Czech Republic (Programme NPU I Project No. LO1602). I want to acknowledge HiLASE, Czech Republic "Guest Expert on Ceramic Processing and Engineering for High Power Lasers" for their work collaborating on these chapters.

I would like to acknowledge NIT Rourkela; the Department of Science and Technology (DST, EEQ/2017/000028, DST/TSG/Ceramic/2011/142-G, SR/FTP/ETA-088/2009), India; the Department of Biotechnology (DBT, BT/PR13466/COE/34/26/2015), India; Board of Research in Nuclear Sciences (BRNS, 2012/34/46/BRNS), India; Korea Research Institute of Science and Standards (KRISS), South Korea; Tata Steel Ltd, Jamshedpur; Tata Krosaki Refractories Ltd, Belpahar, Ants Ceramics, Nashik and Keramic Resource and Industrial Services, Rourkela for their support.

I would like to thank all the researchers for their contributions to scientific society, which are the building blocks of the conceptual knowledge in this book.

Editor

Prof. Debasish Sarkar is currently a Professor at the Department of Ceramic Engineering, National Institute of Technology (NIT), Rourkela, Odisha, India. After completing his undergraduate degree in Ceramic Engineering from the University of Calcutta in 1996, he completed a M.Tech in Metallurgical and Materials Engineering from the Indian Institute of Technology, Kanpur, followed by a PhD in Ceramic Engineering at NIT, Rourkela. As a visiting researcher at the Korea Research Institute of Standards and Science (KRISS), South Korea, he has gained extensive experience in ceramic processing from a wide perspective. As principal investigator, he has managed several research projects relating to refractories and nanomaterials for functional and structural applications. The government of India and businesses like Tata Steel Ltd., Tata Krosaki Refractories Ltd., etc. sponsored most of the research and development fund.

Prof. Sarkar is an internationally recognized expert in academic and industrial research. Several of his works are highlighted in Global Medical Discovery, Canada and have been published as European patents. He was awarded the Materials Research Society (MRSI) of India Medal Award in 2016 for his work on patient-specific orthopedic implants, and this is a symbol of his outstanding career. He is a life member of the American Ceramic Society, the Indian Ceramic Society and the Indian Institute of Metals.

He has a cumulative 23 years of experience and published 65 peer-reviewed international journal papers, 20 papers in conference proceedings, and three chapters in books published byPan Stanford Publisher, Wiley–VCH Verlag GmbH and Elsevier. In 2018 he published *Nanostructured Ceramics: Characterization and Analysis* (CRC Press). He has been a lead inventor on one Korean patent, one European patent and one Indian Patent and co-inventor for two Korean patents. He has been published and reviewed in ACS, Blackwell Publishing Inc, Electrochemical Society, Elsevier, RSC, Springer, etc. He is a member of various editorial boards and technical committees around the globe.

Chapter Contributors

Chapter 1

Mr. Mithugopal Mandal, Senior Business Consultant, Global Consulting Practice, Tata Consultancy Services, Bangalore, India. Mr. Mandal has a B.Tech in Ceramic Engineering from the University of Calcutta, followed by an MBA from IIM, Kolkata. He has extensive experience in process optimization, particularly in the manufacturing industry for the last 22 years.

Chapters 3 and 4

Dr. Samuel Paul David, Senior Researcher, HiLASE Center, Institute of Physics of The Czech Academy of Sciences, Prague, Czech Republic. Dr. David has around three years of teaching and an extensive five years of postdoctoral research experience at the University of Central Florida, USA followed by HiLASE, Czech Republic, in the field of optics and fabrication of ceramics. To his credit, he has published more than 20 publications, including several book chapters. Dr. David also acknowledges the contribution of Dr. Debasish Sarkar, who is also affiliated with HiLASE Center during the work of Chapters 3 and 4.

Chapters 5 and 6

Mr. K. Sarath Chandra, PhD Scholar, Department of Ceramic Engineering, NIT Rourkela, India. Mr. Sarath is extensively involved in the development of a black refractory project for Tata Steel Limited, Jamshedpur.

Chapters 7 and 9

Dr. Partha Saha, Assistant Professor, Department of Ceramic Engineering, NIT Rourkela, India. Dr. Saha has completed his PhD in Metallurgical Engineering from The University of Alabama, Alabama, followed by four years postdoctoral research experience at University of Pittsburgh, Pennsylvania. He has more than seven years of research experience in the

United States and five years teaching experience in India. He has published more than 25 peer-reviewed papers in various high impact factor journals and has 2 US patents.

Chapter 8

Mr. Susanta Basu, Manager, Furnace and Batch House, Emirates Float Glass, Abu Dhabi, United Arab Emirates. Mr. Basu completed a B.Tech in Ceramic Engineering from the University of Calcutta, India and has more than 22 years of experience in the glass industry around the globe and completed the Executive Management program from IIM Lucknow.

Chapter 10

Dr. Sangeeta Adhikari, Postdoctoral Researcher, School of Applied Chemical Engineering, Chonnam National University, Gwangju, South Korea. Dr. Adhikari has a PhD with the Department of Ceramic Engineering, NIT Rourkela. Dr. Adhikari has published more than 25 papers and book chapters and has completed two years of the UGC Kothari Postdoctoral fellowship at IISc, Bangalore, up to 2017, followed by a recent engagement as a KRF fellow in South Korea.

1

Manufacturing Excellence in Ceramic Industry

Mithugopal Mandal and Debasish Sarkar

CONTENTS

1.1 Introduction

The manufacturing sector has a significant impact on a country's GDP, so the manufacturing capabilities of any country decides upon the macro- and microeconomic position as well. As key and diverse industries, ceramics industries (like refractories,

1

cement, tile, glass, etc.) make a substantial contribution toward the overall manufacturing output of the country. It is important to understand the dynamics of conventional ceramic manufacturing and the paradigm shift due to emerging disruptive technologies. This chapter is a brief discussion on how manufacturing excellence philosophies can be adopted in the ceramic manufacturing context with appropriate strategy. A brief note on Industry 4.0 is also provided, as it is being considered as the next big change in manufacturing.

1.2 Inception and Advancement of Ceramic Manufacturing

Is the ceramic industry the oldest one in terms of systematic production serving human need? Or can we call this industry the mother of all other industries? Maybe not so. Maybe textile manufacturing started at the infancy stage of human community. Some scientific research says that initial textile weaving started as early as 100,000 – 500,000 years ago, whereas the earliest earthenware is presumed to be made 24,000 years ago. However, large scale textiles production started not before 50,000–60,000 years ago, when humans first started using sewing needles made of wood or from the stem of plants. Evidence from South Africa, Siberia, Slovenia, Russia, China, Spain and France reinforces this assumption. Many evidence suggests that there has been a continuous progression in the textile industry since its inception. However, structured textile production may have started around 5000 BC, when human started using net gauges, spindle needles and weaving sticks for large-scale production (Figure 1.1).

Still, the ceramics industry is undoubtedly one of the most ancient industries. Clay was the first raw material used to manufacture ceramicware that is predominantly known as earthenware. Subsequently applying heat and adding new substances to increase the strength of the body have been a major change in the manufacturing process. Clay-based fired pottery has benefited human life, starting with storage and preservation or transportation of solid and liquid materials from one place to another. It was around 9000–10000 BC when clay pottery (Figure 1.2) took its position in day-to-day human life.

FIGURE 1.1
Evidence of textile usage in the ancient age.

FIGURE 1.2
Few examples of early day's clay potteries.

In those days or little later, ceramic tiles and bricks were discovered, though in a very initial form. It is not clear when exactly glass ceramics emerged, but it is generally assumed to be around 8000 BC in Greece. Initially it was treated as an overheated glazed material and applied to the surface of clay pottery to provide shine, color and strength; later, glass was recognized as a separate category of material. In a later age, glass technologies opened a remarkably new area for material science and society as well.

Thus, the contribution of the ceramics toward human progress is immense. In the advancement of ceramics, refractories and insulators have been other milestones. Refractory has given an edge to the manufacturing of steel and glass-like materials, which melt in higher temperatures in order make a definite shape of various forms and structures. During the 16th Century, the refractory brought an industrial revolution. Not only did refractories help glass and steel manufacturing but they aided in the manufacturing of cement, coke, chemicals and other high temperature-assisted products. The usage of ceramics as insulation came into the picture during the 19th Century, and they have been widely used in various fields like automobiles, radios, televisions, computers, furnace, etc. More recently, ceramic products have been further classified into traditional and advanced ceramics, where the traditional ceramics include refractories, glass, cement, whiteware, abrasives and conventional coating, whereas the advanced ceramics cover magnetics, dielectrics, photocatalytic, load and impact resistance ceramics, etc. [1, 2].

Evidently, there has been research done to understand the inception of the ancient ceramic production and how the human community first started it, but most of them fail to explain the degree of automation, the production system, mode of flow, etc. It is obvious that the driver of early pottery was the needs of individuals. Mass-scale production was out of reach until the machine replaced some of the manual work. If we look at the early days, people used to manufacture their necessary and utility items, like cookware, cups, saucers, etc., through manually driven processes. Thus, the textiles industry grew alongside the ceramic industry to fulfill the basic needs of life and improve the standard of living. The need for automation arose with the need to increase the production, perfection and quality of the articles being manufactured. Handmade or manually constructed items may not have the same repeatability in terms of quality because of multiple variables involved in manual production.

1.3 Process Design in the Ceramic Industry

Since the beginning, there has been continuous trial and error to improve the product and its quality, so there have been permutations and combinations of the raw materials used for manufacturing the different ceramic wares. For a long time scientists have been asked, "What kind of raw materials did the ancient people use? How did they process it? How did they move to mass-scale production from unit-level production?" Let us understand what raw materials were used, how they were processed and finally how they were given the final coatings before heat treatment and delivered as a finished product. Depending on the nature of the ceramic ware, typical process steps have to be followed in the target of specific size, shape and properties, as illustrated in Figure 1.3.

- Raw material collection, crushing and grinding (more of a manual method)
- Mixing (dry and wet)
- Forming (giving a specific shape, either manually or with a wooden mold)
- Surface coating with some design
- Drying to reduce the moisture
- Firing (to harden the shape and make it unbreakable)
- Machining and finishing (as if required)

If we look at the raw material selection and crushing and grinding process, initially it started with clay, maybe a combination of different natures of clays (plastic and non-plastic clay), a mix and match of different varieties. With time different materials were tried, like sand, quartz, feldspar, etc. Therefore, the mixing and grinding itself became a process which needed automation for the bulk materials.

Next is shaping or forming. This started with a simple handmade process. With the progress of time, demand-specific design, dimension, and thus molds were discovered that sped up the manufacturing process. Let us take the example of firing that may have started with open-air firing and subsequently closed chamber firing, which follows a

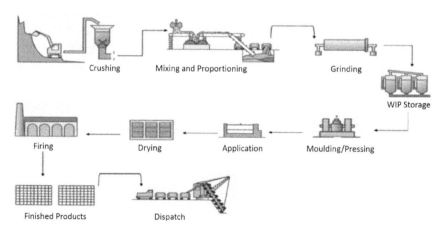

FIGURE 1.3
A common ceramic ware processing protocol continued from early days.

specific firing profile including temperature, time and pressure. With the progress of time and development of a manufacturing system to mass-scale production, multiple players joined in the market so competition among them became the common feature for survival. As competition grows, there is a need for organizational strategy that primarily consists of sustainability in the market and healthy financial practices. Marketing strategy was thought of as a survival weapon for the organization, but that was not sufficient. Companies had to think differently and re-prioritize their focus areas. During the 18th Century, manufacturing strategy was introduced to support the corporate strategy and thus integration started. Moreover, it was thought of as a competitive weapon to outperform the competitors and uniquely position oneself in the marketplace. The following section covers this manufacturing strategy and some guidelines on how it can be best utilized in the ceramic sectors.

1.4 Manufacturing Strategy

The survival of any business depends on consistent profitability that outperforms the competition. Therefore, the utmost approach of any enterprise is to target higher revenues while keeping the cost of production and sales at the lowest possible level. In this circumstance, questions arise: how is the manufacturing strategy designed and how does it really help to address manufacturing challenges and improve the efficiencies? In order to analyze such questions, one should understand and consider the common challenges across all industries:

- Differentiating the product and services in the ever-changing market place
- Delivering greater value to the customers at the lowest possible cost
- Integration among people, process and technologies to improve productivity
- Low capacity utilization with effective line balancing
- Variance in demand forecasting vis-à-vis how actual market demand impacts material procurement
- Identification of the various factors apart from plant layout and process design that result in higher internal delivery lead times
- Non-adherence to a preventive and predictive maintenance schedule results in prolonged breakdown and downtime
- Product design complexity and higher production lead time results in higher production losses
- Moving away from key competence areas because of lack of clarity in manufacturing strategies
- Profitability impact due to inadequate cost consciousness, higher production cost and delayed delivery
- Unavailability of skilled man power, improper allocations of resources and low productivity
- Inadequate response toward the fluctuating market demand attributed to inefficient manufacturing facilities

To achieve the operational excellence in any manufacturing environment, the primary requirement is to design the appropriate manufacturing strategy aligned with the corporate strategy. The strategy is designed to differentiate the firm's production from that of the competition. Thus, it provides specific direction to the manufacturing activities in the medium to long term. A production strategy that does not consider production facilities, including the availability of the production assets, flexibilities and capabilities to accommodate the urgent needs, is not effective enough. Retaining existing customers and acquiring new customers fuels growth of the market demand, which in turn helps to grab the greater market share day-to-day. However, to continuously fulfill needs within the stipulated lead time, provide the asked quantity and meet the expected quality demands, an organization needs to be very effective and efficient. There is no right or wrong manufacturing strategy, just whatever is most suitable to a specific organization. Following are some strategic elements related to manufacturing that you can jot down for future reference.

a. **Focus on competitive strength of production**: Every company has their own strengths and weaknesses. Firms should focus on their key competitive strengths so they can segregate themselves in the marketplace through their unique product attributes and quality. For example, one should not follow the same protocol for the refractory industry as was used in the electronic device business portfolio and strategy.

b. **Higher productivity with low cost**: A high degree of output with desired quality and minimum cost is a primary objective. To have maximum productivity, the company has to ensure the upkeep of assets, train the employees, supply the demanded quality of raw materials and reduce the variability of the production and processes as much as possible. Broadly, cost is attributed to direct sources and indirect sources. Direct costs may typically be material and manpower costs, and indirect costs may be production and sales overheads. Cost reduction has to target both the direct as well as indirect valuation.

c. **Process integration with higher automations**: As a part of the manufacturing strategy, the vertical integrations among the different sections and units are very important. Automations through technology implementations can give an edge over the competitors. The technologies also provide flexibilities in the production system.

d. **Wining product design and customization**: A product design that is compatible with the production line and not complex in nature boosts productivity as well as quality. As the cost of production is directly linked to the product design and configuration, it is paramount to have a right design selected for production. The product should have features that will win customers, be less troublesome in production and, at the same time, should not be easily replicable for competitors. The market continuously changes with preference and tastes. Many customers want something unique, especially their own customized product, so it is inevitable that the trend of customization will continue growing with time. As a part of the manufacturing strategy, the organization must have the flexibility in the production system to accommodate customer-specific customizations. For example, the production line is essential for accommodating and switching over the specific shape of refractories or float glass thickness for the customers' interests.

e. **Consistent quality**: Delivering products with consistently good quality is one of the major elements of the manufacturing strategy. Following the guideline of "first time right" makes a huge impact on the minimization of production loss and the cost of production. To have a consistently good quality product, a company may implement TPM, Six Sigma, ISO-9001, etc.

f. **Efficient usage of resources and capacity**: The organization should have a strategy for utilizing production capacities and resources as much as possible. For an example, if the whiteware shop floor has 10 production lines with different configurations and flexibilities, the sales and operations (including capacity planning) should be handled in such a way that none of the lines remains idle at any point of time. Starting from the raw material sections and up to the final product, the available resources should be efficiently planned for proper utilization. The ceramic industries exist from ancient times and are not overcrowded by modern technologies and automations in specific traditional ceramics. Take the example of the most prevalent industry, the refractory manufacturing sector. Until today, much of the unshaped refractory production was handled manually and by batch process, followed by intermediary inventory is storage and manual quality checking. However, few industries are as well progressed as tile manufacturing, where automation is employed in extensive ways. Several advanced ceramics manufacturers are also adopting automation in order to maintain the precise micron-scale dimension of high-dense to porous 3D prototypes or nanoscale ceramic coatings.

g. **Different manufacturing scenarios**: Although effective management, healthy financial habits, adequate risk management and strong marketing defines the success of companies, effective manufacturing strategy makes a huge difference between the success and failure of any enterprise. Any manufacturing company needs to understand its strength and weakness, and accordingly it should focus on its core strength to sustain and grow in the competitive marketplace. To design an effective manufacturing strategy, it should understand its supply chain in its entirety. Depending on the different nature of the ceramic products, whichever market they serve, like B2B and B2C, the manufacturing process strategy and design changes quite significantly. For example, the refractory serves the B2B market whereas cement serves the B2B as well as B2C market. Likewise, lifestyle products like tiles (wall, floor and synthetic granite), bathtubs and sanitary wares serve either or a combination of both the markets, and thus to fulfill the market demand the production strategy and the design changes accordingly. Most companies focus on their key competitive strength, and thus the major or key components are indigenous and the rest are manufactured by the subsidiaries, third-party manufacturers or contractors.

As the refractory companies have their core competitive strength in the manufacturing of refractories, they do not necessarily engage in processing pure alumina or magnesia extraction, purification, etc. However, this is another sector associated with raw material beneficiation of particle management for ceramic industries. Likewise, tile manufacturing companies focus predominantly on their core strength on processing strategies for different categories of tiles rather than fritz or stains, etc. Depending on the core skill set of a specific manufacturing house, the production unit can be classified into:

- Inhouse manufacturing
- Outhouse manufacturing or third-party manufacturing

1.4.1 Inhouse Manufacturing

Every company has their own key competency and may not be proficient in producing all semi-finished products/components and final assemblies. This practice may not be cost effective as well. Thus, as a part of the manufacturing strategy of the organization, it is decided which of the components/semi-finished they want to produce and which ones to strategically outsource. The manufacturing process design is done in such a way that it produces key components and final assemblies and has the flexibilities to accommodate the outsourced components in the different stages of the manufacturing cycle. Including the finished products, the final assembly and some key components and intermediary products are manufactured inhouse prior to brand the product.

1.4.2 Outhouse Manufacturing (Third-Party Manufacturing/Subcontracting)

As a part of the enterprise manufacturing strategy and the process design, several components are outsourced to authorized vendors to reduce process complexities and minimize the overall production cost. However, it requires systematic and continuous integration with the vendors/outsourced partners to feed into the different stages of the manufacturing. Otherwise, there must be a structured inventory management process either at the manufacturing site or at the vendor site. Storing at the manufacturing site or nearby areas should take into consideration the manufacturing lead time of the components and the lead time of delivering the components to the manufacturing site and the consumption rate/day and safety stock required for the contingency situation. Despite that, there must be proper protocol for quality checks before putting the components in the production line.

Developing a manufacturing strategy primarily depends upon the ability to produce a certain category of the products, which in turn depends on the product design, production layouts, technological flexibilities and supply chain. Moreover, it depends on management's *vision* and *goal* as well. Nevertheless, there may not be any single strategy that suits the company best. Before designing the manufacturing strategy, the company needs to understand its supply and distribution network and cost of sending a product to the end customer. Strategies are designed to extract the best possible outcome of the business. In actuality it is always difficult to decide how much (total quantity) to produce. Which production line to produce? When exactly to produce? That depends on the *push* and *pull* strategy of the market. That is mitigated through effective demand planning, proper capacity management and shop floor planning, etc. In consideration of the push and pull from the market, the production system may be *make to stock* or *make to order*, respectively.

In the push system, the production is done on the basis of the average forecasted market demand and the line capacity. The products are produced and dispatched to warehouses in different locations until they reach maximum capacity and are supplied to the retail shops from the nearby warehouses and end customers. In this system, the primary driver becomes the capacity and the resource utilization to reduce the production cost. In the make to stock situation, there is a systematic production plan which helps to run the production line to its optimal capacity, with less changeover time and better expected quality. In this kind of situation, priority is given to productivity optimization. In the make to order or pull situation, the market demand is recognized first in the form of confirmed sales order or customer-specific demand

and production is planned accordingly. The priority is given to the bulk order and less customized requirement, whether the customized or very specific requirement is dealt separately. Cumulatively, it requires special attention starting from the planning phase itself including set-up time for change over, line balancing and tear down time for next product.

The push system is beneficial in terms of less inventory storage, inventory carrying cost and space utilization. Typically, if we look at any manufacturing scenarios, none of them are completely make to stock or make to order; rather, it is a mix of both. During production planning, the material requirement planning (MRP) is run depending on the incoming sales order and forecasted demand subjected to the past trend, including seasonal increased or decreased demand. Depending on the nature of processes, the production system may be *discrete manufacturing* and *continuous manufacturing*.

Discrete manufacturing is a process where the semi-finished products are identifiable in every stockable stage of the production. Suppose an alumina refractory is manufactured in 10 steps; then, in each and every step, the stock can be identified, measured and reported. Most of the batch-processed manufacturing process is discrete manufacturing. Continuous manufacturing is majorly applicable to the process industries, as for example the chemical processing industries, pharmaceutical industries. In this kind of manufacturing scenario, the identification of the intermediate products is very difficult. The process runs continuously with set production parameters unless there is a specific need for change. Running a different product in the same production line is very difficult since it requires a considerable amount of changeover time, which typically attracts a considerable amount of production losses. Depending on the nature of the production process, the production optimization methodology can be employed.

Most of the refractory companies, sanitary wares, bath tubs, tiles (wall, floor and synthetic granite) and manufacturing systems are examples of discrete manufacturing. However, glass and steel manufacturing can be categorized as continuous manufacturing. Therefore, the Lean implementation approach would be different in these two categories of the industries. By nature of the manufacturing process, discrete manufacturing may have multiple work stations and works in process (WIPs) are stored in the buffer stations. Production may have different batches and batch-wise semi-finished stocks are also stored in the intermediate stations to feed in the next stations. An example of this is in the tile industry: the dried tiles are stored in sufficient quantities to feed in the next stations like the firing process. If the stocks fall below the critical stock, the kiln output may get impacted in either firing strength or defects that eventually demand sorting for quality and reliable product. But in case of the continuous manufacturing, there is very little or no chance to store the WIP in between the work stations. The intermediate products are continuously fed to the next stations automatically and production run uninterruptedly. The continuous manufacturing is less flexible in nature; on the contrary, the discrete manufacturing is more flexible, line balancing, capacity management and customization.

Tile, bathtub and user-friendly lifestyle-based ceramic products have higher technology adoption than their peers like the refractory or cement industries. For the ceramic-based lifestyle product manufacturing process, starting from raw material processing and ending when the finished products are delivered to the market, most of the steps are automated and there has been attempt to reduce manual intervention. Other industries have wide room for the automation enhancement in their manufacturing life cycle.

1.5 Operational Excellence

Let us understand what operational excellence is, and how it is pertinent to ceramic manufacturing industries. Operational excellence is the integration of people, process and technology in a way that provides the maximum productivity per unit time with per excellence quality and assists in obtaining the lowest possible product cost per customer. One question that comes into mind: "Is manufacturing excellence different from industry to industry?" Maybe yes. In general, production motives and objectives are the same or similar but the methodology for implementation may differ from company to company, unit to unit and obviously varies from industry sectors to sectors. In common, the basic philosophy is to reduce the inventory, improve the quality, increase the productivity, improve resource utilization, bring more flexibilities to the production system, improve response time to customers' needs and finally reduce overall cost increasing profit margins, which ultimately increase the stakeholders' value.

It is a strategic agenda for all kinds of manufacturing industries irrespective of the nature of the industry or the products and type of human resource employed and the degree of automation/mechanization is in place. The operational excellence is a methodology that is applicable across all sectors, including manufacturing, irrespective of the push and pull system. But the approach may differ in the perspective of implementation. Since the inception of the production system, the aim has been to increase the productivity without compromising the quality and with minimal effort/unit time. The hunt is still on for most of developing or the developed ceramic companies. In this circumstance, analysis discloses the reasoning of why production output suffers and what are the root causes. Such important operational features are:

- Unexpected down time in the machine
- Lack of proper skillsets among the labor
- Irregular maintenance and poor equipment health
- Poor quality forces to stop the production
- Input output ratio mismatch of the production
- Higher amount of line rejection
- Supply disruption so on and so forth

1.5.1 Toyota Production System (TPS)

In order to obtain and establish the high throughput in the manufacturing industry, a strong management philosophy is essential, and Taiichi Ohno and Eiji Toyoda invented such a hypothesis between 1948 and 1975 for the automotive industries and coined it the Toyota Production System (TPS). It is not just a system or tool; rather, TPS is a broad management philosophy that can be implemented across processing industries. Though many believe that it may initially look like TPS is only meant for Lean manufacturing, waste reduction and zero inventory approach, it is not so. Broadly, it implies maturity in the production system, reduces variability, improves productivity, quality and reduces inventory.

Thus, TPS is a thinking production system because it makes the people motivated to improve the system in all considerations so that it can be made leaner, smarter and highly productive. TPS core people can bring a great change in the production environment.

People are engaged in continuous coaching for a knowledge upgrade in order to change production scenarios as the technology and process changes with time. This innovative approach cannot directly transfer from one industry to another, nor from one plant to another. The TPS can give exceptionally good results in any field in which anybody wants to achieve overall improvement in terms of productivity, quality, safety and reliability. TPS also stands for manufacturing problem solving and provides guidance on how to do it in the production shop floor. Following are eight significant aspects that are essential during analysis and toward solving the problem:

- Define the problem (plan for it).
- Break down the problem into smaller areas to handle it better (plan).
- Identify the root cause of the occurrence (plan).
- Target setting for achievement (plan).
- Select the potential or probable solutions from different countermeasures (plan).
- Execute the plan (do).
- Review the outcomes (check).
- Analyze what is going wrong, adapt, adjust and then repeat the cycle (act).

Irrespective of the production system, one can design the appropriate approach in the target of specific and effective work to that of production environment. It depends on the technology, machine employed for production, process, product designs, shop floor layout, people at work, etc. While discussing the management philosophy in the perspective of high throughput, TPS can be better explained by 13 pillars which are summarized as follows [3]:

1. **Just-in-Time** (Get the Material Only When It Is Required)—A cornerstone of modern manufacturing, just-in-time production was pioneered by Toyota. It consists of a *pull* system (as opposed to push) that provides different processes in the assembly sequence with only the kinds and quantities of items that they need and only when it needs them. It allows cars to be built for optimal efficiency and financial management.

2. **Muda Muri Mura** (Waste, Overproduction and Unevenness/Lean Production)—Muda, Muri and Mura work in tandem to eliminate waste. Muda actually means waste, but in the context of the Toyota Production System, waste is defined as non-value-adding activities such as over-processing. Muda divides waste up into seven categories: transportation, inventory, motion, waiting, over-processing, over-production and defects. Muri means to overburden, and this is avoided by even distribution of production tasks in the assembly processes. Mura means unevenness, which is eliminated in the Toyota Production System by training workers to operate multiple machines, so that there is cohesion between their operations.

3. **Kanban** (Card You Can See)—Immortalized as a flashing signboard, Kanban is a system that conveys information between processes and automatically orders parts as they are used up. Toyota has six rules for the effective application of Kanban: (1) never pass on defective parts; (2) take only what is needed; (3) produce the exact quantity required; (4) level the production; (5) fine-tune production; and (6) stabilize and rationalize the process.

4. **Poka-Yoke** (Error Prevention/Avoid Mistakes)—Put simply, Poka-Yoke means to avoid (yokeru) mistakes (poka). The Toyota Production System employs devices that automatically stop the line if there is an error.

5. **Kaizen** (Continuous Improvement)—Kaizen is at the heart of the Toyota Production System. It serves as a mantra for continuous improvement; the effects of which are far-reaching, from waste elimination to efficiency optimization. Kaizen gives voice to the workforce, empowering individuals to identify areas for improvement and suggest practical solutions.

6. **Andon** (Visual Indicators/Action Point)—Andon is a visual aid that highlights where action is required. Usually activated by a button or pull-chord; production is automatically halted when a member of staff pulls it.

7. **Genchi Genbutsu** (Go and See/Problem Solving)—Genchi Genbutsu is the idea that the best way to solve a problem is to see it for yourself. In the Toyota Production System, managers are present on the factory floor; this immersive approach means they fully understand the working environment and processes and can advise the best possible solution when a problem arises.

8. **Genba** (The Actual Place)—Genba is the physical place where work is done, and its philosophy is that all actions and processes are as transparent as possible. Toyota team members regularly conduct "Genba Walks" on the factory floor in order to identify areas where potential improvements might be made and to better understand the workload of their colleagues.

9. **Hansei** (Self Reflection/Learning from Mistakes)—Hansei is the process of recognizing and learning from mistakes, in order to prevent them from occurring again. Toyota hosts hansei-kai meetings, in which failures experienced during the production process (if there were any) are reflected on and future prevention plans put in place.

10. **Heijunka** (Leveling)—Heijunka means having the correct number of parts required to build a specific number of components (for example, cars) for the smoothest possible production process. Heijunka is important when sequencing production. For instance, if a factory was required to send batches of high spec components down its assembly line at the same time, workers would be required to manage additional build tasks not present in less well-equipped components. Heijunka solves the problem by assembling a mix of models within each batch.

11. **Jidoka** (Humanized Automation)—Designing equipment to detect problems and stop automatically when required is fundamental to Jidoka. Toyota Production System operators can stop production the moment they observe something untoward, preventing the wasted production of defective items.

12. **Konnyaku Stone** (Smoothening)—Konnyaku stone is used to make precision, smooth unpainted body panels and remove imperfections; it is known as *devil's tongue*. Brushing the stone across the surface of a panel will smooth the metal so that it is ready for painting.

13. **Nemawashi** (Dig Around Before Planting /Information Sharing)—Decisions shouldn't be dictated by individuals, they should be made as a team; this is the thinking behind Nemawashi. In the Toyota Production System, information is shared openly with employees in order involve them in decision-making processes and allow them to voice their opinions.

However, an extensive discussion on lean manufacturing, the Kanban system and Just-in-Time (JIT) is necessary to provide insight into the subject, and such philosophy must be discussed chronologically.

1.5.2 Lean Manufacturing

As discussed earlier, the objective of Lean manufacturing is to reduce the waste in the production system. Now, what is waste? Waste can be defined as anything which does not add value to the customer. Waste is something which a customer should not pay for. The following are the seven wastes suggested by the TPS system; these can be either removed or minimized to the greater extent: processing waste, stock in hand, transportation time, waiting time on hand, movement waste, defective products and underutilized workers.

The most deadly waste is overproduction. Overproduction is the unplanned production. Typically, a *push*-based system follows the production without the customer's demand and firm requirement. Should we adopt a *pull*-based system with respect to the customer requirement, so there is less chance of overproduction? Overproduction utilizes excess capacity and resources, which eventually enhance the stock in hand and need to be stored until the requirement comes from the market. Inventory made out of overproduction or process variability consumes the working capital, which does not give immediate returns. At the same time, the transportation waste can be reduced through improved transportation administrative process, better transportation network design, logical lot size planning and proper transportation asset management. Waiting-time waste can be related to either material, machine or human. This has a direct impact on the production cycle time, as it increases with waiting time results decreasing the throughput per unit time. Movement is related to the tools/material movement and their systematic arrangement for judicial use in the different stages. The execution process may delay the process if any tool or material is misplaced from their original position. This is one kind of process inefficiency and it has direct impact on the output. Defective parts need the additional rework time and may attract additional costs for repairs also. The customer lead time may increase for order fulfillment due to loss of time in the defective products. The waste of unutilized human resources is the loss of opportunity for higher production. Thus, it is necessary to utilize the human resources to their best potential in order to add value in both ends.

In the manufacturing cycle, the product passes through several stages of conversion. Though it is known for waste reduction, the heart of the Lean manufacturing is to add value to the customer's requirement [4]. The reduction of wastes in turn reduces manufacturing costs and improves the quality of the products. The quality of the products adds value for customers. Therefore, Lean manufacturing is all about removing the fats from the system and making it slim or slimmer; a diagram is shown in Figure 1.4. Lean manufacturing can be precisely defined as a system and methodology to remove the wastes defined by the TPS system and improve the customers' value. Lean is basically a set of steps followed to remove the non-value-added activities and improve the productivity and quality. The designing of the lean system can be easier if we understand the customers/segments and their preference/choice and what is meant by *value* to them. However, Lean is never a destination; it's all about the journey for the continuous improvement. Technology and processes change continuously, and Lean ideas have to be revalidated in the context of the new system. There are several tools available for Lean operations:

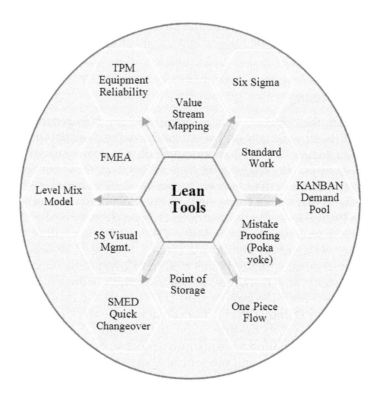

FIGURE 1.4
Lean tools [4].

- Just-in-Time (JIT) production and supply
- Total productive maintenance (TPM)
- 5S Program (seiri, seiton, seiso, seiketsu and shitsuke/*Sort, Set In Order, Shine, Standardize* and *Sustain*)
- Quick changeover techniques (such as Single Minute Exchange of Die techniques)
- Mistake-proofing
- Workspace layout improvements
- Failure mode and effects analysis (FMEA)
- Lean tools

Several ceramic and glass industries have already begun implementing Lean. This is true for both process (continuous) as well as discrete ceramic industries. For example, the Pittsburg Glass Works and Cardinal Glass Manufacturers both took the Lean route to improve the operational performance, reduction of inventories (predominantly WIPs), reduction of order fulfillment lead time and adding overall value for customers.

1.5.3 Value Stream Mapping and Fishbone (Ishikawa) Diagram

The value stream mapping (VSM) methodology helps to map the current product and the information flow of the organization, starting with the raw materials until the finished products are dispatched to the market.

It consists of all the required production stages, inventory storage, WIPs and quality checks, including all the related information. This is to visualize the present state of the process. On the basis of the current state, improvement areas are identified and the future state of the business process is designed (a briefing is given in Figure 1.5). The Fishbone diagram is widely known as the cause–effect diagram. It can be a good starting point for the enterprises seeking for process improvement. The Fishbone diagram primarily identifies the possible root causes of the problem occurring at any stage of the production and logically eliminates them one by one to arrive at the real root cause of the problem. The Ishikawa diagram has been named after its originator Kaoru Ishikawa, the Japanese quality pioneer, and it has been used in multiple manufacturing industries because of its simplicity of usage [5]. The Fishbone diagram is not just a diagram but a way of brainstorming and finding the root cause. It is a methodology to systematically eliminate the less potential causes.

1.5.4 Six Sigma and Lean Six Sigma

Six Sigma is one of the most prominent process improvement tools. Motorola first introduced it in 1986. Within a few years it made rapid progress. In 1995, GE adopted it. Predominantly, Six Sigma made its presence felt in the manufacturing sector and later it was practiced in service sectors and IT companies as well.

Six Sigma is a quality improvement tool that helps to reduce the defects and variabilities in the process and hence it reduces the cost of production and improves customer satisfaction. Over the course of time the Six Sigma ideologies were amalgamated with Lean manufacturing and were known as Lean Six Sigma and was used as an operational excellence tool. Companies like GENPACT, IBM, Accenture and GE have advocated this as a transformational tool to enhance the business growth and efficiencies.

There are two methodologies suggested by the experts:

- DMAIC (Define, Measure, Analyze, Improve and Control) which typically helps to improve the existing business process
- DMADV (Define, Measure, Analyze, Design and Verify) which focuses on new project or product development

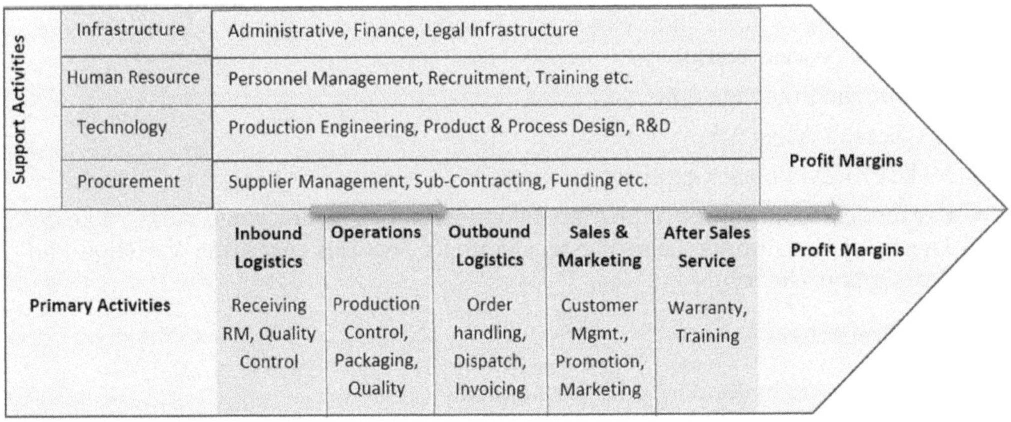

Support Activities	Infrastructure	Administrative, Finance, Legal Infrastructure					
	Human Resource	Personnel Management, Recruitment, Training etc.					
	Technology	Production Engineering, Product & Process Design, R&D					Profit Margins
	Procurement	Supplier Management, Sub-Contracting, Funding etc.					
		Inbound Logistics	Operations	Outbound Logistics	Sales & Marketing	After Sales Service	Profit Margins
Primary Activities		Receiving RM, Quality Control	Production Control, Packaging, Quality	Order handling, Dispatch, Invoicing	Customer Mgmt., Promotion, Marketing	Warranty, Training	

FIGURE 1.5
Value stream mapping.

For Six Sigma implementation, roles are classified as Green Belt (part-time employee, works under the guidance of a Black Belt), Black Belt (dedicated 100% to the project), Master Black Belt (100% dedicated and assists Champions) and Champions (solely responsible for Six Sigma implementation) [5].

1.5.5 KANBAN System

Let's understand what Kanban is. In the Japanese terminology, "Kan" means card and "ban" means signal. So the Kanban is "the visual card system that controls the production and inventory." This is an open-to-all visual system which shows the output vs input required, i.e., production vs withdrawal, giving a clear indication of where and how to manage the inventory efficiently. It recommends storing only the required amount of inventory in each and every stage of the production, thus keeping the inventory as low as possible. Thus, "Kanban is a production scheduling tool which connects the entire manufacturing cycle of a product" [6].

This is a two-card system for Production (P) and Withdrawal (W). The number of cards for P and W is decided by the management and it is gradually reduced to match it to the lowest possible level. The P should match to the short-term demand and W should match accordingly with P.

On the shop floor, the workers pass the card or Kanban to the next team in order to communicate effectively and subsequently the same is passed to their next team, and likewise the chain of communication passes through the shop floor supply chain to external supply chain like respective Suppliers. This depends on the bill of materials (BOM) of the Sub-Assembly/Assembly/Semi-Finished/Finished products. Suppose when a bin of materials being used in a specific operation of a production line is exhausted or finished, a Kanban signal is sent to the raw material section describing what material is needed, the quantity required and the specification of the materials required. By that time, the raw material section might have a new bin ready for supply. Likewise, the process of supply and receive move like a chain. Finally, the requirement goes to the supplier who is also ready for supply with the required amount of material. This triggers the JIT system. As per David Anderson, a pioneer in the Kanban system, following are the 10 properties including five core properties and five emergent properties.

Core Properties:

- Visualize process workflow
- Measure and manage flow
- Restrict the WIPs
- Make process policies explicit
- Use models to recognize improvement opportunities (well-known models like the Theory of Constraints, Lean Waste, Deming's Theories regarding Variation and Real Option Theory)

Emergent Properties:

- Prioritize work by (opportunity) cost of delay
- Optimize value delivery with classes of service
- Spread risk with capacity allocation

- Encourage process innovation
- Manage quantitatively

Historically it took 12 years to flow down Kanban within Toyota, and cumulatively it took 8 years to have 60% of first tier suppliers on Kanban and 20 years to get 98% of first tier suppliers on Kanban.

Only approximately 50% of Toyota suppliers are using Kanban systems internally today, although there is wide scope to implement Kanban in ceramic manufacturing sectors. In order to implement the Kanban system, one has to follow:

- Keep it simple.
- Develop strategy in phases.
  - Utilize stockroom and warehouse on floor.
    - Then eliminate stockroom.
    - Then eliminate warehouse.
- Have supplier continue to play larger role
- Make it very visual. Should not have to count:
  - Label shelves and boxes.
  - Use water levels.
  - Taped/sealed boxes.
  - Color codes, bar codes.
  - Egg carton approach, mistake-proof system.
- Continue to reduce quantity and increase replenishment frequency. Kanban are inventory (evil) and need to be constantly minimized or eliminated.

1.5.6 Just-in-Time (JIT)

The primary objective of JIT is to deliver the right quantity of the products at the right time. JIT is the typically a *pull* system. In this pull system, the production can only be initiated when there is a predefined demand that may be confirmed by sales orders or customer commitment. JIT optimizes the inventory to the minimum, thus it saves the inventory carrying cost, which helps in working capital management. At the same time, it helps in improving quality and plant efficiency [7].

On the contrary, the Material Requirement Planning (MRP) is a *push* system. MRP is a process, through which it is identified how much raw material or components are required for producing the final product to fulfill the market commitment. In the MRP process, the demand considers the forecasted requirement and confirmed customer requirement, expected confirmed demand (may not be the sales order) and un-confirmed demand; the basis may be past trends, some assumption depending on the market intelligence in terms of market research or survey results. Summation of all of them determines the demand and becomes the input for the MRP run process. The MRP run gives the planned quantities of all the required raw materials and the components. It excludes the orders in the pipeline, the warehouse and the in-transit stocks. The requirement is translated in terms of purchase requisition, which in turn is converted to purchase orders (PO). The POs are sent to the respective vendors for delivering the components and the raw materials within the stipulated lead time. The JIT system has the following key features:

a. **Moving Away from the Forecasting-Based System**: Traditionally production is driven by the forecast. The forecast is based on the historical trend of the market for a product or group of products. The JIT suggests to move away from the forecast-based push system. In the pull-based system, there is very little or no reliance on the forecast.

b. **Minimal Storage Cost**: The result of the pull system is the minimal inventory storage and lower carrying cost. The following are the tangible benefits:
   - Ease of space management
   - Less inventory
   - Less handling cost

c. **Limit the Dead and Slow-Moving Stocks**: JIT has a direct impact on the accumulation of dead and slow-moving stocks. In the pull-based system the inventory is never procured on the basis of the presumed demand, which in many instances is proven wrong and the stock gets piled up, so there is the chance of unwanted inventory blocking the money.

d. **Reduced Cost of Production**: The input cost of the materials and the activities required for production is less compared to the conventional production because of the lesser inventory and human costs, and less wastage.

e. **Higher Return on the Capital Asset Ratio**: As the input cost and the investment are less due to less inventory and other associated costs of production, the cost of goods manufactured (COGM) is less. Due to higher productivity the asset efficiency is also greater. As the profit is high, the return on investment is naturally high. The details of advantages and disadvantages of the JIT–Pull and MRP–Push systems have been highlighted in Table 1.1 [7].

Every industry is unique due to its exclusive product and services, manufacturing process, mode of operations, plant layout and demographic diversities. The ceramic industries are no different. Certain industries, like ceramic and vitrified tiles, sanitary wares, etc., have advanced with high-end technologies and modern philosophies of production, but industries like refractory, cement and glass may not be so equipped with sophisticated technologies. If we look back, during the 1990s the ceramic and synthetic tiles industries had end-to-end automations, laser-guided vehicles that did not require human drivers. The final product quality was checked by the automated sorting machine with predefined quality parameters. In comparison, the other industries may not have advanced so much due to its nature of operations. So the adoption of modern philosophies like JIT, Lean, Six Sigma, TPM and TQM into the production is not observed much. But there is room for improvement and scope for implementing Lean, Six Sigma, TPM and TQM in full scale. The ceramic industries predominantly are push-based systems or MRP-based systems, where production is done based on the forecasted demand and kept in store. Depending on the customers' need it is dispatched from the different distribution locations. But it is observed that there is a growing trend of moving away from the traditional production system to advanced production system. It is also found that multiple glass, ceramic and refractory producers are moving toward implementing lean, Six Sigma, TPM and TQM, which is promoting the need to process automation from the production processing as well as the transaction processing perspective. Way back, in 1999/2000, H. R. Johnson, the leader in ceramic lifestyle products implemented the ERP system and had a state-of-the-art

TABLE 1.1

Advantages and Disadvantages of JIT-Pull and MRP-Push Systems

Advantages	Disadvantages
JIT – PULL	
Inventory is always limited as most of the productions are done against some confirmed requirement	Every job is a high priority and high stress one as every requirement is important
Raw Materials are consumed only what is required and so is the case for the time consumption for workers	Possibility of frequent changes because of smaller lot sizes
Quality must be the first priority in anything done on the shop floor	Set-up time is considerably high as every production lot may be different
MRP– PUSH	
Managers have the opportunity to plan properly and control the things	MRP system is prone to accumulation of higher inventories
It helps to achieve the economies of scale in purchasing and production	Chance of generating more scraps than JIT system
Flexibility to prioritize and produce some complex assemblies and components	MRP system management needs dedicated effort

manufacturing system. It started initiatives for implementing the Six Sigma, TPM and TQM in greater scale.

Apart from the above manufacturing process improvement philosophies, there are other process optimizations methodologies like Design of Experiment (DOE) that are widely lauded for improving the manufacturing processes. This is basically a data-driven process optimization approach where we need to identify the critical process parameters and statistically analyze the input vs output. The following section covers the DOE in detail.

1.6 Design of Experiment

Design of Experiment (DOE) implies the systematic planning of an experiment, collection of data and the establishment of the statistical relationship between input parameters and output responses. In a wide utility, this tool is successfully applied in the field of science and engineering for process optimization and development, process management and data validation tests. R. A. Fischer first introduced both experimental design in agricultural research and the basic principle of factorial design to analyze the relevant data, coined as Analysis of Variance (ANOVA) [8]. Critical interaction between mathematical and statistical techniques including regression, ANOVA, non-linear optimization and desirability function leads to optimized manufacturing processes and quality characteristics. ANOVA is capable of assisting and identifying the effect of each factor (input) against the objective function. Factorial design, Taguchi's method and Response Surface Method (RSM) are the most effective tools for the composition and process optimization among other DOEs.

1.6.1 Factorial Method

Factorial design is used for conducting experiments as it allows study of interactions between factors. The phenomenology *interaction* is the prime driving force for many processes. However, some significant interfaces may not be exposed in the absence of adequate factorial design of experiment. The full factorial experiment determines the responses in all probable combinations for factors and their levels to accomplish the process optimization. Herein, each experimental condition is designated as *run* and the measured response is called *observation*, while factorial design can be run on two-level, three-level and multi-level factorial. The entire set of runs is known as the *design*. In consideration of full factorial design, all possible combinations of the factor levels are encountered in order to achieve more reliable data, but conducting such a run is expensive and sometimes prohibited.

1.6.2 Taguchi Method

Conventional experimental design techniques have limitations in an industrial environment. In order to overcome such a situation, Dr. Genichi Taguchi developed a method known as *orthogonal array design* that encourages a new dimension of conventional experimental design. The prime objective of the Taguchi method is to evaluate the optimal settings of input parameters without consideration of uncontrollable or noise factors. Factor refers to an input variable during the experiment. In brief, the Taguchi DOE method is described by $L_a b_c$, where L_a is the orthogonal arrays of variables or deign matrix, and b_c is the levels of variables and c the numbers of variables. It conducts the balanced (orthogonal) experimental combinations that lead more effectively than a fractional factorial design, and governs a mean performance characteristic value close to the target value rather than a certain specific limit value, which results in an improved process performance. Employing this tool, industries are effectively minimizing the product development cycle time for both design and production, thus minimizing costs. However, this method is not a universal approach, since it does not assume interaction between factors that is non-scientific when it is really a significant issue, like high-temperature ceramic processing. In actuality, interactions imply a much large number of experiments for same number of control factors. Thus, the major disadvantage of the Taguchi method is that it yields only relative data that do not exactly indicate which parameter has the highest influence on the performance characteristic value. Herein, the orthogonal arrays do not test all variable combinations, and thus the reliability is criticized when all interactions and relationships are not encountered. Limitations also exists in dynamic analyses such as simulation study, as it deals with designing quality rather than modification of poor quality, and thus it is an effective tool for the early stages of process development. However, it is a regular and unique tool in automobile, electronics and other processing industries to develop high-quality and low-cost products [9].

1.6.3 Response Surface Method (RSM)

Response surface method (RSM) is an ideal mix of both mathematical and statistical methods in order to achieve improved method and process optimization. In contrast to the Taguchi method, the RSM is an excellent analyzing tool to determine the relationships and influences of input parameters on responses during extensive interaction between factors. G.E.P Box and K. B. Wilson first introduced the first-degree polynomial model in RSM and used a set of designed experiments to obtain an optimal response [10]. RSM explores the relationships between several independent variables and one or more response variables; the response variables can be graphically viewed as a function of the process variables (or independent variables). Furthermore, the RSM is constructed to check the accuracy of model that uses the build time as a function of the process variables and other parameters [11]. Asiabanpour et al. developed the regression model to describe the relationship between the factors and the composite desirability [12]. In consideration of that, it improves the analyst's understanding of the sensitivity between independent and dependent variables, as it is an experimental strategy and has been employed in the industry for research and development, with considerable success in a wide variety of situations to obtain solutions for complicated problems. The following two designs are widely used for fitting a quadratic model in RSM.

a. **Central Composite Designs**: Central Composite Designs (CCDs), also known as Box-Wilson designs [13], are appropriate for calibrating the full quadratic models described in Response Surface Models. There are three types of CCDs, *viz.* circumscribed, inscribed and faced. The geometry of CCD's is shown in the Figure 1.6.

Each design consists of a factorial design (the corners of a cube) together with *center* and *star* points that allow estimation of second-order effects. For a full quadratic model with n factors, CCDs have enough design points to estimate the $(n+2)(n+1)/2$ coefficients in a full quadratic model with n factors.

The type of CCD used (the position of the factorial and star points) is determined by the number of factors and the desired properties of the design. Table 1.2 summarizes some important properties of CCDs. A design is rotatable if the prediction variance depends only on the distance of the design point from the center of the design.

b. **Box-Behnken Designs**: Figure 1.7 illustrates the Box-Behnken designs that are used to calibrate full quadratic models. These are for a small number of factors (four or less), rotatable and require fewer runs than CCDs. By avoiding the corners these designs allow researchers to work around extreme factor combinations like an inscribed CCD [14].

Some extensions of RSM deal with the multiple response problem. Multiple response variables create difficulty, as the optimal variable for one response may not be optimal for others. Significant criticisms of RSM include the fact that in most of the cases the optimization is done with a model for which the coefficients

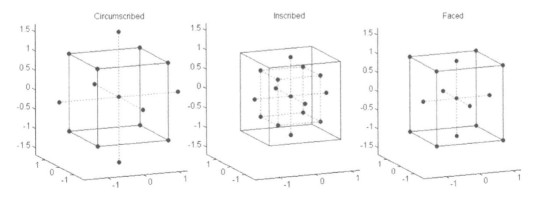

FIGURE 1.6
Circumscribed, inscribed and faced designs.

TABLE 1.2

Comparison of CCD's

Design	Rotatable	Factor Levels	Uses Points Outside ±1	Accuracy of Estimates
Circumscribed (CCC)	Yes	5	Yes	Good over entire design space
Inscribed (CCI)	Yes	5	No	Good over central subset of design space
Faced (CCF)	No	3	No	Fair over entire design space; poor for pure quadratic coefficients

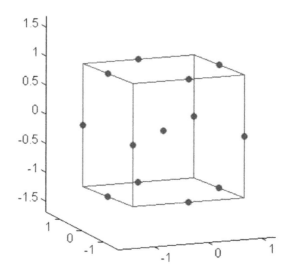

FIGURE 1.7
Box-Behnken design.

are estimated and not known. Thus, an optimum value may only look optimal, but is far from the real value because of variability in the coefficients.

Thus, RSM is a set of techniques that encompass: [15] (i) the designing of a set of experiments for adequate and reliable measurement of the true mean response of interest, (ii) determining the mathematical model with the best fits, (iii) finding the optimal set of experimental factors that produce maximum or minimum value of response and (iv) representing the direct and interactive effects of process variables on the output responses through two-dimensional and three-dimensional graphs. The accuracy and effectiveness of an experimental program depends on careful planning and execution of the experimental procedures.

1.6.4 Other Methodologies

DOE is a method that revolves around experimentation for optimization and modeling. The other methods that are commonly used for this purpose are analytic methods and Artificial Intelligence (AI)-based techniques. Some of the commonly used analytical methods are exact solution and numerical solution, while Artificial Neural Network (ANN) and fuzzy logic are widely used AI techniques. Population-based optimization techniques such as Genetic Algorithms (GA) and Particle Swarm Optimization (PSO) in combination with DOE are also used to solve complex problems.

1.6.5 Advantages and Disadvantages of DOE

DOE became a more widely used modeling technique superseding its predecessor, the One-Factor-At-Time (OFAT) technique. One of the main advantages of DOE is that it shows the relationship between parameters and responses. In other words, DOE shows the interaction between variables which in turn allows us to focus on controlling important parameters to obtain the best responses. DOE can also provide us with the most optimal setting of parametric values to find the best possible output characteristics. Besides, the mathematical model generated can be used to predict the possible output response, based on the input values. Another main reason for using DOE is savings in time and cost in terms of experimentation. DOE can determine the number of experiments or the number of runs before the actual experimentation is done. DOE allows the handling of experimental errors while still continuing with the analysis. *DOE is good when it comes to predicting linear behavior; however, when it comes to nonlinear behavior, DOE does not always give the best results.* DOE is extremely helpful in discovering the key variables influencing the quality characteristics of interest in the process.

1.6.6 DOE in Ceramic Research and Development

In the conventional approaches to develop new ceramics, the researchers use trial-and-error or one-time-variable method to optimize the process variables. Newer computational approaches are currently being investigated to establish Process–Structure–Properties (PSP) linkages of engineering materials [16]. While such an approach is relatively well explored in the case of metallic materials (dual phase steels or steels with non-metallic inclusions) [17], such statistical modeling-based analytical approaches are not well-explored in case of ceramics. It has been experimentally well demonstrated in ceramic systems that the interplay among the process variables (sintering temperature and sintering time) and material variables (sinter-aid and second phase addition) has a profound influence on both sinter density and grain size [18]. The complex interaction among these variables is difficult to be assessed using conventional experimental module by varying one parameter, while keeping other parameters constant. Several existing quantitative modeling methodologies that consider factor and level to define the design parameters are commonly used, including full factorial design, fractional factorial design, Box-Behnken, Response Surface Method (RSM) and Taguchi statistical modules.

In factorial design approach, the level is restricted to two-level for less than or equal to four number of factors, which is also an expensive aspect to run the exponential number of sample size. For example, an attempt was made to prepare pure hydroxyapatite particle suspension from base precursors by this method and the utility of this statistical tool was explored [19]. Using the same model, full factorial central composite design circumscribes the fabrication method of reaction-bonded ceramic microparts by low-pressure injection molding. The powder volume content, Zr/Si ratio, binder composition and processing temperature on flowability of the feedstock were described as input process variables, whereas dynamic viscosity and yield point were considered as output responses [20].

Fractional factorial design is reported as an alternative to the full factorial design approach, which reduces the number of runs but loses information due to the lack of effect on two-factor interactions [21]. For example, various responses, such as green density, specific surface area and mean granule diameter of spray granulated 3YTZP-Al_2O_3 (10 to 95%) composite particle preparation methodology are optimized through input variables of solid content of Al_2O_3, spraying pressure and temperature [22]. As another alternative,

Box-Behnken designs are considered as a response surface design, specially prepared and restricted to only three levels, coded as -1, 0 and +1. For example, the phase purity, crystallinity, crystallite size, lattice parameters and particle size of hydroxyapatite were optimized from input variables such as reaction synthesis temperature, stirring speed, ripening time, acid addition rate, initial calcium concentration and atmospheric environment [23].

Among the different statistical modeling approaches, RSM comprises effective interaction among the design factors on the resultant response [24, 25]. Such an approach is not considered in other statistical methods. The name of such a model is justified as the model predictions provide a precise map, leading to successful optimization of the composition, process, grain size, mechanical properties, etc. In a recent article, the optimization of relative density and bending strength of MgO-PSZ was pursued using three independent factors, namely heating rate, sintering temperature and holding time by RSM approach [26]. In another study, sintering temperature and time-based optimization of the elastic modulus and fracture toughness were reported for zirconia-toughened bioglass ceramics [27]. In the RSM approach, the interactive effect among process variables on the output parameter can be assayed. Such an interaction effect however cannot be captured using Taguchi method. For example, the mechanical behavior and failure of SiC-Y_2O_3 composite were analyzed using the Taguchi method in reference to processing parameters, volume percentage of particulate, temperature, time and load [28]. Recently, the Taguchi method has been used to statistically assess the machinability of AISI 4340 steel with Zirconia Toughened Alumina (ZTA) ceramic inserts, and optimized the cutting speed, depth cut and feed rate [29]. In consideration of advantages of RSM, a recent article demonstrates the statistical modeling approach with the predictive capability of sinter density and grain size as a central theme in the development of next-generation ceramics. Such a computationally intensive method can be equally significant, if the predicted process conditions adapted experimentally to develop complex-shaped ceramics with variable sizes. Using ZrO_2-toughened Al_2O_3 (ZTA) as a model system, the adopted RSM approach has been used to quantitatively predict the independent and interactive role of process and material variables on sinter density and grain size [30].

1.7 Smart Manufacturing and Industry 4.0

The forces of change in research and development has been inexorable, and thus technology once again has brought its unprecedented change on to industries as a whole. With its staggering march, the message is clear to everybody - Change or Perish. Industries, especially the manufacturing industries, have come a long way through various transitions and different phases of evolutions, and there is further scope in contemplation of different production and quality control strategies.

> We have come to an age where Smart Manufacturing is all about, fully-integrated, collaborative manufacturing systems that respond in real time to meet changing demands and conditions in the factory, in the supply network, and in customer needs.

The primary drive has been to increase the speed of the productions, increase throughput and improve the quality of the output to the highest possible level. Ultimately, it has always been to serve the customer better within the commitment date and with unquestionable quality. Therefore, the need for automation has been perennial to avoid manual intervention and bringing the repeatability for the large-scale production. Without automation it was probably never a close possibility. Due to their history, ceramic industries are relatively less crowded by the technology compared to other industries like automobile and high-tech electronics and more driven by labor clubbed with machine power. If we look at the Industrial Revolution there have been four stages of evolution since the first Industrial Revolution.

England, where the first Industrial Revolution was initiated in the middle of the 18th Century and where the steam engine was first invented. It was a complete paradigm shift from the conventional complete labor-based manufacturing to the first step of automation where the water- and steam-powered machine was developed to aid the workers. With increased production capability, the manufacturing set up moved from a cottage-based industry to a more organized and structured operation where they started serving customers instead of only their known circles/neighbors. In addition, the organized market and customer segment started forming. This phase of the evolution can be termed as *Industry 1.0*. This was just a stepping stone for the organized production and market formation.

The second Industrial Revolution took another 100 years to happen: it took place in Europe and the United States during the second half of 19th Century. This revolution is known for mass production and the replacement of the steam power-based engine by electrical energy, which became the main source of power. In order to meet growing demand, several technologies in industry and mechanization have been developed, such as the assembly line with automatic operations, allowing the increasing of productivity. The revolution came in the form of machines designed for mass production and quick delivery to the market. This period also saw the development of a number of management programs that made it possible to increase the efficiency and effectiveness of the manufacturing facilities. Divisions of labor, where each worker does a part of the total job, increased productivity. Mass production of goods using assembly lines became commonplace. American mechanical engineer Fredrick Taylor introduced approaches of studying jobs of optimize worker and workplace methods. Lastly, Just-in-Time and the Lean manufacturing principle further refined the way that manufacturing companies could improve their quality and output. This phase of industrialization is known as *Industry 2.0*.

Humanity's quest is relentless. The aim has always been to bring more automation across production lines and remove production constraints. In last few decades of the 20th Century invention and manufacture of electronic devices, such as the transistor and later the integrated circuit chips, made it possible to more fully automate the individual machines to supplement or replace operators. The third Industrial Revolution (*Industry 3.0*) happened with the invention of the integrated circuit (microchip), which triggered a new milestone. The use of electronics and Information Technology in order to achieve further automation in production is the key feature of this revolution that emerged in the last few years of 20th Century in many industrialized countries around the world. This period also spawned the development of multiple software applications to capitalize on the potential of hardware for better automation.

Integrated systems like Material Requirement Planning (MRP) were supported by the Enterprise Resource Planning (ERP) tools that enabled humans to plan, schedule and track product flows through the factory. Pressure to reduce costs caused many manufacturers to move component and assembly operations to low-cost countries. The extended geographic dispersion resulted in the formalization of the concept of supply chain management. The increasing of productivity is the core of every industrial revolution. The first three industrial revolutions had a strong impact in industrial processes, allowing a productivity and efficiency increase through the use of disruptive technological developments, such as the steam engine, electricity or digital technology. *Industry 4.0*, which may eventually represent a fourth industrial revolution, is a complex technological system that has been widely discussed and researched, having a great influence in the industrial sector, since it introduces relevant advancements that are related with smart and future factories.

1.7.1 Inception of Smart Manufacturing

The idea of Industry 4.0, or Smart Manufacturing, has been as disruptive as the invention of electricity or the first electronic transistor. An idea about Industry 4.0 is illustrated in Figure 1.8, to describe how modern manufacturing, automations and data are linked together to create a Cyber Physical System [31]. Still, it is in the nascent stage of transformation. Over the past several years, Smart Manufacturing has been a hot topic of conversation among manufacturing experts, strategists and thought leaders. Still it is not clear to many what exactly Smart Manufacturing is. Is it additional automations? Is it all about employing robots in the shop floor? Is it another big movement to reduce the manpower? Is it more of an integrated manufacturing? However, despite its recent coverage in the press and in journal articles, many in the front lines of manufacturing aren't quite sure what Smart Manufacturing entails, why its importance or how it's even relevant to their organization.

The Smart Manufacturing Leadership Coalition (SMLC) and the National Institute of Standards and Technology (NIST) have come up with the widely quoted definition of the Smart Manufacturing. The SMLC definition states, "Smart Manufacturing is the ability to solve existing and future problems via an open infrastructure that allows solutions to be implemented at the speed of business while creating advantaged value" [32].

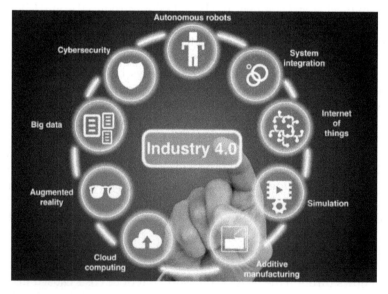

FIGURE 1.8
Smart Manufacturing and Industry 4.0, a Cyber Physical System [31].

According to NIST, Smart Manufacturing are systems that are "fully-integrated, collaborative manufacturing systems that respond in real time to meet changing demands and conditions in the factory, in the supply network, and in customer needs" [32].

Smart Manufacturing (along with Digital Transformation) is expected to be the next big disruptive idea for all the manufacturing sectors, be they small medium enterprises (SME) or large-scale manufacturers. For an example, in the recent years the mobile has become an inseparable device from our lives due to its contribution in our day-to-day activities beyond communication. Is it not the device that books us a cab? Is it not the device that captures, stores and delivers our critical information? It is just like a miniature of the computer muscled by multiple technologies and apps. The advent of cloud technologies in combination with the Industrial Internet of Things (IIoT) has made it more powerful. It is networked and integrated into our life, as depicted in Figure 1.9 [33].

The advancement of chip-making technologies, the mobile is becoming more and more powerful like never before. The smart phone is becoming smarter. Now mobile cloud service providers help us store almost unlimited data, be it personal or official.

Smart Manufacturing utilizes all of these components, addressing the complexities of security, interoperability and intellectual property for manufacturing. The next industrial revolution is upon us, as Industry 4.0 brings in a new wave of connected manufacturers and smart factories. Industry 4.0 is a current trend in manufacturing that involves a combination of cyber-physical systems, automation and the Internet of Things (IoT), which together create a smart factory. Industry 4.0 manufacturers worldwide are connecting their machines to the cloud and developing their very own Industrial Internet of Things (IIoT). In doing so, they are scratching the surface of untapped potential, which promises exponential growth and enormous scalability for their businesses.

In order to perform maintenance and repairs on connected machines, technicians must offer a level of technical expertise, in addition to their foundational mechanical knowledge, in order to keep up with the accelerated service demand that comes with IoT connectivity. Because of this, manufacturing customers turn to our crowd service model to alleviate the pressure that comes with servicing IoT devices.

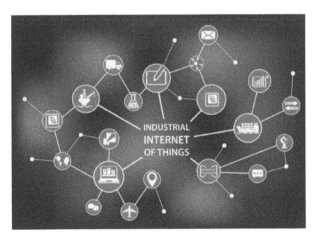

FIGURE 1.9
Industrial Internet of Things (IIoT) [33].

Digitizing manufacturing processes is not as simple as connecting devices to Wi-Fi. For one, the manufacturing industry is historically known for a mechanical marriage of oil and steel to make moving metal parts—not so much cloud computing or cyber-physical systems. With that, a manufacturing organization's upgrade to Industry 4.0 can require a full-fledged paradigm shift—from factory floor workers to C-suite decision-makers—to instill organizational change and company-wide rethinking of existing processes. While machine-to-machine and human-to-machine connectivity are the paramount focuses of Industry 4.0, the true underlying benefit of Industry 4.0 resides in the machine-to-business connectivity, which we call - "machine-as-a-service". [34]

In the 21st Century, Industry 4.0 connects the internet of things with manufacturing techniques to enable systems to share the information, analyses it and use it to guide intelligent actions. It also incorporates cutting edge technologies including additive manufacturing, robotics, artificial intelligence and other cognitive technologies, advanced materials and augmented reality, according to the article Industry 4.0 and Manufacturing Eco-system by Deloitte University Press. [35]

The development of new technology has been a primary driver of the movement to Industry 4.0. Some of the programs first developed during later stages of the 20th Century, such as manufacturing execution system (MES), shop floor control and product life cycle management (PLM), were farsighted concepts that lacked the technology needed to make their complete implementation possible. Now Industry 4.0 can help these programs reach their full potential.

This new approach will bring together the digital and physical worlds through cyber-physical system (CPS) technology, embracing a set of future industrial developments that will allow the improvement of productivity and efficiency among the companies that are adopting this new manufacturing paradigm. Industry 4.0 holds huge potential and will provide a set of economic and social opportunities through the paradigm shift regarding work organization, business models and production technology that eventually can be employed in ceramic industries. The influence of Industry 4.0 is being researched by academics and companies in recent years, which has resulted in an increasing number of publications about this topic. However, this concept intends to highlight the new industrial environment and the involved technological advancements is not always consensual, as well as its potential consequences in industry and manufacturing, which are not yet clearly defined.

Given the historical pedigree of the manufacturing industry, Industry 4.0 adoption will not be an overnight transformation. However, manufacturers today are partnering with service platforms and providers to catalyze digital transformation. There will be a shift from the journeyman engineers who are good with their hands to the service technicians who use digitally available information to bridge the skills gap and open new business channels—in not only maintenance and repair but also new business and sales.

Ultimately, the customer always comes first, and Industry 4.0 is one step closer to achieving the pinnacle of customer satisfaction through this machine-as-a-service philosophy. Digital transformation is not just limited to machine connectivity; rather, it involves a comprehensive approach, from product to service to sales, to achieve greater stability and adaptability business-wide. Indeed, an integrated approach is key to implementing widespread adoption of Industry 4.0, setting the foundation for the next industrial revolution.

1.7.2 Why Smart Manufacturing is Called Smart

Is "smart manufacturing" a term just randomly chosen? Is it a real paradigm shift from the conventional manufacturing process? Looking at the focus from the big enterprises and attention from the chief experience officers (CXOs), apparently, it is evident that this shift of the production dynamics is remarkably different from the earlier few. A few decades back, many used to ask the same question about Lean manufacturing. But nowadays, management people know what Lean manufacturing means and what the guiding principles are. Smart manufacturing is an endeavor to enhance connectivity across the entire manufacturing value chain to a new height.

To adapt with the new wave of Industry 4.0, what should manufacturers do now? Companies need to strategize and create their own roadmaps for the digital journey. The following are some guiding principles for the smart manufacturing initiatives:

- Minimize manual data entry at each and every step of the manufacturing process.
- Integrate the entire value chain including inbound and outbound supply chain, manufacturing units, product lines and machines, and extract the information for real-time decision making
- Leverage the acquired data from different process steps and apply the advanced analytics and machine learning algorithms to recommend process adjustments at different levels of the enterprise from controls to operation.
- Adopt machine-to-machine (M2M), application-to-application (A2A) and business-to-business (B2B) integration that will enable multi-vendor hardware and software plug-and-play solutions with open integration platforms to the Internet.
- Develop a culture of educating the workforce beyond the conventional functional boundaries like production, general automation and IT. Now manufacturing personnel need to understand advanced IT and vice versa. Workers have to learn to configure and maintain smarter machines and robots.

Over the next few years, smart manufacturing will evolve to new levels of connectivity. Revolutionary productivity gains are expected from the resulting integrated value chain processes. These are exciting times for the manufacturing sector.

1.7.3 Need of Smart Manufacturing

Do we need smart manufacturing? Maybe all the manufacturers in small-scale to large-scale industries would respond with a big "yes." But why do we need it? What is the impact on the manufacturing value chain? Will it serve my customer better? Will costs be optimized? Will the production be more integrated? Will my machine be maintained predictively? Will the down time of production be far less? Will the quality ultimately be better or the best?

Maybe yes will be the answer to all these questions. But the following areas are expected to improve:

- On-time delivery to customers
- Inbound supply
- Product and service quality and scrap reduction
- Capex and inventory position

- Logistics and transportation
- Overall productivity
- Utilization of manpower

As per the Capgemini Smart Factory Survey by the Capgemini Digital Transformation Institute, all of the above parameters can improve at least tenfold, if not more. The majority of the big manufacturers are now focusing on smart factory initiatives, especially the manufacturing leaders from the United States, France, Germany and the UK [36].

The initiatives are not confined to any specific sector or segment but widespread across aerospace, automotive, energy and utilities, pharma, life sciences, biotech industrial products, etc.

1.7.4 Does Smart Manufacturing Mean Only Robots?

Many think that it's all about robots and the replacement of human forces. The day might come when humans have to fight for jobs with bots. The factory may be placed under the custody of the robots [37]. Is it? It's not so. It's not the replacement of the human forces, but complementing them in many places where human cannot execute the job due to adverse climate in the factory, like blast furnace repair in the running condition, the wearing out of glass tank refractories, etc. Automation in factories isn't new. Today though, the disruptive force of digital transformation is taking manufacturing far beyond automation. Industry 4.0, mass customization and advances in tech like 3D printing and nanomaterials have placed humanity at the cusp of several game changers when it comes to this $11.6 trillion industry. For more on 3D printing, see Chapter 9.

Robotic automation began back in the 1800s with mechanized cotton spinners, steam power and the arrival of the First Industrial Revolution. By the 1930s, the automotive industry was leading the Second Industrial Revolution of mass production, paving the way for the digital control systems of the 1970s. In the 1980s, car makers became intensive adopters of industrial robots, at which point computers and automation were embodying the Third Industrial Revolution.

Jump forward to more recent milestones, and Foxconn in China was running up to 10 automated production lines in some of its factories at the end of 2016 in the second of its three-phase full automation plan. Also, in 2016, Adidas unveiled its first fully robot-built sneaker, one of 500 planned prototypes for its new factory in Germany. Though we're not quite there yet, the arrival of lights-out manufacturing is a case of when, not if.

Automation is certainly not new, but digital transformation is so much more than robots assembling parts—it's destined to disrupt every link in the manufacturing value chain and lead us into the Fourth Industrial Revolution: the cyber-physical age (Figure 1.10). As data takes center stage, connectivity, cloud, big data, IoT, AI and virtualization will act in concert to create a new business paradigm [38].

But there's a problem: manufacturing enterprises have been sluggish when it comes to embracing digitalization. Why the slow start?

One major issue is outdated legacy infrastructure. The complexity of virtualizing the production environment is exacerbated by IT systems that were developed before the cloud, inexpensive storage and ubiquitous connectivity came along. Going fully digital is also risky. Shutting down an assembly line to fix a software or network failure could be cripplingly expensive for a manufacturer. Connectivity requirements in smart manufacturing are also very high, often to the tune of sub-millisecond latency and data rates of 10 Gbps, as in the case of machine vision and cooperative robots. Fortunately, that's what the

FIGURE 1.10
Robotic automations [37].

latest wireless network solutions deliver: high-bandwidth, low-latency and reliable connections that can cut costs by up to 50% and energy consumption by 10%.

Equally significant, though, is the skills gap that exists in data analytics, a central facet of manufacturing and the source of insight into processes, faults, consumer habits and much more. Many companies aren't all that clear on how and where to deploy analytics solutions or how to use the huge volumes of data generated by sensors or systems. Moreover, McKinsey estimates that there will soon be a *shortage of around 1.5 million analytics experts in the United States alone*. While *Forbes* writer Meta S. Brown questions the McKinsey stats and analysis, she also identifies the human factor as an issue: "Managers who have trouble finding analytics talent have usually not given enough thought to their business goals" [39].

Moreover, in a survey by Tata Consulting [38] about big data analytics in manufacturing, the top problem identified by enterprises was building trust between data scientists and functional managers, which in turn creates a gap between data insights and how and which business strategies are executed. Of the 17 categories surveyed, the second biggest problem was determining what data to use for which business decisions and the third was the inability to handle the volume and velocity of data generated by sensors. Simply put, manufacturers can't and aren't making the most out of the data they have access to.

The complexity of the manufacturing industry means that no coherent industry-wide digital transformation strategy exists, with individual enterprises digitalizing at different rates and in different directions. Moreover, many companies lack the agility to quickly shift from traditional goals like Lean manufacturing. Indeed, the Tata Consulting survey [38] found that the top three benefits of data analytics for manufacturers are still in line with the old-school aim of optimizing processes: tracking product defects and quality, supply planning and identifying manufacturing process defects. Considering such operations, ceramic process industries can boost up their production and quality efficiency.

Reflecting the industry's commitment to Lean processes, manufacturers have been relatively fast movers in analytics, smart sensors and Industrial IoT (IIoT). That's all well and good, but the productivity gains from Six Sigma and Lean manufacturing have petered out over the last five or so years, because processes have become as optimized as they can be.

1.7.5 Possibility of Mass Customization

Those who work in the factory today are very much aware that any customization requires special planning for shop floor machine parameter setting and, after production, the output quality check. The digitally operated factory may have immense flexibility in customizing the products for the customers who prefer to have their own impression or expressions.

Smart robotics and machine learning will help achieve the desired mass customization. ABB, a leader in digital tech and robotics, is working with Huawei to combine wireless tech, smart sensors and smart components to solve manufacturing challenges. According to Joni Rautavuori, president of ABB Robotics and Applications, "The development that is happening on smart components and sensors makes it possible to use machine learning to develop new ways of programming robots" [37]. This increases the potential for adaptive programming, which in turn helps enable mass customization.

But it comes in last in one of the Tata Consulting surveys [38]. However, given shifting consumer expectations, it's quite probable that this will change for many products.

In March 2017, the tech magazine *Manufacturing Business Technology* reported that manufacturing is the second most hacked industry after healthcare, in large part because of inadequate investment in security. Although cyberattacks cost businesses $400 billion every year, there is a scarcity of experienced cyber-security analytics professional. *Forbes* cites the nonprofit information security advocacy group ISACA, which "predicts there will be a global shortage of two million cyber security professionals by 2019" [40].

The transition to Industry 4.0 is creating larger attack possibilities due to more complex networks, a vast number of connected IIoT devices, and big data processed in the cloud. Many companies lack a robust end to end (E2E) Information Security Solution that protects against attacks from a hacker's armory, including server, client, web and software, etc. Equally, the R&D related data, IPR and sensitive data requires a network solution that separates the R&D intranet from the office extranet.

Over the next decade, smart manufacturing will extend past individual factories to connect groups of factories and the manufacturing industry with other verticals. The convergence of manufacturing and services will continue with the cloud model based on IoT and data insights. Thus, the services that manufacturers will require and deliver as a result of the products they make will increase, many of which will be driven by data insights and consumer demand.

In the B2C space, consumers in emerging economies will become a dominant market player, while demand in developed countries will stay stagnant. However, customization—both in products and after-sales services—is likely to increase.

Upcoming technologies, including 3D printing, will evolve from prototyping to a viable means of mass production in the 2020s. Advances in 3D printing will enhance parts design, manufacturing processes, and printing technology. At the same time, the use of nanomaterials, which we're seeing today in products like clothing, sports goods and electronics, will expand into an industry worth $170 billion a year. Coupled with improvements in robotics and AI, new areas of demand will emerge.

1.7.6 Impacts of Industry 4.0

It is undeniably true that the Industry 4.0 will impact many areas of the entire business value chain starting from order fulfillment to the customer to the sourcing of the raw materials and the final components. In the downstream, the following areas are expected to be affected:

- Customer commitment and order fulfillment lead time
- Visibility of the order processing at the shop floor and direct interactions with the customers
- Service delivery operating models for after markets

In the production units and the shop floor, drastic changes are expected in operations, maintenance and resource planning. Many things won't operate the way they do today. The new production is certainly going to impact the following areas:

- Product lifecycles
- Predictive maintenance and machine Safety
- Reliability and continuous productivity
- Workers' education and skills
- Capacity planning and line balancing
- Final product inspection and delivery to market

In the upstream, there are changes in the sourcing and procurement cycle, as it needs continuous and real-time feed to the manufacturing system. Impacted areas would be:

- Logistic management, vehicle scheduling and transportation root management
- Domestic procurement for materials and services
- Import of the materials and services

Other factors in terms of safety, security and macro industry perspectives are expected to change:

- Environment, health and safety.
- IT security: companies like Symantec, Cisco and Penta have already begun to address the issue of IoT security.
- Socio-economic factors.
- Industry demonstration: to help industry understand the impact of Industry 4.0, Cincinnati Mayor John Cranley signed a proclamation in 2013 to state "Cincinnati to be Industry 4.0 Demonstration City."
- An article published in February 2016 suggests that Industry 4.0 may have a beneficial effect for emerging economies such as India [41].

1.7.7 Technology Roadmap for Industry 4.0

It's a radical change to the business cutting across all sections. Technology change is at the core of the fourth industrial revolution. Be it IoT/IIoT, cloud, big data analytics, artificial intelligence/machine learning, cyber security or adaptive robotics, all are the key elements of the large-scale Digital Transformation. To effectively adopt it, there should be a complete deployment planning with appropriate investment justification.

From both strategic and technological perspectives, the Industry 4.0 roadmap visualizes every further step on the route toward an entirely digital enterprise. In order to achieve success in the digital transformation process, it is necessary to prepare the technology roadmap in the most accurate way. In today's business, Industry 4.0 is driven by digital transformation in vertical/horizontal value chains and product/service offerings of the companies.

Depending on the nature of the industry, its operating model, product or service, the market they serve, level of automations required, skill set of the resources, the strategic as well as tactical planning for the enterprise may differ. But to be a part of the revolution there must be first a first step, which is nothing but strategic-level planning.

Enterprise must focus on the following before jumping into the Digital Transformation;

- Amount of investment required and the stakeholders willingness to support both from financially and morally.
- What is the current operating model and state of automations in terms of machine and information flow?
- Product or service they want to offer the customers and degree of automation required.
- Plans for re-skilling of the resources for the effective operations of the production system.
- The technology support system in case of breakdown or failure in operations.
- System security, especially cyber security management.
- Identifying the weak linkages of the business value chain and strengthening it [42].

1.8 Concluding Remarks

Regardless of the nature of the products, manufacturing infrastructure, automations and demographic diversities, continuous improvement is the primary objective for every industry's sustainability. Refractories, cements, glass, tiles, insulators and other ceramic wares have a lot of scope for improvements in term of process maturity, product quality, inventory turnover, supply chain integration and cost proposition. Simultaneously, the robustness of the DOE approach can be adapted in order to perform in-depth research and development and fabricate ceramic prototypes. Adoption of continuous Improvement tools like Lean, Just-in-Time, Kaizen and Six Sigma can facilitate the integration of process, people and technologies. Moreover, as this is the age of disruptive technologies, enterprise can strategize and leverage the benefit of the technology deployment. Industry 4.0 is the buzzword for all and is going to be the transformer for tomorrow.

References

1. Ancient Pottery (from 18000 BCE). http://www.visual-arts-cork.com/pottery.htm.
2. Sarkar, D. 2018. *Nanostructured Ceramics: Characterization and Analysis*. CRC Press: Boca Raton, FL.
3. Monden, Y. 2011. *Toyota Production System: An Integrated Approach to Just-In-Time*, 4th Edition. Productivity Press: New York.

4. Shingo, S. 2017. *Fundamental Principles of Lean Manufacturing*. Productivity Press: London.

5. Adams, C. W., Gupta, P., and Wilson, C. E. 2003. *Six Sigma Deployment*. Butterworth-Heinemann: Burlington, MA. ISBN: 0-7506-7523-3. OCLC 50693105.

6. Anderson, D. J., and Reinertsen, D. G. 2010. *Kanban: Successful Evolutionary Change for Your Technology Business*, Blue Hole Press: Seattle, WA.

7. Rios, R., and Ríos-Solís, Y. A. 2012. *Just-In-Time Systems*. Springer-Verlag: New York.

8. Fisher, R. A. 1992. *Statistical Methods for Research Workers*. Oliver and Boyd: Edinburgh.

9. Antony, J. 2003. *Design of Experiments for Engineers and Scientists*. Elsevier Science & Technology Books: Amsterdam, the netherlands.

10. Box, G. E. P., and Wilson, K. B. 1951. On the experimental attainment of optimum conditions. *Journal of the Royal Statistical Society: Series B* 13(1):1–38.

11. Mc Clurkin, J. E., and Rosen, D. W. 2002. Computer-aided build style decision support for stereo lithography. *Rapid Prototyping Journal* 4(1):4–9.

12. Asiabanpour, B., Khoshnevis, B., and Palmer, K. 2006. Development of a rapid prototyping system using response surface methodology. *Journal of Quality and Reliability Engineering International* 22(8):919–937.

13. Box, G. E. P., Hunter, W. H., and Hunter, J. S. 1978. *Statistics for Experiments*. John Wiley and Sons: New York, pp. 112–115.

14. Khuri, A. I., and Cornell, J. A. 1996. *Response Surfaces; Design and Analysis*. Marcel Dekker: New York.

15. Montgomery, D. C., and Peck, E. A. 1992. *Introduction to Linear Regression Analysis*. Wiley: New York.

16. Panchal, J. H., Kalidindi, S. R., and McDowell, D. L. 2013. Key computational modelling issues in integrated computational materials engineering. *Computer-Aided Design* 45(1):4–25.

17. Gupta, A., Cecen, A., Goyal, S., Singh, A. K., and Kalidindi, S. R. 2015. Structure-property linkages using a data science approach: Application to a non-metallic inclusion/steel composite system. *Acta Materialia* 91:239–254.

18. Basu, B., and Balani, K. 2011. *Advanced Structural Ceramics*. John Wiley & Sons, New York.

19. Jaworski, R., Pierlot, C., Pawlowski, L., Bigan, M., and Martel, M. 2009. Design of the synthesis of fine hydroxyapatite powder for suspension plasma spraying. *Surface and Coatings Technology* 203(15):2092–2097.

20. Schlechtriemen, N., Binder, J. R., Knitter, R., and Haußelt, J. 2009. Optimization of feedstock properties for reaction-bonded net-shape Zr ceramics by design of experiments. *Ceramics International* 36:223–229.

21. Rekab, K., and Shaikh, M. 2005. *Statistical Design of Experiments with Engineering Applications*, 4th edition. Chapman and Hall: Boca Raton, FL; New York.

22. Zárate, J., Juárez, H., Contreras, M. E., and Pérez, R. 2005. Experimental design and results from the preparation of precursory powders of ZrO2 (3%Y2O3)/(10–95)% Al2O3 composite. *Powder Technology* 159(3):135–141.

23. Kehoe, S., and Stokes, J. 2011. Box-Behnken design of experiments investigation of hydroxyapatite synthesis for orthopedic applications. *Journal of Materials Engineering and Performance* 20(2):306–316.

24. Mason, R. L., Gunst, R. F., and Hess, J. L. 2003. *Statistical Design and Analysis of Experiments*, 2nd Edition. John Wiley & Sons, New York.

25. Pan, J., and Cocks, A. C. F. 1994. A constitutive model for stage 2 sintering of fine grained materials –I. Grain- boundaries act as perfect sources and sinks for vacancies. *Acta Metallurgica et Materialia* 42(4):1215–1222.

26. Li, J., Peng, J., Guo, S., and Zhang, L. 2013. Application of response surface methodology (RSM) for optimization of sintering process for the preparation of magnesia partially stabilized zirconia (Mg-PSZ) using natural baddeleyite as starting material. *Ceramics International* 39(1):197–202.

27. Fernandez, C., Verne, E., Vogel, J., and Carl, G. 2013. Optimisation of the synthesis of glass-ceramic matrix biocomposites by the response surface methodology. *Journal of the European Ceramic Society* 23: 1031–1038.

28. Mohanty, D., Sil, A., and Maiti, K. 2011. Development of input output relationships for self-healing Al2O3/SiC ceramic composites with Y2O3 additives using design of experiments. *Ceramics International* 37(6):1985–1992.
29. Mandal, N., Doloi, B., Mondal, B. and Das, R. 2011. Optimization of flank wear using zirconia toughened alumina (ZTA)cutting tool: Taguchi method and regression analysis. *Measurement* 44(10):2149–2155.
30. Sarkar, D., Reddy, B. S., and Basu, B. 2018. Implementing statistical modeling approach toward development of ultrafine grained bioceramics: Case of ZrO2-toughened Al2O3. *Journal of the American Ceramic Society* 101(3):1333–1343.
31. McKinsey & Company. 2016. Industry 4.0 After the Initial Hype, McKinsey Digital, New York.
32. Brand, S. 2019. Don't Get Left Behind, SMART Manufacturing is the Future for Small and Medium-Sized Manufacturers, CMTC Manufacturing Blog. https://www.cmtc.com/blog/dont-get-left-behind-smart-manufacturing-is-the-future-for-small-and-medium-sized-manufacturers
33. McKinsey & Company. Digital 4.0 Model Factories.
34. Grenacher, M. 2018. *Industry 4.0, The Smart Factory And Machines-As-A-Service*. Forbes.
35. Zhong, R. Y., Xu, X., Klotz, E., and Newman, S. T. 2017. Intelligent Manufacturing in the Context of Industry 4.0: A Review. Science Direct, Amsterdam, the netherlands.
36. Trstenjaka, M., and Cosica, P. 2017. Process Planning in Industry 4.0 Environment. Science Direct, Amsterdam, the netherlands.
37. Maidment, G. *Smart Manufacturing, More than Just Robots*. Huawei.
38. Tata Consulting. https://sites.tcs.com/big-data-study/manufacturing-big-data-benefits-challenges/.
39. Brown, M. S. 2016. *What Analytics Talent Shortage? How To Get And Keep The Talent You Need*. Forbes.
40. ISACA Annual Report 2017. https://sqps.onstreamsecure.com/origin/Isaca/Annual-Reports/2017-isaca-annual-report.pdf.
41. Deloitte University Press. The Smart Factory, Responsive, Adaptive, Connected Manufacturing, A Deloitte Series on Industry 4.0, Digital Manufacturing Enterprises, and Digital Supply Networks. Westlake, Texas.
42. Wikipedia, Industry 4.0. Last modified: March 20, 2016. https://en.wikipedia.org/wiki/Industry_4.0.

2

Particle Management

Debasish Sarkar

CONTENTS

2.1 Introduction

In the beginning, a common question comes to mind on the necessity of a topic like *particle processing*. Usually, processing defines a protocol to make a product starting from raw materials; obviously the foremost step is particle management to fulfill the desired properties of ceramics. The ceramic processing follows a different class of techniques on the basis of different solid loading, such as the pressing of dry powders, extrusion of semi-dry mass, casting of viscous slurry, suspension for drug delivery media, etc. A different class of densification, product shape and size, complexity and cost manipulation of ceramic components are common practice, and most of the traditional and advanced ceramics begin with particles although a few exceptions, such as laser ablation, lithography, etc., do not demand particle processing. Thus, ceramic manufacturing protocol accepts either naturally occurring rocks and minerals or synthetic raw materials as starting resources that consist of different degrees of purity, particle size, particle size distribution, homogeneity as well as heterogeneity. In this context, the particle processing is essential and discussed in three modes, predominately *particle generation and characterization, particle mixing* and *particle scaling up*, in the view of either traditional or advanced ceramic processing.

In principle, the material possesses two classes of properties: *intrinsic* and *extrinsic* properties. Intrinsic properties are inherent to the material and depend on atomic scale structure (atomic bonding, crystal structure); however, they may alter depending on the presence of inclusions during processing (for example, a porous body reduces the elastic modulus). Typical intrinsic properties, including hardness, brittleness, melting point, chemical inertness and others, are predominate criteria to encounter wide horizon of insulators to conductors. On the other hand, extrinsic properties depend on a microstructure comprised of grains, void space, porosity and cracks. Several engineering properties depend on the microstructure that rely on the starting particles and processing conditions, and thus the combination of different properties plays a vital role in defining the behavior of materials where new classifications are incubated as advanced ceramics. It is certainly difficult to distinguish between traditional ceramics (since earliest civilization, low-cost ceramics) and advanced ceramics (recently classified, technical ceramics), but in this book we considered this classification for convenience.

While processing controls the ceramic properties, such processing further can be classified into two modes, first-stage *particle processing* and second-stage *ceramic processing*. A classic relation has been noticed within properties, microstructure and chemical composition that eventually depends on both the particle processing and ceramic processing.

Figure 2.1 illustrates how ceramic product characteristics depend on both intrinsic and extrinsic properties; while intrinsic properties are difficult to modify, the only option that remains is to synchronize the properties through the tailoring of extrinsic properties that

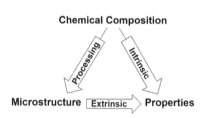

FIGURE 2.1
Relationship between chemical composition, microstructure and properties.

eventually depends on microstructure and thus processing. In the perspective of product development, the first stage of management is particle processing and this is discussed in this chapter, and Chapters 3–10 describes the different ceramic processing protocols and their industrial practices for both traditional and advanced ceramics.

2.2 Particle Size Reduction

Crushing and grinding are the basic comminution processes to reduce coarse mineral ore or calcined raw materials by four to nine times through impact by a harder material as compared to feed. The basic purpose is to reduce the particle size in different gradients in order to obtain a proper mixing for adequate packing and reactivity during the final stage of processing. In terms of size reduction, coarse range is termed *crushing* and fine range is *grinding*. Before the period of the First Industrial Revolution, muscle power was employed for this purpose, but with the advancement of technology, several industrial-scale instruments have been developed as per required feed and output size. Detailed information about different types of crushers available with their particle hardness and size reduction ratio is given in Table 2.1.

Several governing parameters are essential during the *selection of equipment* for effective particle reduction from natural raw materials. These are:

  i. Frequency of operation; either batch-wise or continuous operation
  ii. Quantity of material and instrument capacity
  iii. Feeding method and rate
  iv. Impact liner should possess high abrasion resistance in contact with feeding material
  v. Size range of feed material and targeted output size

Along with the equipment selection, *operational features* are also critical to monitor effective particle reduction. The most important are listed below:

  i. Energy input rate should be maintained at as minimal levels as possible
  ii. Operate economically under minimum supervision and maintenance
  iii. Passing should not choke or jam during continuous operation
  iv. Effective production of desired shape and size
  v. Consistent output quality and minimum wastage

Crushing in general is an energy-consuming process, and it depends on the physical properties of the material, the amount of material to be crushed and the input energy. The crushing forces need to be so intense that it has to cross the elastic limit of the material; thus, they are most energy efficient with brittle materials. Depending on the feed and output size of material, crushing and grinding can be broadly classified into three categories as primary crushing, secondary crushing and fine grinding or milling. Categorization of such instrumentation is illustrated in Figure 2.2.

TABLE 2.1

Commonly Used Crushers and Their Input/Output Parameters

Crusher Name		Material Hardness	Type of Milling	Reduction Ratio
Jaw Crusher	Blake	Soft to very hard	Dry	4 to 9:1
	Overhead pivot		Dry	4 to 9:1
	Overhead eccentric		Dry	4 to 9:1
Gyratory Crusher	Standard	Soft to very hard	Dry or Wet	4 to 6:1
	Attrition		Dry or Wet	2 to 5:1
Impact Crusher	Horizontal Impactor	Soft to medium hard	Dry or Wet	10 to 25:1
	Vertical: shoe and anvil	Medium to very hard	Dry or Wet	8 to 12:1
	Vertical: autogenous	Soft to very hard	Dry or Wet	5 to 10:1
Roll Crusher	Single Roll	Soft to hard	Dry	7:1
	Double Roll		Dry	3:1
Compound Crusher		Medium to very hard	Dry or Wet	3 to 5:1
Mineral Sizers		Soft to hard	Dry or Wet	2 to 5:1
Crusher Buckets		Soft to very hard	Dry or Wet	3 to 5:1

Note: For Example, 4 to 9:1 Denote the Size Reduced by 4 to 9 Times, in the Other Way a 20 cm Lump Can Be Reduced in the Range of 20/9 cm to 20/4 cm Using A Jaw Crusher [1].

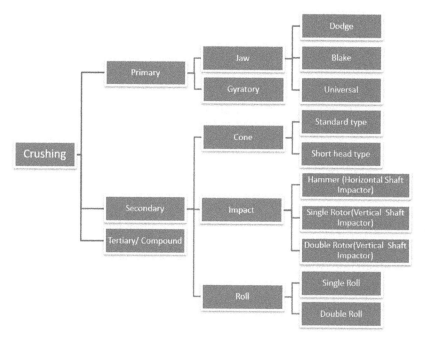

FIGURE 2.2
Hierarchy of different classes of crushers to reduce the particle size.

A primary crusher receives raw material directly and usually breaks the material by means of impact, compression and attrition. The energy requirement for a primary crusher is due to the high reduction ratio for starting large size raw material. Jaw crushers and gyratory crushers come under the category of primary crushers.

2.2.1 Jaw Crusher

Jaw crusher is one of the main types of primary crushers; it uses compressive force for breaking particles. Compressive force is developed between two vertical jaws, one of them fixed and the other one movable, termed as a swing jaw. The materials are entrapped between the upper parts of the jaws and dropped down toward narrower space for crushing until they are small enough to pass through the narrow opening. The movement of swing jaw is preferentially fixed as per the required size reduction ratio as depicted in Figure 2.3a. The gap between the jaws are conical in shape with a wider opening at the top for feeding that implies the output size becomes smaller compared to feed size; herein, the force required to crush is provided by flywheel to the swing jaw.

Depending on the pivoting of the swing jaw, jaw crushers are divided in to three types: blake crusher, dodge crusher and universal crusher. Blake crushers have the swing jaw pivoted at the top position; thus, they have a fixed feeding cross-section and variable discharge cross-section. Blake crushers are divided into two types, single toggle (Sandvik) and double toggle (Pennsylvania) crusher. Single toggle crushers have a compact design compared to double toggles, as they have swing jaws suspended in the eccentric shaft. In a single toggle crusher, the swing jaw moves in a combined elliptical path due to movement by toggle plate and eccentric, whereas in a double toggle crusher, it is only horizontal motion controlled by a pitman.

Dodge crushers have the swing jaw pivoted at the bottom position, thus they have a variable feeding cross-section and fixed discharge cross-section; this sometime leads to the choking of the crusher. But universal crushers maintain the pivoting at an intermediate position. It is essential to calculate the crusher capacity in order to maintain the effective output, and it primarily depends on the amount of material in the opening between two jaws and the characteristics of the feeding material.

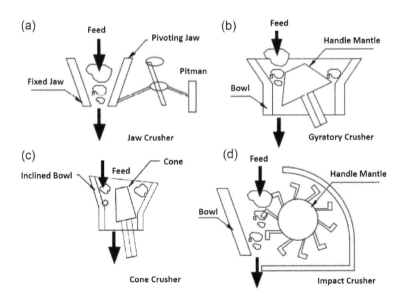

FIGURE 2.3
Cross-sectional view of several types of crushers: (a) jaw crusher, (b) gyratory crusher, (c) cone crusher, and (d) impact crusher. (a) and (b) are primary types, (c) and (d) are secondary type crushers.

The theoretical capacity of a jaw crusher can be defined as:

$$Q = [\text{B. S. s. cot(a) k}] \cdot [60\,\text{N}] \; \text{m}^3/\text{h} \tag{2.1}$$

Where Q = volume throughput, m³/h; B = inner width of crusher, m; S = open side setting, m; s = jaw throw, m; a = angle of nip; N = speed of crusher, RPM, k = material constant (1.5–2.0) varies with respect to the feeding method and characteristics of crushed materials. First part depicts the works out volume of one motion of the jaw, multiplies of RPM and 60 provides volume/hour. However, on-site discrepancy noticed because of inaccurate input of "s."

2.2.2 Gyratory Crusher

The working principle of a gyratory crusher is same as a jaw crusher. Thus, it uses compressive force in a cone for size reduction. It consists of a movable conical head and outer concave surface in which both of the surfaces are made of harder materials, typically Mn-steel, as illustrated in Figure 2.3b. In this crusher, the moving cone is placed in the center and the particle crushed until it is small enough to pass through the cavity within the cone and liner. Although gyratory is one of the main types of primary crusher, it can also be used as a secondary crusher by adjusting the gap between two surfaces on the demand of feed and output size. It facilitates a complete revolution during crushing, whereas the jaw crusher facilitates a forward stroke of the jaw only, and this mode of operation ensures that gyratory crusher has more efficiency than the jaw crusher.

The secondary crusher reduces the particle size further after performing the primary crushing approximately up to 1mm in particle diameter. It is further classified into three types: *cone crusher, impact crusher* and *roll crusher* (as described in Figure 2.2).

2.2.3 Cone Crusher

Cone crushers can be used as both a secondary and tertiary crusher depending on the design parameters and can eventually be capable of generating uniformly crushed fine materials. A cone crusher operates similarly to a gyratory crusher, with some differences such as less steepness of chamber, smaller opening, smaller cone size and more rotation speed, and it produces more uniform sized particles, as shown in Figure 2.3c. Depending on the cavity size, there are two types of cone crushers, standard type (secondary) and short head type (tertiary). Standard-type crushers are used for intermediate crushing and short head-type crushers are used for finer crushing. The cone crusher is further divided into four types, namely compound cone crusher, spring cone crusher and hydraulic cone crusher (single cylinder and multi-cylinder); different working modules are categorized based on the input and output size and parameters of the feed.

2.2.4 Impact Crusher

Compressive force is employed in the jaw, gyratory and cone crusher, but this class of crusher predominately works under impact to feed material. It is assembled with a reasonable structure as depicted in Figure 2.3d. They demand low maintenance without machine parts damaging and better productivity, leading to fascinating choices for crushing and grinding. There are two types of impact crushers: horizontal shaft impact crusher (HSI) or hammer crusher and vertical shaft impact crusher (VSI). Hammer crushers break the material with the impact of hammers attached to the roller inside a shell or bowl and used for breaking relatively softer materials compared to VSI crushers. VSI crushers follow a different approach: they contain a wear-resistant high-speed rotor which throws the feed material into the walls of the crushing chamber. Herein, velocity is the governing factor to crush the materials; the final discharge size can be controlled by adjusting the velocity of the rotor and the distance which material travels for impact. VSI crushers provide relatively uniform particle size as compared to HSI crushers.

2.2.5 Roll Crusher

The roll crusher is an effective choice for intermediate range of crushing, and popular in mining sectors. In brief, the shells are made up of either manganese or chrome alloy steel depending on the hardness range of the material to be crushed. It demands more maintenance and less productivity in terms of volume compared to the cone crusher; however, it is capable of producing a narrow particle size distribution and needs less maintenance if the feed is less abrasive. Roll crushers are sometimes used as movable crushers attached to a crane, termed as a bucket crusher. Roll crushers are of two types: *single roll crusher* and *double roll crusher.*

2.2.5.1 Single Roll Crusher

Single roll crushers are comprised of a single cylinder to crush the material by compressing it against the stationary outer shell with the rotation of a smooth or teeth-equipped roll and it produces relatively coarser particles. It is designed for long life and heavy-duty applications. The gap distance between the outer shell and teeth of the roller decides the final grain size and can be further adjusted to achieve the desired output. It also expedites an auto-toggle mechanism that opens the hinge to pass unbreakable hard materials and return to its original position to follow the continuous operation. In accordance with the choice and demand of the output, the roll crusher may be smooth or have teeth with having different geometry, like trapezoidal, slugger, etc.

2.2.5.2 Double Roll Crusher

The double roll crusher is essentially designed to be two cylinders and it is widely used for secondary and fine crushing. Two rollers rotate in opposite circular directions driven by a motor through a pulley or gears, and feed falls in between two rollers and naturally passes downward after being crushed enough. The gap between two rollers can be manipulated to obtain different size output. It has a self-protection mechanism and is a good choice to crush for medium hardness of feeding material. When an unexpected hard and unbreakable or bulk material is fed between the two rollers, the rollers widen automatically under the effects of hydraulic pressure and springs allowing the foreign material to drain and roll back to continue the crushing operation. The theoretical capacity of a double rolling crusher is [2]:

$$Q \text{ (theoretical)} = 60\pi D_1 D_3 bN\rho \tag{2.2}$$

Where b is the width of the roller face, N is the number of revolutions per minute (rpm) and ρ is the density of material in Kg/m^3. D_1 (roller diameter) and D_3 (gap between two rollers) are illustrated below in the Figure 2.4. Actual capacity remains around 10–30% of theoretical capacity. So,

$$Q_{actual} = (0.1 \text{ to } 0.3) * Q_{theoretical} \qquad (2.3)$$

Despite these crushing mechanisms, several tailor-made crushers and milling operations are well accepted for the comminution of the mineral ore or calcined raw materials to obtain different grades of materials starting from a few millimeters to micron particles. A brief of such a disintegration process is shown in Figure 2.5 [3]. The comminution process describes a protocol wherein the disintegration process goes through series of crushing

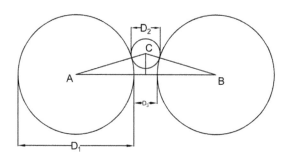

FIGURE 2.4
Front view of a double roll crusher. Schematic view of feed with diameter of D_2, where $D_2 > D_3$ for effective crushing.

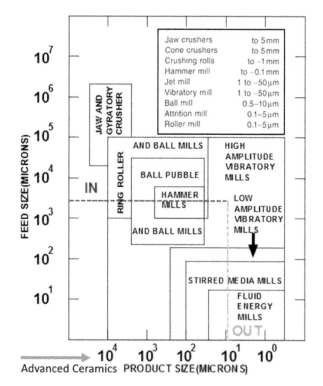

FIGURE 2.5
A brief on the different particle processing protocols based on feed size and output size. More fine particles are eventually considered for advanced ceramics [3]. One can pick up the specific process sequence to achieve the desired particle size. For example, if feed size varies 1–10 mm, it is necessary to follow the different stage of operations denoted by the dotted line (IN) to obtain 10 μm (OUT) particles.

operations, which is followed by a size classifier segregation of the coarser particles and a return back to the operation to complete the process as desired. This is a continuous process, and preferably ball milling is the intermediate step and the last step is vibratory milling (grind in mill) to obtain fine particles up to a few microns in size.

However, several synthetic methods are available to obtain particles the size of microns and nanometers, as discussed in Section 2.5. Usually, the grinding is done in two modes: dry grinding and wet grinding. When water content is <1% in the system it is referred as dry grinding, and when water content is >1% it is referred to as wet grinding. However, in the presence of high water content (in the range of ~34wt%), the system forms a slurry; thus, the grinding effect is not predominate in this circumstance.

Unless the product needed is coarser, it is impossible to complete in a single operation. To increase the efficiency of a grinding it can be processed two parts, the first part consisting of particle sizing, separation and impurity removal, followed by grinding. During particle purification, particles separate in different size ranges and gangues (i.e., unwanted impurities) are removed using methods such as screening, floatation, magnetic separation (dry drum and wet drum), electro static separation and classifiers. Eventually, grinding equipment used for industrial application are mostly the tumbling mill type; details are discussed in Section 2.3.

2.3 Milling

Milling is an important unit operation in particle processing that includes the particle size reduction, separation, sizing or classifying of the resultant materials. The basic privilege of milling is that it increases the surface area to enhance the surface energy and ensure a high degree of reactivity during high-temperature operations. In the early days, the milling operation was done by manual effort but it had limitations, and thus automation, including improved machineries, are being developed and employed to provide attrition, compression and impact force during operation. In brief, the attrition refers to particle-to-particle collision or particle shearing; compression is the application of pressure on material, causing fracture; and impact is either a medium or the particles themselves providing impact or bombardment to cause fracture. Process mechanism depicts interactive motion in between two solid surfaces such as vial wall and grinding media using a shear or impact. Milling using impact is known as tumbling mill, which makes a medium tumble and fracture into either lumps or powdered materials. To produce finer particles, the most used milling type is the tumbling mill, which is again classified as ball mill, rod mill, tube mill and pebble mill depending on the medium used. Among these tumble milling units, the ball mill is the most used milling machine in ceramic industries. Despite the selection of unit operations, it is important to remember that the process becomes more costly if feed material hardness is excessive than that of the shell wall or grinding media, and thus effective grinding media and grinding protocol selections are a important criteria. Grinding or milling method may be dry or wet milling, as highlighted earlier. Dry milling does not demand an additional drying process, particles neither form hard agglomerates nor react with liquid, there is less wearing of the mill lining, it is easier to optimize and it can be started or stopped at any time. Disadvantages include the fact that milled powders are prone to get stuck to the walls of the mill and grinding media. On the other hand, wet milling has several advantages including low power consumption for required output, no dust problem or loss, higher rotational speed, better homogenization, narrow particle size distribution and smaller particle size achievable, but it needs additional drying process if one would like to use it for compaction. The specific gravity of media plays an important role as it decides the impact efficiency and results in the fine grinding of total mass. Thus, the selection of grinding media and type of milling are important aspects with respect to feeding material hardness, size and output particle size, and the resultant mode of utility of particles. For example, a dry milling is beneficial for refractory brick manufacturing, whereas homogenous slurry for slip is essential for whiteware industry; thus, wet milling can be advantageous over dry milling.

Prior to particle processing for a specific use, it is essential to store the different graded particles; thus, particle storage and discharge are an important aspect to reduce the hazardous environment in industry and maintain continuous production. The raw material storage capacity depends on the bulk density of the material. Usually, the solid is defined as numerous dry or wet solid particles ranging from coarse particles to fine powders that are being handled in a large quantity. This bulk quantity of materials is stored either outside, which is known as bulk storage, or in vessels like bins, silos, elevators or process vessels, known as bin storage. Details of this are discussed in Section 2.4.

2.3.1 Ball Milling

The ball mill, a type of fine grinder, is one of the most used milling machines. It works on the principle of impact, i.e., the size of particles is reduced by the impact of milling media. A ball mill consists of a slightly inclined or horizontal hollow cylinder rotating about its axis, partially filled with material and grinding media. The grinding media is of a spherical shape or balls made up of steel, ceramic, stone or rubber depending upon the material being grinded. The inner surface of the cylinder is lined with an abrasion-resistant lining; frequently rubber lining is employed to reduce the wear. In order to produce finer particles, ball mills are more suitable as compared to other tumbling mills, as ball mills can produce finer particles ranging from 10 μm to a fraction of a micrometer [4].

Usually, the ball mill operates with a loading range of 30–50% of the chamber volume and maintains the length by diameter ratio in the range of 1.5 to 2. In a simple operation, the mill is loaded with both feed and grinding media through one end, and the entire mix rotates at a speed sufficiently high so that the beads can tumble, then finally discharge at the other end. The efficiency of the mill depends on several factors like speed of rotation, degree of filling, hardness of feed, geometry of feed and the size and shape of shell and media [5]. During long-term operation, it is essential to remember that repeated use of grinding media may promote impurities through wear and tear, and thus replacing of grinding media further increases the output efficiency. Sometimes, same grinding media and feed composition reduce the impurities; however, it is a costly affair. In common practice, the ball mill runs at a low speed; with heavier loads it is even less than 60 rpm. Larger size grinding media predominately contributes potential energy at low speed, whereas smaller size grinding media expedites the kinetic energy, resulting in effective grinding of the feed. Thus, the larger grinding media cannot reduce further after attaining an optimum fine particle and it is necessary to the second stage of milling through smaller grinding media to obtain submicron to nanoscale particles. However, grinding media must be denser than feed and larger in size compared to the highest particle size present in the feed.

The rotational speed of a ball mill plays a crucial role as it affects the resultant movement of the particles inside the cavity. During a high-speed operation, it follows three stages of the grinding process, termed as cascading, cataracting and centrifuging at low, intermediate and higher speed of rotation, respectively. Three stages are pictorially illustrated in Figure 2.6.

In both lower and higher speeds, no free fall occurs that eventually minimizes the impact force for effective grinding. However, an intermediate speed, referred to as *cataracting*, promotes effective grinding, but after a critical speed, the centrifuging action starts causing the powders to stick to the wall, and no longer causes grinding anymore. So, the critical speed (η_c) of rotation is essential to calculate and can be defined as:

$$\eta_c = \frac{1}{2\pi}\sqrt{\frac{g}{R}} \tag{2.4}$$

Where g is the acceleration due to gravity and R is the radius of the ball mill. Another type of ball mill consists of an agitator is present to mix and rotate the material instead of the rotation of a drum to obtain effective milling. At critical speed, the material is about to complete rotation or start cascading, so ball mills are used at ~75% of η_c to avoid materials going to the top of the mill [4].

2.3.2 High-Energy Milling

High-energy milling depicts a system working through high-energy impact during continuous operation, and thus the feed material experiences excessive strain and eventually produces fine materials, up to nanoscale particles. Fine milling, which produces particles sized below 100 µm, is usually done by conventional ball milling, but very fine milling (below 10µm) is done by high-energy milling. High-energy milling cannot be executed in conventional milling machines, so several types of high-energy milling machines are available for the same, for example, planetary mill, vibratory mill, high-energy rotating mill and attritors. Important parameters involved in high-energy milling are container composition and hardness, time, speed, temperature, grinding media composition, density and size, packing ratio, atmosphere and milling control agent/lubricant [6]. Discrepancies in these parameters can cause impurities and contaminations, and thus several steps are needed to obtain pure materials. For an example, removal of contaminated WC-Co (grinding media) from ground particles can be done through wet chemical treatment that consists of 5% (v/v) 15.4M HNO_3 in 30% H_2O_2 solution at 95°C for an hour followed by repeated ethanol washing [7]. A certain amount of contaminated steel can be removed through acid leaching, etc.

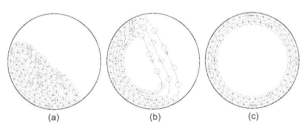

(a) (b) (c)

FIGURE 2.6
Different stages of the grinding process in ball milling: (a) cascading, (b) cataracting and (c) centrifuging [6].

In consideration of operational parameters, one has to remember the following points: [8–11]

i. Time optimization is the most important aspect to obtain the desired product without any contamination during high-energy milling.

ii. Milling speed needs to be just enough for causing the desired impact and energy.

iii. Density of the grinding media must be appropriately higher than the feed, again more energy can be obtained using different radius grinding balls.

iv. Milling atmosphere is crucial in controlling the contamination, specifically for materials prone to oxidation. Using milling atmosphere, the desirable product can be achieved such as nitrides and hydrides in N_2 and H_2 atmospheres, respectively.

v. The milling temperature has an effect on processes such as alloying, welding, intermetallic, solutions; whereas low-temperature milling, termed as cryo-milling, reduces welding probability due to an increase in brittleness.

Using high-energy milling technology, nanocrystalline powders, amorphous powders, nano-composites, intermetallic and ceramic powders are manufactured. A wide range of ceramic powders are produced using high-energy milling whereas it has a constraint of minimum particle size being in the range of 25 to 50 nm[12]. Over time the high-energy milling process has tremendous potential for advanced material processing that is yet to be used in industry scale manufacturing. For that purpose, extensive research is in progress, demanding extensive modeling based on the existing data to predict the quality of the output material.

2.4 Storage and Transportation

While setting up a ceramic processing unit, space administration is a critical aspect in order to maintain continuous and uninterrupted production, and thus one has to design and select effective raw material storage and transport management after examining different crushing, grinding and milling operations. Conventionally, the industry comes across four different types of materials that require storage, such as gases, liquids, semi-liquids and solids, and their consecutive transport systems. Gases are stored in tight containers and are usually transported using pneumatic conveying, while liquids and semi-liquids are stored in tight or open containers and are usually conveyed using a hydraulic system. Solids remain the major class of materials used in industries stored in bulk manner and transported using conveyors or elevators. A well-developed storage and transportation system improves the efficiency of production, reduces indirect costs with a reduction in damage during storage and movement, optimizes proper space utilization and in turn reduces the overall cost of production. If the solid material is non-reactive in atmosphere contact, then it is preferred to store it in an open space on ground or in storage bins, however it is necessary to protect from contamination and segregation of stored materials. A dumped conical-shaped heap follows the conical angle Ø with respect to the horizontal plane and is known as the *angle of response* that eventually determines the flowability of the material.

Thus, a higher angle of response implies a larger height of the pile and more solids can be accommodated in a smaller area. For a smoother surface, the angle is less, i.e., more area is required for the storage of material. Despite such a class of storing process, solids that are either valuable or hazardous to environment, like rock salt, crushed or milled raw materials and gunpowder etc., are stored in bins or silos. These storage containers are usually cylindrical or cuboid in shape and have a larger volume that is maintained either by wider shape and shorter height or vice versa. Storage containers have two parts, the bottom vertical zone of the vessel is cylindrical or parallel, and the bottom converging portion is the hopper; charged feed is put through the opening at the top and is usually discharged through the bottom opening. Common storage protocols and their different parameters are listed in Table 2.2.

Feed input and output both experience some degree of friction during operation and this results in the interlocking of the solid particles that eventually minimize the flow next to the container inner wall that finally prefers spontaneous discharge through the bottom part. The hopper slope is maintained at more than 50° with respect to the horizontal surface and the corners should be rounded to ensure easy movement of

TABLE 2.2

Common Stored Bulk Materials and Their Classification as Per IS: 8730: 1997 [14]

Classifying Parameter	Classifications	Defining Limits
Lump Size	Dust (Clay)	Up to 0.05 mm
	Powdery Material (Sand)	0.05 to 0.50 mm
	Granules	0.5 to 10 mm
	Small Lumps	10 to 60 mm
	Medium Lumps	60 to 200 mm
	Large Lumps	200 to 500 mm
	Very large Lump	Greater than 500 mm
Bulk Density	Light	Up to 0.6 t/m^3
	Medium	0.6 to 1.6 t/m^3
	Heavy	1.6 to 2.0 t/m^3
	Very heavy	2.0 to 4.0 t/m^3
Flowability (Angle of Repose)	Very Free Flowing	0° to 20°
	Free Flowing	20° to 30°
	Average Flowing	30° to 40°
	Sluggish	More than 40°
Abrasiveness	Non-Abrasive	NA
	Abrasive	
	Very Abrasive	
	Very Sharp	
Miscellaneous Parameters	Develops Fluid	
	Contains Explosive Dust	
	Sticky	
	Contaminating	
	Degradable	
	Emitting Harm Fumes	
	Highly Corrosive	
	Hygroscopic	
	Oil/Chemical Containing	
	Very Light/Fluffy	
	Packs under pressure	
	Elevated Temperature	

materials toward the central outlet. In order to store for a longer period of time, moisture content, temperature of bulk solid and atmosphere need to be well maintained. During discharge, three types of flow are typically observed in a container: (i) funnel flow follows the first-in-last-out principle, (ii) mass flow follows the first-in-first-out principle and (iii) expanded flow is a mixture of both type of flows. Design of a typical bin or silo and three types of material flows are pictorially presented in Figure 2.7. Problems with the flow of solids include the segregation of solids, flushing or uncontrollable speed flow and formation of an arc at opening due to interlocking leading to restrict the material flow.

An effective storage system is an essential component to promote regular production; however, transportation of different grades of particles is another critical issue, and thus the ceramic industry has to take care of it seriously. Transportation of solids is done mostly using belt conveyors, screw conveyors and bucket elevators. Screw conveyors can be operated for both short- and long-distance movements, where solids are loaded onto a belt surface through a feed hopper and at the end of the belt they are discharged over the drive pulley. The conveyor belts can be operated under flat or troughed conditions created using the idlers beneath the belt, usually the idlers avoid sagging of the belt during operation. The capacity (Q) of a belt conveyor can be described as:

$$Q = AV\rho_b \tag{2.5}$$

Where A is cross-sectional area, V is linear speed and ρ_b is bulk density of solid.

Screw conveyors are one of the oldest types of conveying methods adopted for short-distance movement. It essentially consists of a screw with a spiral flight mounted on a shaft that is placed parallel to the bottom surface of a U-shaped trough. The shaft is supported at both ends and pushes the solid material forward with the help of a motor attached to

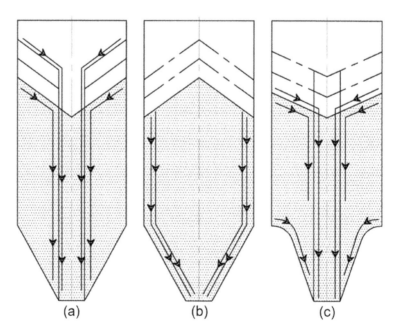

(a)   (b)   (c)

FIGURE 2.7
Schematic of a silo and three types of material flow: (a) funnel flow, (b) mass flow and (c) expanded flow [3].

it. The maximum particle size that can be transported depends on the clearance between screw flight and the surface of the trough. The capacity (Q) of the screw conveyor can be expressed as

$$Q = C_i\left(\frac{\pi}{4}D_s^2\right)P\left(\frac{N}{60}\right)\rho_s C_f \qquad (2.6)$$

Where C_i is the correction factor for inclination, D_s is the screw diameter excluding the shaft diameter, P is pitch, ρ_s is the density of the solid, C_f is the filling coefficient and N is shaft rpm.

Bucket elevators are only used for the vertical transportation of solids. It essentially consists of several buckets attached to a continuous double strand chain passing over two spokes. Materials are fed directly into the buckets, transported and discharged to a hopper as the buckets turn over the upper spoke.

2.5 Solid-State Reaction

Solid-state reaction is the intimate reaction between two solids to develop another solid with a new composition. Solids do not react at room temperature, hence high temperature is required to initiate and complete the reaction. Chemical reaction in between solids is assumed to be three models: (i) transport of oxygen molecules; (ii) transportation of oxygen ions and (iii) counter diffusion of cations, nothing but solid–solid interaction. In this circumstance, the first is not valid in case of ideal contact between two solid interfaces, the second is invalid as the mobility of oxygen ions is very restricted and the process becomes very slow. The schematic of solid–solid interaction is represented in Figure 2.8.

Important factors influencing the solid-state reaction are the contact area of reactants, reactiveness of reactants and rate of nucleation of the final product; obviously these depend on the starting particle size [14]. Starting reagents with a large surface area and reactiveness are preferred to maximize the contact, thus there is a rapid reaction rate between the reactants. High purity, apposite stoichiometric composition of starting particles and their effective mixing facilitate intimate interaction and final stage of reaction. Frequently, mixing is performed through grinding to achieve a homogeneous mixture and this is pressed into pallets to obtain better contact. While high temperature calcination or sintering is the major step to achieve the desired phase or product,

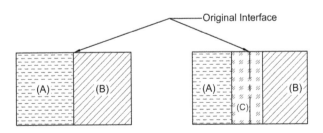

FIGURE 2.8
Schematic representation of solid–solid interaction.

the reaction temperature is typically maintained by two-thirds of the melting point of the lower melting reactant. Definite control over the reaction rate promote different categorized particle sizes, and thus the diffusion rate needs to be accelerated by one of two ways: either by increasing the temperature or by introducing defects by adding starting reagents that decompose during the reaction, such as carbonates and nitrates. The final step of temperature selection is essential to control (either by thermal analysis or phase diagram) to ensure the completion of the reaction. The advantages of solid-state reaction include the direct reaction of solids to produce a final product; a solid–solid reaction is simpler to perform as large volume of raw materials are readily available at low cost; and a large range of compounds can be made using this method. Restrictions of this method include chances of contamination and non-homogeneity due to high temperature; slow reaction rate or restricted diffusion rate; some loss of reactants; careful synchronization to obtain the desired microstructure; product may decompose at a higher temperature; difficulty to obtain stable products at an intermediate temperature; difficulty in incorporating highly volatile ions; crucible may react at high temperature, etc.

Hence, an advanced method has been developed to prepare ceramic materials having a single phase and a fine particle size known as the precursor calcination. It has been observed that calcining the precursor mixture at a reduced pressure helps in reducing the reaction temperature and several steps are associated with the conventional solid-state reaction [15]. The drawbacks of this reaction include the difficulty in getting the reactants with comparable water solubility and precipitation rate to be different for the reactants. Due to the limitations with most combination of ions, it becomes unfeasible to find companionable reagents and further maintaining exact stoichiometric ratios. However, a suitable material combination can highlight this as an effective and easy method as it lowers the reaction temperature, produces final product with high surface area, removes impurities and feasibly stabilizes metastable phases.

2.6 Wet Chemical Synthesis

Wet chemical synthesis is a common protocol to make synthetic and high-purity nanoscale raw materials that sometimes are essential to manufacture advanced ceramics as well as partial additions to enhance the electrical, optical, mechanical, thermal and thermo-mechanical properties of traditional ceramics. Although we have already discussed large-scale particle processing either through the crushing–grinding of natural minerals or mechanically induced high-energy milling performs to obtain submicron to nanoscale particles., in the same time, the direct solid–solid interaction during calcination at elevated temperature promotes a specific product having definite phase. In one step ahead, the wet chemical process leads a bottom-up approach to control the size, shape and composition of a material to produce the desired product either at laboratory- or industry-scale. In ceramic processing, chemical processes are extensively used through either single step or multiple steps, and each step is termed as unit operations of reaction. Thus, each of the unit operations is sequential and these processes can be easily illustrated using block diagrams or flow charts [16]. Despite several wet chemical processes, a few common and industrially viable powder synthesis processes, including the auto combustion, hydrothermal, co-precipitation and sol-gel processes, are discussed in detail.

2.6.1 Auto Combustion

Auto combustion is a process in which self-heating or self-propagating combustion takes place through an exothermic reaction assisted by a mixture of an oxidizer and a fuel. Typically used oxidizers include metal nitrates, and fuels are urea or glycine. The reaction kinetics and performance are affected by several parameters such as combustion temperature, composition of mixture, oxidizer to fuel ratio, etc. With advantages such as less reaction time, low energy consumption, low-cost precursors and simple experimental setups, it produces ultra-fine, pure phase nanoscale particles with a narrow size distribution; thus, it is one of the extensively used processes to ensure formation of homogeneous pure nanopowders [17–19]. Addition of such nanoscale particles in bulk ceramics facilitate improved properties compare to their only bulk counterpart, and thus this powder processing technique is a potential candidate for the advanced ceramic processing.

Herein, the precursor processed either at room or elevated temperature and auto combustion in high temperature are the two major steps. The first step is the formation of highly viscous liquid or gel that ensures proper mixing, and the second step is the ignition of the formed gel; this includes heat of combustion generation and gas/fume formation [20]. Usually this process is accompanied with sol-gel method, as the prepared gel of the oxidizer and fuel in the first step is heated at a predetermined temperature (say 450 °C) to initiate the combustion. A solid mass consisting of porous powders is obtained that is further annealed to obtain dense and homogeneous nanopowders. The primary characteristics of powders depend on the second step of the reaction where the temperature has a great effect on powder characteristics. With an increase in temperature, the grain size may enhance along with hard agglomeration as secondary particles consist of primary particle clusters.

2.6.2 Hydrothermal

Hydrothermal synthesis protocol produces solid products by crystallization from an aqueous solution at high temperature and pressure, where the resultant product is insoluble under ambient conditions. It works over a range of temperatures and pressures and the crystal formation depends on the solubility limit of solute in water. The process is carried out inside a pressure vessel resistant to high temperature and pressure known as autoclaves, in which solids are deposited from the mixed solution during the hydrothermal process. This process is one of the frequently used methods to produce large and pure single crystals. Advantages include the production of phases with high vapor pressure or probable temperature-assisted transformation, narrow particle size distribution and better control over crystal composition, but it is difficult to control the growth pattern of a single crystal. Another drawback is the separation of particles without agglomeration, and it can be avoided by steric stabilization or using a freeze-drying technique [21]. Mineralizers are used in order to increase the solubility of salts and morphology controlling species are used to control the morphology in the hydrothermal process that eventually enhances the process output and different morphologies like cuboid, fiber, plate shaped, etc [22]. It is worth mentioning that the synthesis kinetics and resultant crystallinity depend on the process temperature, pressure, pH, time and solute concentration. Substantial advancement includes the use of additional management, such as ultrasonic, microwave, mechanochemical and electrochemical, introduced to control the reaction medium during hydrothermal synthesis. This process is used to develop some class of synthetic zeolite, diamond, quartz, gems and other crystals, and it is obvious that these materials have economic importance. Depending on the process parameter, this can be further categorized into three types:

2.6.2.1 Difference in Temperature Method

This is the most commonly used method for growing crystal. Herein, the super saturation is achieved by the difference of the temperature zone. A temperature gradient is maintained over the length of the autoclave, and the solute dissolves in the hotter zone. Eventually the saturated solution is transported and crystallizes in the cooler zone.

2.6.2.2 Reduction in Temperature Method

Unlike the previous method, this process does not have any temperature gradient or zones. The supersaturation and crystallization are achieved by gradually decreasing the temperature of the complete system. However, this is a rare method due to the limited growth by the seed crystal and the difficulty in controlling this crystal growth.

2.6.2.3 Metastable Phase Method

This method is based on the solubility difference in between the product to be grown and starting material. Components in the starting material are unstable at higher temperatures and thus the solubility of metastable phase is more comparable to that of stable phase. As the metastable phase dissolves, the stable phase is likely to crystallize. This method is usually combined with any one of the above said methods.

2.6.3 Co-Precipitation

Co-precipitation is a process in which precipitation of more than one solid simultaneously takes place through dissolving soluble salts of desired cations or fractional precipitation of a specific ions [23]. Herein, the solid products are formed starting from liquid solution and the nucleation is initiated by controlling the pressure, temperature and pH. Some of the universally used salts in co-precipitation include chlorides, nitrates, oxalates and hydroxides. This method is an exceptional choice where excessive control over purity and stoichiometry is desired. Despite process control parameters, solute concentration and precursor purity leads to synchronize the particle size distribution; as we know, impurity facilitates heterogonous growth and results in wider size distribution [15]. This process develops a substantial amount of solid precipitation and immediately after it has to dry and calcinate at a particular temperature to obtain the pure and desired phase. Calcination temperature for co-precipitation is comparatively lower, and sometimes it may produce hard agglomerates that need milling to obtain the desired particle size; otherwise, this behavior assists the formation of intragranular pores after powder compaction, which is followed by sintering [24].

Steps involved in a typical co-precipitation process are shown in Figure 2.9a. Co-precipitation can be classified into three categories in terms of mechanisms, such as inclusion, occlusion and adsorption. Inclusion takes place when an impurity with similar ionic charge and radius of carrier occupies a lattice site in the crystal structure and results in a crystallographic defect. However, an adsorbate is one type of impurity weakly bound at the surfaces of precipitate. The term *occlusion* is used when an adsorbed impurity is

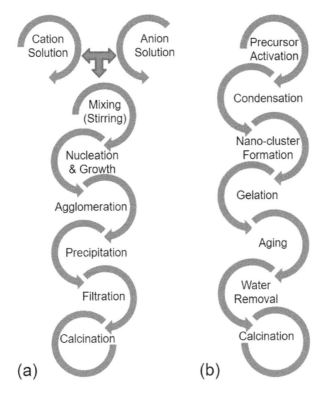

FIGURE 2.9
Flow diagram of a typical (a) co-precipitation method and (b) sol-gel method.

physically trapped inside the crystal during the crystal growth [26]. Apart from application in chemical processing, co-precipitation is extensively used for the synthesis of functional nanoparticles, radiochemistry and it also has the potential for use in several environmental issues such as waste water management, radionuclide transportation and metal contaminant migration, etc. [25, 26]. Some classic theory and YAG particle synthesis for optical application is discussed in Chapter 3.

2.6.4 Sol-Gel

Sol-gel is part of the accepted new class of material processing methods that has emerged over the last few decades to deliver the desired functions and capabilities with advancement [27, 28]. This process is a low-temperature method with fine control over the final chemical composition of the products. The sol-gel process comprises of producing both oxide ceramics and glasses from the homogenous solution where new small molecules develop and gradually form the gel of polymers. The basic steps of a sol-gel process are illustrated in Figure 2.9b.

A typical inorganic salt or metal oxide precursor undergoes hydrolysis and partial condensation to form a colloid, solid particles ranging from 1 nm to 1 micron. Upon destabilization of sol particles and further condensation, it coagulates to form a three-dimensional gel material. The pH of the system affects the rate of hydrolysis and condensation. For example, hydrolysis occurs faster in acidic conditions and the gel is less branched, whereas in basic condensation conditions, the resultant gel is prominent and forms excellent network [29]. The amount of time taken by the sol to form an interlinked gel structure is known as gelation time. Herein, the viscous solution transforms into an elastic gel, capable of sustaining elastic stress of en bloc mass. The type of metal oxide bond and rate of hydrolysis–condensation determine the gel structure. The gelation (gel forming) process is again affected by the temperature, type of solvent and zeta potential. Zeta potential is defined as the effective potential difference between the surface charge of solid particles and that of the solvent. High zeta potential reduces the gelation rate because of the high degree of repulsive states. Thus, high gelation rate is expected near to isoelectric point where the zeta potential is zero. Gelation followed by aging is an important step in order to complete the conversion to the desired density of the entire mass. Aging may include processes such as polycondensation, syneresis, coarsening and phase transformation. In the final stage, it is important to complete the removal of the solvent through elevated temperatures. The final step is associated with apposite heating or firing, depending on the parameters of both linear and cross-linked polymers present in the system. The heating comprises of three stages: during the removal of adsorbed water it is 100 to 200^{0}C, during the decomposition of organic residuals it is 200 to 600^{0}C and finally the formation of corresponding metal oxides is done at high temperatures. The solid structure size varies from discrete colloids to polymer networks [29, 30]. Several types of products obtained using the sol-gel process include dense coating film, dense ceramic, aerogel and ceramic fibers.

2.7 Flame Synthesis

Flame synthesis is the most used versatile and favorable method for the production of a large quantity of nanoparticles; production using flame synthesis reaches thousands of tons per day. This process is popular owing to its simple single step operation; however, its very rapid conversion process limits control over the product properties. Thus, mastering the principles of flame synthesis is a key task to obtain the desired characteristics in the product [31]. The chemical conversion process consumes energy from the flame and liquid precursor which transforms into atomic clusters that are assisted by surface growth to develop nanoparticles. Advantages such as high purity, low-cost synthesis, continuous process, flexibility to produce wide range of products and scalability up to industry-scale makes one go for flame synthesis over other processes [32, 33]. By sacrificing product purity to some extent, one can use additives to control the extent of agglomeration, primary particle size and crystallinity [34].

Evaluation of flame-assisted reaction kinetics is a complex process as this reaction involves the mixture of fuel and oxidizer that eventually produces water, CO_2 and saturated or unsaturated hydrocarbons. Flames are of three types: premixed, partially premixed and non-premixed. In premixed flames, the mixture is completely mixed before coming out of the burner, partially premixed flames have the fuel and oxidizer partially mixed and non-premixed flames have two separate burners providing oxidizer and fuel directly at the reaction site, also known as diffusion flames. Diffusion flames are divided into three types according to the orientation of reactant nozzles, such as co-flow diffusion, counter flow diffusion and inverse diffusion [35].

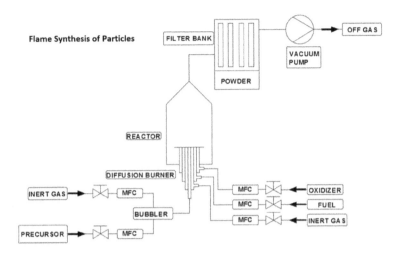

FIGURE 2.10
Schematic illustration of flame synthesis process [38].

Typical flame synthesis equipment consists of a burner or diffusion burner through which supply of materials is enabled, a reactor where the reaction takes place, a filter bank to retain the product formed and a vacuum pump to remove the gas generated; such an arrangement is shown in Figure 2.10. The process initiated through the burner consists of several nozzles through controllers to release the precursor, fuel, oxidizer and inert gas. A multi-flame controller is used mostly to manipulate the flame both inside and outside of the ring. A reaction takes place inside the reactor and a vacuum pump is fixed that evacuates the formed gases from the top of the reactor. The product gases pass through a filter chamber, which retains the desired product formed and leaves off the gas. The rapid shifting of heating and cooling during the process enables the product to obtain special morphology, composition, structure and high purity. This protocol is commercially used for manufacturing unique products such as oxide particles, carbon black, printing ink, and optical fibers and carbon nanotubes through additional controlling arrangements with a conventional flame synthesis unit [36, 37].

2.8 Spray Pyrolysis

Spray pyrolysis is a popular industrial technique where precursor solution is sprayed using atomizers to produce droplets that eventually experience evaporation, condensation, drying and thermolysis in a heated chamber to form fine particles. This technique has several advantages, such as the requirement of simple and inexpensive equipment, better reproducibility, high growth rate, potential for growth spreading over a large area, easy addition of doping material and chemical homogeneity of the product. This versatile method has widespread applications, like producing dense and porous particles, non-oxide ceramics, metals, composites, ceramic coating, solar cell and powders and it is even used in the glass industry and when producing multi-layered films [38, 39].

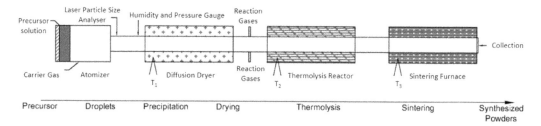

FIGURE 2.11
Illustration of different steps involved in powder synthesis by spray pyrolysis method.

A typical schematic on the process involved in the generation of particles is shown in Figure 2.11.

The desired structure and properties of the product can be tailored by controlling the type of precursor characteristics that comprise the type of solute; concentration of solute; and additives used to influence the physical, chemical and thermal properties of precursor solution. Although several types of precursors are used for different types of application, aqueous solutions are the most widely used precursors in spray pyrolysis [40]. A solute with a high degree of solubility, and thus concentration, increase the process yield and control the particle density. Droplet formation through atomization depends on several factors such as solution viscosity, density and surface tension of precursors. During atomization, part of the droplets coalesce, deflect and reflect, which eventually decreases process efficiency [34]. The selection of atomizer and spraying process parameters are a crucial stage in obtaining the desired droplet size. In this process, a different class of atomizers are used, known as the pressure atomizer by compressed air, electrostatic atomizers under electric field, nebulizers and ultrasonic or ultrasound atomizers [7, 41]. With atomizers, the size of droplets and the speed of droplets vary, which is critical as these two things decide the heating rate and time of transportation. A typical laser particle analyzer is introduced to estimate the atomization efficiency process prior to completion of precipitation of particles. During the process of condensation and evaporation, several steps of reaction occur, including the collision of liquid droplets, evaporation of liquid droplets, diffusion and evaporation of the solvent from the surface of the droplet, change in droplet temperature and shrinkage. Coagulation is possible only when the liquid phase is present in the droplets and the type of product varies with the type of reaction and distance from the surface at which the reaction takes place. The relation of the mean concentration of dissolved salt with the radius of the droplet can be described as [42]:

$$C_m = C_0 \left(\frac{R_0}{R} \right)^3 \tag{2.7}$$

Where C_m and C_0 are the mean and initial concentrations of precursors, R and R_0 represent the mean and initial droplet radius, respectively.

Precipitation occurs only at the parts where concentration is greater than the equilibrium saturation and, accordingly, two types of precipitation can be noticed: volume precipitation and surface precipitation. Volume precipitation refers to the complete droplet precipitation and it produces a solid particle, whereas in surface precipitation the center concentration is lower, hence it produces porous or broken particles. For fine grain ceramic applications, it is essential to understand the effect of precursor properties and precipitation characteristics to synchronize the resultant microstructure of particles. A rapid evaporation leads to the creation of surface precipitation; thus, it is suggested avoiding unless it is essentially required to make hollow particles. Drying and sintering are the last steps of spray pyrolysis and thus in situ sintering is preferred to maximize the benefits of this process. A low temperature promotes porous particles, but a relatively high temperature is required to synthesize fine and dense particles.

This versatile method can be used to produce widespread products like dense and porous particles, non-oxide ceramics, metals, composites, ceramic coatings, powders, multi-layered films, solid, hollow, porous and fibrous material with a variation in precursor properties, solution properties and process parameters [42, 43]. Spray pyrolysis has many opportunities to produce powders with advanced and novel properties.

2.9 Particle Size and Packing

This section deals with the importance of particle size analysis and details packing in the perspective of ceramic processing, as it has great influence on the resultant properties. Adequate knowledge of the grading of coarse particle and particle size distribution of fines facilitates to make the desired microstructure consist of either a very dense or porous matrix. For example, a high-density MgO–C brick matrix performs better in a steel ladle but a porous working lining in tundish is desirable during steel casting. Therefore, we have to have an understanding of the strategies to obtain the desired microstructure from processed particles in the range of a few millimeters to microns. At the same time it is also essential to obtain a narrow size distribution of particles for functional application (e.g., DSSC) in comparison to making it highly dense and compact. Thus, particle size distribution varies with respect to the mode of application. It is noticed that the performance of concrete is greatly affected by the type and degree of packing of its constituents. Thus, knowledge about the concept of particle packing and its influence on concrete performance is required to enable a mixture designer to select from a wide range of cement replacement materials [4]. Using particle size distribution and packing, we can minimize the volume occupied by powder, leading to an increase in strength, green density and a decrease in firing/drying shrinkage.

A conventional classification of particles in the perspective of particle size and their respective range is shown in Table 2.3. Upon maximizing the packing efficiency of a material, one can maximize solid loading for high-density slip casting. This is also applicable for the bulk storage and transportation of raw materials in ceramic processing. Simultaneously any model in ceramic processing which supports maximizing the packing efficiency should automatically include the ability to flow from storage with less external aid [44].

TABLE 2.3

Size Range for Particles in Ceramic Processing

Type of Particle	Size Range
Aggregate	>1 mm
Granule	100 μm–1 mm
Coarse Particle	1 μm–100 μm
Colloidal Particle	1 nm–1 μm

2.10 Sieve Analysis

Sieve analysis is the most commonly used and oldest method for particle sizing to separate particles of different size ranges. Sieves are constructed of metal wires with an opening size ranging from 20 micron to 10 mm, recognized in terms of mesh number and aperture size. Wire mesh consists of square apertures and is defined in terms of wire diameter and number of wires per unit dimension; mesh size is defined as the number of apertures available per square inch. The range can be extended up to five microns using a special type of metal sieve and 125 mm in the upper limit using the punching technique. Particle sizes are identified in terms of whether they pass through a specific mesh or not, and a typical schematic of mesh is represented in Figure 2.12.

The parameters governing the sieving process are interrelated by the following equations:

$$M = \frac{1}{a+w} \text{ or } a = \frac{1}{M} - w \tag{2.8}$$

$$A = \frac{a^2}{\left(a+w\right)^2} = (Ma)^2 \tag{2.9}$$

Where M is mesh number, a is aperture width, w is wire diameter and A is the available opening area. Previously, the reference value was considered to be 75 micron; however, recently the International Organization for Standardization (ISO) modified the old system and 45 micron was established for determining the apertures [4].

2.11 Particle Size Analysis

Particle size distribution is statistical data representing the relative amount of mass or volume of particles present corresponding to the respective size or range of size or in other words, it is the quantity of material in terms of the function of size. Various types of distribution can be represented, subject to property used as a basis for measurement. Particle size analysis is typically assessed in terms of the quantity of individual particles and the

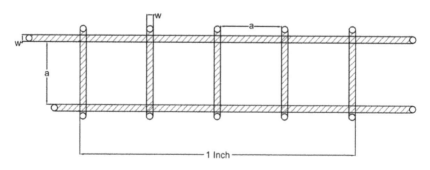

FIGURE 2.12
Dimensions of woven wire in a mesh [4].

TABLE 2.4

Common Methods Used for Measurement of Particle Size and Their Size Ranges [4, 46].

Characterization Technique	Working Principle	Size Range (μm)
Optical Microscopy	Particle Counting	>1
Ultrasonic Attenuation Spectroscopy	Ultrasonic	0.05–10
Optical Centrifugal Sedimentation	Sedimentation	0.01–30
X-Ray Centrifugal Sedimentation	Sedimentation	0.01–100
Scanning Electron Microscopy	Particle counting	>0.1
Transmission Electron Microscopy	Particle counting	>0.001
Sieving	Sieving	20–100,000
Laser Light Diffraction	Electromagnetic wave interaction and scattering	0.04–1000
Light Scattering Intensity		0.1–1000
Brownian motion		0.005–1
X-Ray Line Broadening		< 0.1
BET Absorption	Surface area	NA
Electroacoustic Spectroscopy	Ultrasonic	0.1–10
Micro-electrophoresis	Electromagnetic wave interaction and scattering	0.1–1

quantity of particles in each pre-determined size fraction is used to obtain the particle size distribution [41]. Prior to a discussion of such particle size distribution analysis, one can pick up the different characterization protocols, apart from sieve analysis, used to collect the particle size data (see Table 2.4).

Weibull statistical analysis provides three important parameters to identify the particle size characteristics and can be estimated by the Equation 2.10:

$$f(x) = 1 - \exp\left[-\left(\frac{x-\gamma}{\beta}\right)^{\alpha}\right] \tag{2.10}$$

Here, x, α, β and γ are all positive, and $f(x)$ is the cumulative undersize percent of particle size x present in the distribution. Three parameters α, β and γ represent shape, scale and location, respectively. The high-value shape parameter that indicates narrow size distribution, the scale parameter that describes the size of most of the particles in the system and the location parameters depict the presence of minimum particle size present in the system. After obtaining the particle size distribution data, it is necessary to evaluate all parameters and justify the narrow size particle processing conditions [15].

In another end, a relatively wide range of particle size demands to make dense body, and thus discrete model prefers to fabricate refractory bricks. However, a continuous particle size distribution described by *Furnas model, Andreasen model* and the modified Andreasen model known as *Dinger and Funk model* is in forefront to design the concrete and castable [45]. The discrete model explains that the particle size distribution does not have a continuous distribution and comprises discrete sizes of coarse, medium and fine grades materials. Thus, each class of particles occupy the maximum volume available only. In actuality, after the mixing of different graded materials, the discrete particle fraction exhibit a somewhat continuous model. These mixed particles facilitate free flow during compaction; however, continuous particle size distribution accommodate all probable voids and eventually promote high-density packing. The continuous model assumes to have particles of all possible sizes present in the particle distribution system; it is a type of discrete model consist of an adjacent class of particles diameter ratio near to 1. A predefined continuous particle size distribution obtained from different particle sizes leads to less friction and particles follow roll-over flowability without additional compressive force, and thus among the three continuous model, the most acceptable is the *Dinger and Funk model* within a confined particle size horizon considered and expressed in Equation 2.11.

$$\frac{\text{CPFT}}{100} = \frac{\left(D^n - D_S^n\right)}{\left(D_L^n - D_S^n\right)} \tag{2.11}$$

Where, CPFT is the (cumulative percent finer than) cumulative information about particle size distribution, D is any particle size in the distribution between D_L and D_S, D_L is the largest particle size in the system, D_S is the smallest particle size in the distribution, and n is an exponent that relates the slope of the distribution. Simulation provides a preferable $n = 0.37$ in order to obtain maximum packing efficiency for any size distribution that consists of spherical particles [46]. Thus, the systematic particle size fraction expedites the flow behavior and consistency of the castable.

Furthermore, the particle size and their distribution are among the governing factors influencing the rheology and processing ability of a suspension as well [47]. In the case of suspension-driven systems, disastrous dilatant rheology may be observed with ideal packing and high solid loading. To avoid the same, a wise alternative needs to be considered without compromising rheology and optimized solid loading. Flow behavior can be improved in two ways: either with a wide particle size distribution leading high packing density to consist of more inter-particle interactions, thus reducing the largest particles; or with a narrow particle size distribution width causing less-frequent inter-particle contact [48, 49]. This is the cause behind industrial dispersions, i.e., paints and inks are not made up of mono-sized particles, rather a mixture of multi-sized particles. Coarser particles expedite the shear thickening behavior of suspension; however, their compaction efficiency can be enhanced with an extended particle size distribution [50, 51]. Despite traditional ceramics, particle size distribution has substantial control over the electrical and thermal properties, mechanical strength and density of the end product. If the input particle size and their distribution are not properly regulated, it can cause a high rate of rejection, leading to production loss.

2.12 Particle Packing and Density

The previous discussion emphasizes that particle packing is an important phenomenon to comprehend the performance of the resultant product, and that it is related to the density of the particles that eventually assist in controlling the particle mixing behavior and particle size increment through granulation. In this context, herein, we first discuss particle packing followed by the consequent effect of particle density. Particle packing can be measured in terms of packing density (PD), defined as the ratio of volume of solids to the total volume of arrangement. Total volume of arrangement includes the volume of solids and voids. Thus,

$$\text{Packing Density} = \frac{\text{Solid volume}}{\text{Total volume}}$$

$$PD = \frac{V_S}{V_T} = \frac{V_S}{V_S + V_V} = 1 - VD \tag{2.12}$$

Where V_s = Solid Volume, V_V = Void Volume, V_T = Total Volume and VD = Void Density.

Particle packing can be categorized as regular and random packing arrangement. Regular packing follows the specific regular packing arrangement of particles and random packing does not follow such incidence. Let's consider, the single layer (2D) is arranged by the equal spherical particles and follows regular packing in order to develop two classes of packing (square and rhombohedral packing), as illustrated in Figures 2.13a and b, respectively.

Using 2D square and rhombic arrangement, there are five classes of arrangement predicted in 3D space with each having a different degree of coordination number and theoretical PD. For example, cubical with coordination number (CN) 6 exhibit theoretical PD 52.4%; orthorhombic arrangement with CN 12, PD 60.5%; tetragonal with CN 10, PD 69.8%; and both pyramidal and tetrahedral having the same CN 12, PD 74%.

Despite regular packing and their definite coordination number, random packing and particle morphology plays a crucial role and a characteristic representation on the packing density variation with respect to particle irregularities and spherical roundness, which is

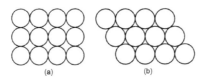

(a)   (b)

FIGURE 2.13
Pictorial representation of (a) square and (b) rhombic arrangement. Each sphere has identical diameter. A significant void within particles is noticed.

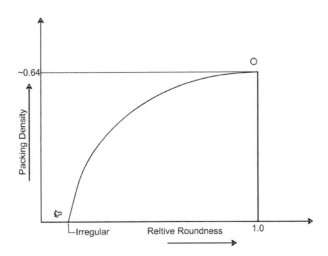

FIGURE 2.14
Influence of particle morphology (irregular to spherical) on the packing density [4].

represented in Figure 2.14. Thus, the pouring of mono-size metallic spheres in a container followed by mechanical agitation, one could attain 60 to 64% packing density [52]. Herein, maximum mechanical packing in the rhombic pattern is considered to be the theoretical density (100%) of the particulate material. To decrease the void density during such mechanical packing, we have to move toward multiple sized particles or at least a binary mixture consist of spherical particles in order to achieve efficient packing. Hence, a binary particle mixture that consists of secondary particles small enough to fit among the voids, and modified packing density can be recalculated as follows: let's consider the first stage of random packing of mono-size particles is 64%; this means it consists of 36% void. In an attempt to fill-up this void, consider another size of relatively smaller particle, so the new packing density becomes 87% (64% + 64% of 36%). Thus, we need a smaller particle size to improve the packing density more than 87%, and it improves to 95.32% (87% + 64% of 13%) of theoretical density (attained after mechanical agitation known as tapping) for a ternary system. This calculation for a ternary system can be represented by Equation 2.12.

$$\text{Packing density for ternary mixture} \left(\text{PF}_{\text{Mix}}\right) = \text{PF}_c + \left(1 - \text{PF}_c\right)\text{PF}_m$$
$$+ \left(1 - \text{PF}_c\right)\left(1 - \text{PF}_m\right)\text{PF}_f$$

(2.13)

Where PF_c, PF_m and PF_f are the packing fraction of coarse, medium and fine particles, respectively, in a ternary mixture, which resembles the discrete model. However, this model may differ to some extent for ceramic particles as it has rare sphericity and deformation. Thus, it is noticed that the density increases when more particles are accommodated in the same volume. Despite these incidences, binders help in more dense packing by increasing the ease of particle movement, but if the binder thickness is higher it causes a negative effect on the packing density.

In order to define and analyze the competitive experimental data and theoretical packing model, it is important to encounter other additional effects such as the wall effect, the loosening effect and dilatant flow. Two types of wall effect are predominant: one is when the particle diameter is similar to that of the container; to avoid this, the ratio of the container diameter needs to be very high (at least >10); and the other is when an isolated coarse particle remains in a matrix of fine particles that eventually increase the void volume in packing. The loosening effect is observed while a fine particle is in a matrix of coarse particles but the size is greater than the void size; it hinders the packing of coarse particles. Dilatant flow occurs due to particle–particle interactions, which is disastrous and at any cost it must be avoided during processing [4, 45, 53]. Hence, an optimum ratio is required in a multi-particle model to obtain dense packing. For example, experimental data established that a quaternary packing of metallic spherical particles achieved 95.1% of theoretical density with diameter ratios 1:7:38:316 and volume compositions 6.1:10.2:23:60.7%, respectively [54].

Particle packing density leads to high green density during compaction. Prior to ceramic processing, starting from particle mixing to sintered product, research laboratories often evaluate the packing efficiency of starting particles through mechanical agitation known as tapping that provides a clear sense of how to obtain a highly dense product from particular graded materials. Conventionally, packing can be divided into loose packing and dense packing. Loose packing comprises bulk density or pour density; however, dense packing results in tapping and is referred to as tap density. Bulk density or apparent density is defined as volume occupied by the unit mass of the material; herein the volume includes particle volume, inter-particle void volume and internal void volume [55]. Bulk

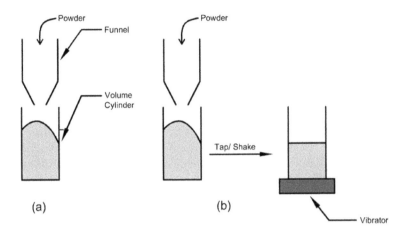

FIGURE 2.15
Schematic representation of the measuring (a) bulk density and (b) tap density. During bulk density measurement, average of upper and lower meniscus is recommended.

density (weight/volume) is measured by pouring a known quantity of mass into a measuring container, and it may increase if the material has a settling tendency. Tap density or true density is defined as the final volume occupied by unit mass after specified amount of compaction, herein, the compaction is done by several procedures either mechanically or manually. A pictorial presentation of measuring bulk density and tap density is shown in Figure 2.15.

The inter-particulate interaction influences the density, interaction at interface and bulk properties by observing bulk and tap density. One can examine the particle's ability to settle and flow behavior with the help of the Hausner ratio (H) and compressibility index or Carr index (C). Both are defined as the $H = \rho_T/\rho_B$ and $C = 100\,(1 - \rho_B/\rho_T)$, where ρ_T is tap density and ρ_B is bulk density of the material, respectively. Although these values are criticized, it is widely used in industries to get a quick understanding on the flowability of particles. The value depends on both of the density measuring protocols. Thus, a poor flowable material enhances the difference between bulk density and tap density, and eventually increases the H as well as C. An optimal H value (greater than 1.33 and lower than 1.18) comprises the poor and good flowability, respectively. In this context, C>25 represents poor flowability and C<15 for good flowability of particles [54, 55]. One can synchronize the resultant microstructure and porosity by starting from the particle packing density analysis.

2.13 Particle Mixing

The prime and foremost unit-processing method is the particle mixing done prior to the consolidation of the body; however, there are only a few advance-processing protocols that are exceptions where direct particle mixing is not required. Adequate solid or wet mixing are used to obtain the desired dry mixer, semi-dry mixer or slurry followed by the processing of different ceramics, like refractory brick by pressing, granules by nodulizer, long ceramic tube by injection molding and large-shaped sanitary ware by slip casting. High temperature sintering is a common process to end the ceramic product prior to finishing, but improper mixing is one of the root causes of developing a non-uniform microstructure and final stage of defects in a sintered body. Mixing is performed to increase the chemical and physical homogeneity of the mixture, and the status of mixing is defined as *mixedness* that is evaluated in terms of physical and chemical analysis. A homogeneous mixing is obtained when the composition does not vary throughout the mixture or else it is an inhomogeneous or heterogeneous mixture. In this context, the basic principle is discussed to apprehend the philosophy of particle mixing phenomenon. Several types of ceramic particle mixing are highlighted in Chapters 6 and 7. Conventionally, three types of successive mechanisms are noticed during mixing: (i) a convection involving the movement of particles from one region to another, (ii) the shear involving interaction and deformation of shape and (iii) diffusion involving microscopic interchange of particles. Because of such incidence, several are results found in mix, and a typical two-dimensional particle arrangements is illustrated in the Figure 2.16. In this process, two different (black and white square) materials become a homogenized mixture, and the reader should not misinterpret this as the formation of compound; this difference is analogous to composite and solid solution, respectively.

Industrial-scale mixing typically follows either *batch mixing* or *continuous mixing*. The type of product and feed input decide whether the unit operation should be adapted to batch or continuous mixing. However, before selecting the mixing protocol and relevant equipment, several factors should be considered, including the processing environment, energy consumption, moisture content, size, shape, consistency and flow behavior of the product, feed rate and properties of material and degree of homogeneity required. For example, continuous vibration can segregate and settle the finer particles from wide particle range distribution that eventually affects the packing density, resulting in a different class of microstructure from targeted appearance and properties.

Among a different class of milling techniques, the pug mill, roller mill, ball mill, etc., are the most commonly used equipment in the ceramic industry. Depending on the consistency, different types of aids are used during the process of mixing, such as binders and plasticizers or dry powder or plastic materials and liquids and deflocculants for slurry making. In the case of dry powders, binders provide the strength to the green body and the percentage of binders would preferably be below 3wt%. Deflocculants improve particle dispersion and stability in the slurry. Recently, the mechano-chemical synthesis has been considered to be an advanced mixing protocol where the intimate mixing and chemical reaction occurs with the heat generation during high-energy mechanical processing, which results in a new class of nano-scale materials. Thus, particle mixing initiates ceramic processing, where particle size enlargement is the immediate consequence and also has a significant effect on the targeted properties of the final product.

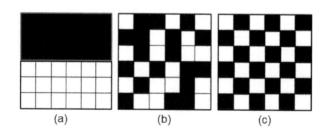

(a)   (b)   (c)

FIGURE 2.16
Particle mixing arrangement representation in two-dimensional, (a) completely segregated, (b) completely random and (c) completely dispersed [3].

2.14 Particle Scaling-Up

Particle scaling-up or enlargement is defined as a reverse step of particle disintegration, where particle mixing is taking place with the desired composition, particle granulometry and binder that eventually promotes the development of larger bodies from smaller particles. Different types of methods are used for particle size enlargement; a few common protocols are compaction, agglomeration, globulation, heat bonding, pelletization, crystallization and flocculation. Depending upon the material and desired output, consideration of single or multiple processing is common practice in an industrial environment. Particles combine due to the presence of adhesive forces; adhesion of any particles can be observed under high pressure, but in the presence of binders or liquids it can initiate the adhesive force under low-pressure conditions. Depending on the inter-particle distance and process involvement, three types of forces govern particle mixing: dispersion forces when the inter-particle distance is less than 10^{-3}Å; electrostatic forces as a result of friction or presence of opposite charges; and particle-solvent bridges that only act in the presence of liquid which increases the strength of granules.

A brief of different particle enlargement processes is here, although several ceramic processing protocols are discussed in different chapters:

i. Compaction: This process involves the formation of granules under pressure or compaction. The material is compressed or extruded through molds and cut into the desired shape as required.

ii. Agglomeration or powder layering: This is an old and commonly used process performed with or without binders through tumbling or agitation. Resultant particle size is controlled in consideration of feed size and rate, binder addition rate and time of process. Gradual powder agglomeration is the result of the solid–liquid bridging force of binders.

iii. Globulation: Globulation or droplet formation is associated with processes like spray drying and spray congealing or prilling. Spray drying includes the spraying of a solution or suspension toward a hot air stream to produce spherical particles, and prilling consists of the dissolution of material into a hot melt that eventually spray below melting point to produce pellets.

iv. Heat bonding: This is an excellent protocol when the industry demands permanent bonding and appreciable strength in enhanced particles. It may follow heating in tumbling to produce hard-bonded spherical particles and appreciable heat treatment or sintering after the formation of large particles.

v. Crystallization: Crystallization consists of super saturation, nucleation and crystal growth. Super saturation is caused by increasing the solubility with the help of external factors, followed by nucleation that implies the formation of a new phase or crystal from the solution and diffusion-assisted crystal growth.

Several factors need to be considered for selecting the most suitable method for particle size enhancement, including quantity of product over time, resultant particle size and distribution of enlarged particles, feed characteristics, morphological characteristics of product, space requirement, dry or wet method, environmental issues, etc [56–58]. Advantages of particle size enlargement are its high density and dust-free material, better stability, compact structure and easiness to handle and transport.

2.15 Conclusions

A wide array of science and technology is involved in engineering particle processing for different classes of applications. Ceramic processing starts with both natural minerals and synthetic mass, but it is important to process those starting raw materials into the desired particle morphology that comprises a range of a few millimeters to nanometer. In order to fulfill such a class of particles, different crushing, grinding and milling systems are common, and adapted comminution is used to fulfill the targeted particles size. A few classical methods comprehends with basic principles of physics and chemistry, which form a framework for understanding how the process variables influence the characteristics of the powders. Practically, the methods vary considerably in the quality of the powder produced and in the cost of production, but sometimes it is important to justify the production cost. Particle size analysis provides an insight into the coarse-to-fine distribution for different classes of traditional and advanced ceramics. A quick look at the statistical analysis and compressibility index illustrates how to pick up the best particle grading to obtain a high compact body. However, several ceramic manufacturing protocols enthusiastically ask for homogenous particle mixing in the target of consistent properties and performance of bulk body. Despite several well-known mixing techniques, herein the basic mixing philosophy has been highlighted. Moreover, the particle size scaling protocols to control the properties of different ceramics have been discussed.

References

1. Duroudier, J. D. 2016. *Size Reduction of Divided Solids*. Elsevier: London.
2. Egbe, E. A. P., and Olugboji, O. A. 2016. Design, fabrication and testing of a double roll crusher. *International Journal of Engineering Trends and Technology (IJETT)* 35(11).
3. Reed, J. S. 2007. Principles of ceramics processing. In: C. B. Carter and M. G. Norton (Eds.), *Ceramic Materials: Science and Engineering*. Springer, John Wiley & sons, 1995.
4. Rahaman, M. N. 2003. *Ceramic Processing and Sintering*. CRC Press: Boca Raton, FL.
5. Beddow, J. K. 1980. *Particulate Science and Technology*. Chemical Publishing Co: New York.
6. Bernotat, S., and Schönert, K. 1998. *Ullmann's Encyclopedia of Industrial Chemistry*, Vol. B 2. VCH verlagsges: Weinheim, 1998. Size Reduction.
7. Alves, A. K., Bergmann, C. P., and Berutti, F. A. *Novel Synthesis and Characterization of Nanostructured Materials, Engineering Materials*. Springer: Berlin.
8. Wollmershausher, J. A., Drazin, J., Kazerooni, D. A., Feigelson, B. N., and Gorzkowski III, E. P. 2018. Nanocrystalline alpha alumina and method for making the same. US2018/0111841 A1.
9. Calka, A., and Williams, J. S. 1992. Synthesis of nitrides by mechanical alloying. *Materials Science Forum* 88–90:787–794.
10. Chen, Y., Williams, J. S. 1996. Hydriding reactions induced by ball milling. *Materials Science Forum* 225–227:881–888.
11. Hwang, S. J., Nash, P., Dollar, M., and Dymek, S. 1990. Microstructure and mechanical properties of mechanically alloyed NiAl. *MRS Proceedings* 213:661.
12. Perez, R. J., Huang, B.-L., Crawford, P. J., Sharif, A. A., Lavernia, E. J. 1996. Synthesis of nanocrystalline Fe-B-Si powders. *Nanostructured Materials* 7, 1–2: 47–56.
13. Kong, L. B., Zhu, W., and Tan, O. K. 2000. Preparation and characterization of $Pb(Zr_{0.52}Ti_{0.48})O_3$ ceramics from high-energy ball milling powders. *Materials Letters* 42:232–239.

14. IS 8730:1997. 1997. Classification and Codification of Bulk Materials for Continuous Material Handling Equipment. Bureau of Indian Standards: Manak Bhawan, India.
15. Sarkar, D. 2019. *Nanostructured Ceramics*. CRC Press: Boca Raton, FL.
16. Balachandran, U., Poeppel, R. B., Emerson, J. E., and Johnson, S. A. 1992. Calcination and solid state reaction of ceramic-forming components to provide single-phase superconducting materials having fine particle size. US5086034A.
17. Lee, B., and Komarneni, S. 2005. *Chemical Processing of Ceramics*, 2nd Edition. CRC Press: Boca Raton, FL.
18. Vajargah, S. H., Hosseini, H. M., and Nemati, Z. A. 2007. Preparation and characterization of yttrium iron garnet (YIG) nanocrystalline powders by auto-combustion of nitrate-citrate gel. *Journal of Alloys and Compounds* 430:339–343.
19. Adhikari, S., Sarkar, D., and Madras, G. 2015. Highly efficient WO3–ZnO mixed oxides for photocatalysis. *RSC Advances* 5:11895–11904.
20. Toksha, B. G., Shirsath, S. E., Patange, S. M., and Jadhav, K. M. 2008. Structural investigations and magnetic properties of cobalt ferrite nanoparticles prepared by sol–gel auto combustion method. *Solid State Communications* 147:479–483.
21. Varma, A., Mukasyan, A. S., Rogachev, A. S., and Manukyan, K. V. 2016. Solution Combustion Synthesis of Nanoscale Materials. *Chem. Rev.* 116(23):14493–14586.
22. Adair, J. H., and Suvacib, E. 2001. Submicron electroceramic powders by hydrothermal synthesis. *Encyclopedia of Materials: Science and Technology*.
23. Adhikari, S., and Sarkar, D. 2014. High efficient electrochromic WO3 nanofibers. *Electrochimica Acta* 138:115–123.
24. Chandra, K. S., Monalisa, M., Ulahannan, G., Sarkar, D., and Maiti, H. S. 2017. Preparation of YAG nanopowder by different routes and evaluation of their characteristics including transparency after sintering. *Journal of the Australian Ceramic Society* 53:751.
25. Harvey, D. 2000. *Modern Analytical Chemistry*. McGraw-Hill: New York.
26. Lu, A.-H., Salabas, E. L., and Schüth, F. 2007. Magnetic nanoparticles: Synthesis, protection, functionalization, and application. *Angewandte Chemie International Edition* 46:1222–1244.
27. Zhu, C., Martin, S., Ford, R., and Nuhfer, N. 2003. Experimental and modeling studies of co-precipitation as an attenuation mechanism for radionuclides, metals and metalloid mobility. *Geophysical Research Abstracts* 5:06552.
28. Roy, R. 1969. Gel route to homogeneous glass preparation. *Journal of the American Ceramic Society* 52:344.
29. Brinker, C. J., Keefer, K. D., Schaefer, D. W., and Ashley, C. S. 1982. Sol-gel transition in simple silicates. *Journal of Non-Crystalline Solids* 48:47.
30. Farrauto, R. J., and Bartholomew, C. H. 1997. *Fundamentals of Industrial Catalytic Processes*. Blackie Academic & Professional: London.
31. Klein, L. C., and Garvey, G. J. 1980. Kinetics of the sol-gel transition. *Journal of Non-Crystalline Solids* 38:45.
32. Kammler, H. K., Mädler, L., and Pratsinis, S. E. 2001. Flame synthesis of nanoparticles. *Chemical Engineering & Technology* 24:6.
33. Stark, W. J., and Pratsinis, S. E. 2002. Aerosol flame reactors for manufacture of nanoparticles. *Powder Technology* 126(2):103–108.
34. Wegner, K., and Pratsinis, S. E. 2003. Scale up of nanoparticle, synthesis in diffusion flame reactors. *Chemical Engineering Science* 58(20):4581–4589.
35. Pratsinis, S. E. 1998. Flame aerosol synthesis of ceramic powders. *Progress in Energy and Combustion Science* 24(3):197–219.
36. Gore, J. P., and Sane, A. *Flame Synthesis of Carbon Nanotubes*. Purdue University: West Lafayette, IN.
37. Kumar, B. M., and Bhattacharya, S. S. 2012. Flame synthesis and characterization of nanocrystalline titania powders. Processing and Application of Ceramics 6(3):165–171.
38. Donnet, J., Bansal, R. C., and Wang, M. J. 1993. *Carbon Black*. Marcel Dekker: New York.

39. Bautista, J. R., Walker, K. L., and Atkins, R. M. 1990. Modeling heat and mass transfer in optical waveguide manufacture. *Chemical Engineering Progress* 86:47.
40. Hill, J. E., and Chamberlin, R. R. 1964. US Patent 3,148,084.
41. Perednis, D. 2003. Thin film deposition by spray pyrolysis and the application in solid oxide fuel cells. PhD Thesis, Swiss Federal Institute of Technology Zurich, Zurich, Switzerland.
42. Messing, G. L., Zhang, S. C., Jayanthi, G. V. 1993. Ceramic Powder Synthesis by Spray Pyrolysis. *Journal of the American Ceramic Society* 76(11):2707–2726.
43. Perednis, D., and Gauckler, L. J. 2005. Thin film deposition using spray pyrolysis. *Journal of Electroceramics* 14:103–111.
44. Hill, J. E., and Chamberlin, R. R. 1964. US Patent 3,148,084A.
45. Funk J. E., and Dinger, D. R. 1994. Fundamentals of particle packing, monodisperse spheres. In: *Predictive Process Control of Crowded Particulate Suspensions*. Springer: Boston, MA.
46. Jillavenkatesa, A., Dapkunas, S. J., and Lum, L. S. H. 2001. Particle size characterization. *Materials Science and Engineering Laboratory*. Special Publication: 960–961.
47. Dinger, D. R., Funk, J. E. 1992. Particle packing III – Discrete vs continuous particle sizes. *Interceram* 41(5):332–335.
48. Wachtman, J. B. 1989. *Wachtman, Materials and Equipment – Whitewares*. John Wiley & Sons: New York.
49. Cheng, D. C.-H., Kruszewski, A. P., Senior, J. R. 1990. The effect of particle size distribution on the rheology of an industrial suspension. *Journal of Mathematical Sciences* 25:353–373.
50. Funk, J. E. 1981. Coal water slurry and method for its preparation. U.S. Patent 4,282,006.
51. Dinger, D. R., Funk, Jr. J. E., and Funk, Sr. J. E. 1982. Rheology of a high solids coal-water mixture. *Proceedings of the 4th International Symposium on Coal Slurry Combustion*, Orlando, FL.
52. Olhero, S. M., and Ferreira, J. M. F. 2004. Influence of particle size distribution on rheology and particle packing of silica-based suspensions. *Powder Technology* 139:69–75.
53. Smith, P. A., and Haber, R. A. 1995. Effect of particle packing on the filtration and rheology behaviour of extended size distribution alumina suspensions. *Journal of the American Ceramic Society* 78(7):1737–1744.
54. McGeary, R. K. 1961. Mechanical packing of Spherical particles. *Journal of the American Ceramic Society* 44(10):513–522.
55. Buckman, H. O., and Brady, N. C. 1960. *The Nature and Property of Soils - A College Text of Edaphology*, 6th Edition. Macmillan Publishers: New York. p. 50.
56. Beddow, J. K. 1995. Professor Dr. Henry H. Hausner, 1900–1995. *Particle & Particle Systems Characterization* 12:213.
57. Kanig, J. L., Lachman, L., and Lieberman, H. A. 1986. *The Theory and Practice of Industrial Pharmacy*, 3rd Edition, Lea & Febiger: Philadelphia, PA.
58. Capes, C. E. 1980. *Particle size enlargement*, Elsevier Scientific Publishing Company: Amsterdam, the netherlands.

3

Transparent Ceramics

Samuel Paul David and Debasish Sarkar

CONTENTS

3.1 Introduction

For several centuries, ceramic materials have been widely associated with building materials, whitewares and refractories. This notion has been changed over the past several decades, as non-metallic, inorganic ceramic materials have found applications in several areas of modern technology including electronics, medicine, energy, lasers and so on. Polycrystalline ceramic materials are generally opaque due to the scattering of incident light because of voids, pores, grain boundaries and birefringence and were believed to be impossible to make transparent. Conventionally, glasses, polymers and single crystals were used for optical applications that require high transparency. Glasses and polymers suffer from poor mechanical and thermal properties, whereas single crystals with good thermal properties require an expensive and time-consuming manufacturing process. Ceramics that are transparent to electromagnetic radiation gained popularity over the last fifty years, after the first use of translucent α-Al_2O_3 [1] envelopes in sodium-vapor lamps that require materials with enhanced thermal and mechanical properties. Even though translucent ceramics are continuously used for such an application, improvement in transparency was made over a few decades with advanced technology to address other technological applications. In a physical sense, at a macroscopic scale, translucent materials do not need to follow Snell's law, but successful transparent materials should obey it. To develop such transparent ceramics, scattering has to be completely eliminated by avoiding pores, voids and secondary phases along with thin grain boundaries. Scattering due to birefringence can be avoided in cubic materials that limit the transparent ceramics to materials with a preferential cubic structure such as Y_2O_3, YAG, $MgAl_2O_4$, CaF_2 and so on. Yttrium aluminum garnet (YAG) gained more interest in its transparent ceramic form because it does not show any birefringence effect along the grain boundaries due to its cubic symmetry along with its excellent thermal and mechanical properties. When doped with laser active ions, polycrystalline YAG performs as one of the best materials for solid-state lasers.

In addition to cubic ceramics, several non-cubic materials such as Al_2O_3, lead zircontae titanate (PLZT) and strontium fluoroapatite (S-FAP) have also been made transparent in their polycrystalline forms. Some of the important transparent ceramics with their structure and applications are shown in Table 3.1. Developing such ceramics with optical qualities similar to that of a single crystal involves the improvement of ceramic technology at each processing step, including powder processing, powder compaction, shape forming, sintering and polishing. The general processes involved in the fabrication of transparent ceramics is shown in Figure 3.1.

With a better theoretical understanding of the entire process, it has been found that the crucial underlying importance is the choice of starting powders and the forming of a green body, which have an impact on the further sintering mechanisms to achieve full density [2]. Hence, on an industrial scale, much attention has been given to powder processing and green body formation. With new sintering technologies, several other materials, which otherwise could not be developed into polycrystalline transparent material, have been made as transparent ceramics. This chapter focuses on the inherent difficulties and procedures such as high-precision stoichiometry powder processing, green body formation and sintering in fabricating transparent ceramics. It also enlightens some of the important applications of transparent ceramics in various fields.

TABLE 3.1

Few Transparent Ceramics with Their Crystal Structure and Their Major Applications

Material	Crystal Structure	Applications
α-Al_2O_3	Trigonal	Envelopes for high-pressure lamps, armor windows
Yttrium aluminum garnet	Cubic	Solid-state laser devices
Y_2O_3, Sc_2O_3	Cubic	Solid-state laser devices
MgO	Cubic	IR windows, Sensor envelopes
$MgAl_2O_4$, AlON	Cubic	Lenses, missile domes, armor windows
GYGAG: Ce, Lu_2O_3: Ce	Cubic	Scintillators

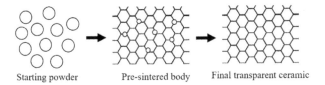

Starting powder · Pre-sintered body · Final transparent ceramic

FIGURE 3.1
Schematic of the ceramic powder consolidation starting from powder to intermediate pre-sintered compact followed by inclusion-free transparent ceramics.

3.2 Single Crystal Vis-À-Vis Polycrystalline Transparent Ceramics

Materials with an optical transmission over a wide range of electromagnetic (EM) radiation are desirable, and conventionally glasses, polymers and alkali fluorides are used in the visible range of the EM radiation. These materials have low thermal conductivity and also exhibit strong absorption in the infrared (IR) region that limits their optical window. Single crystals show better thermal conductivity and superior transmission range as well. However, crystal growth technologies require expensive crucibles to develop materials with a very high melting point that is limited to smaller dimensions of resultant components. For applications such as armor windows or high-power lasers that require very large dimensions, the single crystals are generally bonded with high precision that adds to the fabrication cost. Thus, polycrystalline transparent ceramics have been found to be an alternative to single crystals with superior characteristics, at a relatively lower cost and a faster production time. Polycrystalline transparent ceramics are considered to be both passive as well as active; the former type represents ceramics that are simply transparent to EM radiation, such as IR windows, whereas the latter encompasses functional properties in addition to the transmission of EM radiation, such as scintillators [2]. Nevertheless, passive transparent ceramics such as $MgAl_2O_4$ and AlON exhibit good mechanical strength with high hardness and corrosion resistance that makes them useful in missile domes, military vehicle windows and scratch-resistant lenses. Active transparent ceramics such as YAG are used as solid-state laser hosts and radiation detectors.

To give a better insight into the need for transparent ceramics over conventional crystalline media, a brief look into the development differences between YAG single crystals and YAG ceramics for laser applications is provided. Rare earth-doped yttrium aluminum garnet is still the most promising laser-gain media of last 50 years after the first laser demonstration by Geusic et al. in Bell Laboratories in 1964 [3]. Single crystals of YAG are generally grown by Czochralski or Bridgman methods using expensive iridium crucible because of the high melting point of YAG (~1940°C). The growth period usually demands several days to grow to a large size (growth rate ~1 mm/h) crystals (>10 cm) which are essential to build high-power solid-state lasers. Adding to the woes, thermal stress caused during the growth process creates stress-induced birefringence in the crystal, as in Figure 3.2, that limits the usable material to less than 60% of the grown boule.

In addition, the segregation coefficient $\left(k=(1/g)(C_x/C_o)\right)$, (where C_x is the concentration of dopant ion, C_o is the initial concentration of the melt, g is the solidification ratio) challenges the crystal uniformity over a long range. The difference in ionic radii between the dopant ion and host ion limits the dopant concentration and can also cause an uncontrolled gradient in doping concentration in the grown crystals. If the value of k is 1, all the

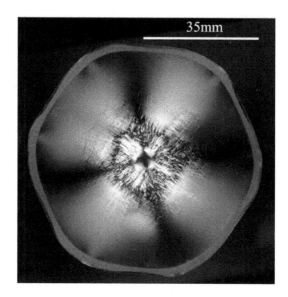

FIGURE 3.2
Striations observed in Nd:YAG crystal boule. It is a cross-polar image of a 1 at% Nd:YAG boule showing striations [4].

Y^{3+} ions can be substituted by the dopant ion. The ionic radius of the Nd^{3+} ion (0.98 Å) is bigger than the Y^{3+} ion (0.90 Å) in the host. This leads to a segregation coefficient value of the Nd ion in the YAG crystal to be 0.18, which limits its concentration to 1~2 at.% in the grown crystal. The concentration of Nd^{3+} ions in the crystal boule is much less than that of melt as the Nd^{3+} ions are retained in the melt. For crystals grown from pulling technique such as the Czochralski method, this leads to a variation in concentration throughout the crystal boule. An Nd:YAG crystal with a length of 3–8 cm can have a dopant variation of 0.05–0.10 wt% of Nd_2O_3 along the length of the crystal, only.

Even though a $Dy:CaF_2$ ceramic rod (2.2 cm × 0.32 cm) was first used to demonstrate lasing (lasing threshold = 24.6 J) at the temperature of liquid nitrogen in 1964, [5] it was not until 1995 when Ikesue et al. achieved a slope efficiency of 28% using Nd:YAG ceramics similar to that of a Nd:YAG single crystal. This breakthrough achievement facilitates exponential growth of transparent ceramics for laser applications [6]. Ceramic technology offers multiple advantages over conventional single crystals to fabricate large size gain media which are essential for kilowatt (kW) level solid-state lasers. This technology also offers other unique potentials to make transparent ceramics with a high dopant concentration, complex doping profile and even in complex shapes and geometries that otherwise are not easily possible with single crystals.

Similar to laser applications, large-sized transparent ceramics gained popularity in the defense applications. Surmet Corporation in the United States has made progress in developing transparent AlON ceramics for use in armor windows (3 ft × 3 ft), infrared heat-seeking missile domes, infrared windows and even to protect against improvised explosive devices (IEDs) [7]. Laboratories such as Fraunhofer IKTS in Germany and Naval Research Lab (NRL) in the United States concentrate on fabricating large-sized $MgAl_2O_4$ spinel ceramic as lightweight windows and shields for similar armor applications because of its hardness, scratch resistance properties along with a wide optical transmission range (0.2–5 µm) [8]. Thus, transparent ceramics find smart consideration for various applications over single crystals.

3.3 Ideal Transparent Ceramics – Challenges

Polycrystalline ceramics can be said to be transparent when the incoming light leaves material without any losses such as absorption or scattering. Ceramics are different from single crystals in their microscopic level with the presence of grains and grain boundaries. The presence of defects or secondary phases along the grain boundaries, voids and pores causes scattering of incoming light because of the difference in refractive indices and hence leads to the degradation of the optical quality of the end product. Different scattering sources are shown in Figure 3.3 and the elimination of all these sources is essential if we target transparent ceramics. Apart from the level of transparency, the size of grains also plays a major role. In the case of YAG, grain sizes do not seem to affect transmission property, whereas fine grain-sized ceramics are preferable to high-power lasers because of their better mechanical robustness over coarse-grained ceramics [9].

However, if the grain sizes are very small (~8 μm), the volume fraction of porosity was found to increase, which resulted in poor transmittance and lasing performance of Nd:YAG transparent ceramics [10]. For scintillation applications, it has been found that an increase in grain size improves the overall light yield due to defect emission along grain boundaries [11]. In the case of birefringent material, such as polycrystalline transparent alumina with grain sizes of less than 1 μm, transmission increases with a decrease in grain size as per the equation derived based on the Rayleight-Gans-Debye theory [12]. Hence, *real in-line transmission* (RIT) becomes,

$$\text{RIT} = (1 - R_s)\exp\left(\frac{-3\pi^2 r \Delta n^2 d}{\lambda_0^2}\right) \tag{3.1}$$

Where $(1 - R_s)$ is the maximum theoretical transmission given by $2n/(n^2 + 1)$, n is the average refractive index, Δn is the birefringence, r is the radius of the spherical grain, d is the sample thickness and λ_0 the incident wavelength. For spinel ceramics, maximum transmission was observed for materials with an average grain size of 40 μm [13]. PLZT transparent ceramics show the effect on electro-optic properties, a lowering of the Curie point is noticed with the increase in grain size [14]. Transparent electro-optic ceramics such as lead zirconate stannate titanate shows improved permittivity when the grain size is controlled to 1 μm by pressure-assisted sintering mechanisms [15].

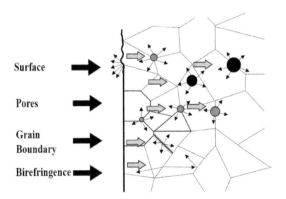

FIGURE 3.3
Probable light scattering sources in consolidate polycrystalline ceramics.

In addition to their optical quality, these ceramics should possess a very good mechanical strength to handle rough conditions such as battlefields for armor windows or mechanical stress induced in laser ceramics under high-power pumping. These mechanical properties in turn depend on grain size and the state of grain boundaries. The Hall–Petch relationship, $H_v \propto 1/\sqrt{D}$ where H_v is Vicker's hardness and D is the average grain size, predicts that the smaller the grain size, the higher the strength ($H_v \approx 3\sigma$) and hardness for the same material in comparison to a larger grain size [16]. The prediction has been proved to enhance the strength in nanocrystalline $MgAl_2O_4$ spinel ceramics by 50% for armor windows when the grain size is near to 30 nm [17]. Interestingly, an inverse Hall–Petch effect was observed in these ceramics when the grain size was further reduced to 17 nm. Such fine-grained ceramics have a large volume fraction of grain boundaries (web of grains), which is weaker than the grains, and this reduces the hardness value from 20 GPa to 17.6 GPa [18]. As can be seen, grain sizes play a major role by affecting different properties of transparent ceramics for different applications. Even for ceramics that show superior characteristics when the grain sizes are reduced, there is a threshold limit for the size, below which the performance goes down, as mentioned earlier for Nd:YAG and spinel ceramics, as well. Ideally, fine grain sizes are preferable for optical ceramics fabricated for laser applications.

Active transparent ceramics for applications, such as lasers or scintillators, need to satisfy a higher degree of optical quality compared to transparent windows. The transparency phenomenon sometimes overestimates the value by considering the scattered radiation over a wide angle and this causes a decrease in transparency as the material thickness is increased. Transparent ceramics for certain applications such as lasers cannot tolerate any atomic-level defects or voids, especially when the task is to fabricate laser materials of large dimensions. Inhomogeneities or the presence of any defects can cause bulk scattering as the light travels through the media that make the material useless for laser applications. For example, to fabricate a YAG ceramic by the solid-state reaction method, a stoichiometric ratio of Y_2O_3:Al_2O_3 should be 3:5; the YAG phase is a line compound in this phase diagram (Figure 3.4) and it exists to only within ~0.2 mol% in Al_2O_3 composition direction [19]. Because of very low solubility (<0.2 mol%), even a small excess of Al_2O_3 results in nanograins of Al_2O_3 inside a YAG grain matrix and causes significant light

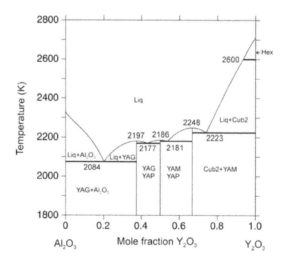

FIGURE 3.4
Al_2O_3–Y_2O_3 phase diagram. YAG corresponds to 37.5 mol% of Y_2O_3 [20].

scattering whereas Y_2O_3 has a relatively high solubility (2–10 mol%) and does not form immediate secondary phases.

However, an excess of Y_2O_3 results in nanoparticles along the grain boundaries which can result in opaque layer with an excess accumulation of such particles. Therefore, any deviation from the ratio of 3:5 results in the precipitation of secondary phases (YAM and YAP) that are detrimental for laser quality transparent ceramics. Starting from the precise weighing of powders, moisture absorption of sub-micron powders have a high surface area that expedites non-stoichiometry and hence pose serious problems in fabricating such optical quality ceramics consistently. Several techniques, such as X-ray diffraction, optical spectroscopy, THz spectroscopy and, recently, laser induced breakdown spectroscopy (LIBS), have been adopted to investigate the non-stoichiometry in such sensitive compounds. The LIBS technique offers unique advantages for analyzing the samples at each processing steps without any loss in the materials and at a much quicker time, even with real-time *in situ* analysis [21]. Considering these factors, a simple optical transparency cannot determine the validity of transparent ceramics for certain applications. Ceramic technology still remains a "black art" even more than twenty years after the first ceramic laser demonstration. This limits only very few suppliers providing bulk ceramic gain media for high-power solid-state lasers. For instance, major high-power lasers built from transparent ceramics use ceramic media from Konoshima Chemicals, Japan. Recently, a kilowatt average power (100 J at 10 Hz) diode pumped solid-state laser was built using such Yb:YAG transparent ceramic slabs, which is a first of its kind in the world [22].

3.4 Role of Powder Processing in Transparent Ceramics

Typical fabrication of transparent ceramics requires three major steps: synthesis of ultra-fine powders (<100 nm), consolidation of powders into green bodies (60~70% density) and the sintering process (>99.99% density). As mentioned in Section 3.3, the functionalities of final transparent ceramics are dependent on the chemical composition and the microstructure that is eventually monitored by the purity, morphology, particle size and agglomeration of starting powders, powder consolidation, sintering atmospheric condition and the sintering profile. Polydisperse powders with a wide range of sizes cause non-uniform microstructure compared to that of ceramics prepared from a narrow range of particle size distribution. Nano-sized powders with a narrow size distribution are highly preferred for transparent ceramics. Smaller particles have high surface energy and the densification rate significantly increases with a decrease in particle size below 1 μm [23]. When packed, due to high surface energy, smaller particles make more contact and require less diffusion distance between the particles that enhance the sintering process even at a relatively lower sintering temperature. The physical and chemical properties of powders are influenced by the preparation methods adopted to produce the chemicals. There are several methods available for producing starting powders for transparent ceramics. These methods are widely classified into mechanical-, chemical- and vapor-phase methods. Mechanical methods are widely popular for traditional ceramics where a certain amount of impurity (≤1 wt%), either from raw materials or grinding media, has no significant influence on the resultant applications. Coarse grains are initially obtained from natural minerals, which then undergo comminution and milling processes to achieve fine-size powders as shown in Figure 3.5 [24]. Details of such processing techniques have been described in detail in Chapter 2.

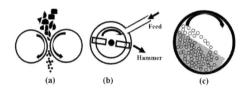

FIGURE 3.5
Different comminution processes: a) roll crushing, b) hammer milling, c) ball milling.

Even though vapor-phase methods yield nano-sized powders, the process can produce less quantity that makes the process become expensive. Chemical methods have been extensively studied and improved over 40 years. Because of the better control of the synthesis conditions to tune powder properties, chemical methods are widely employed to synthesize powders for advanced ceramics. In this context, the chemical methods are widely classified into solid-state reaction and wet chemical synthesis routes.

3.4.1 Solid-State Reaction

Solid-state reaction has been practiced in industries for several decades, and this involves at least one of the reactants to be a solid to form a compound through diffusion either in solid–solid or solid–liquid interface. In solid–solid reactants, there are three possible reactions: chemical decomposition which liberates gaseous product, a simple chemical reaction of two solids to give a solid product and chemical reduction by the exchange of cations and anions between reactants to produce a product. Chemical decomposition is an endothermic reaction mainly controlled by the reaction kinetics that are decided by reaction temperature, reaction time, particle size, surrounding atmosphere and reactant quantity. By controlling the rate of decomposition, which is governed by the Arrhenius equation ($K = A*exp(-Q/RT)$), where K is the rate constant, Q is the activation energy, T is the absolute temperature, R is the gas constant and A is the pre-exponential coefficient, desirable properties of the end product can be obtained, including the microstructure and morphology, by an effective control of the reaction temperature. Even though it is not widely used for synthesizing chemicals for transparent ceramics, however, precursor carbonates, nitrates or hydroxides synthesized by chemical routes undergo decomposition at a higher temperature to end up in oxide products. Most of the simple or complex oxides are produced by the chemical reaction of two or more reactants. The driving force for such solid–solid reactions is the difference between the Gibbs free energy of reactants and products. These reactions are exothermic and occur with the diffusion and counterdiffusion of anions and cations, respectively, between the reactants maintaining charge neutrality. Because of different diffusivity of ions, the diffusion paths are different for anions and cations. For instance, spinel is well studied to understand such mechanism. Oxygen anion (O^{2-}) has a smaller diffusivity (in the order of 10^{-18} m^2/s) value because of its larger size compared to cations (~10^{-8} m^2/s for Al) that causes the major diffusion by the counterdiffusion of cations without affecting the electroneutrality. The reaction rate of such a complex oxide formed by reactants A and B, is determined by the Carter equation,

$$\left[1+(Z-1)\alpha\right]^{\frac{2}{3}}+(Z-1)(1-\alpha)^{\frac{2}{3}} = Z+(1-Z)\frac{Kt}{r^2}$$

(3.2)

Where Z is the volume product formed by the reactant product A, α is the fraction of the reacted volume, K is the rate constant, t is the time and r is the initial radius of the reactant particle. If the reactant particles are smaller in size, the diffusion distance is reduced, which leads to an increase in the reaction rate. Detailed diffusion mechanisms on spinel have been studied by Rahaman [23]. Solid-state reaction offers an advantage of synthesizing large quantity of powders with a lower cost. The first transparent ceramics of Nd:YAG for lasers were fabricated based on the solid-state reaction of Y_2O_3 and Al_2O_3 as per the following chemical reaction:

$$3Y_2O_3 + 5Al_2O_3 \rightarrow 2Y_3Al_5O_{12} \qquad (3.3)$$

When these two reactants are heated together, first a monoclinic phase YAM ($Y_4Al_2O_9$) appears above a temperature of 900°C, then a perovskite phase YAP ($YAlO_3$) forms by further reaction with Al_2O_3 above a temperature of 1100°C. When the temperature is further increased above 1400°C, the cubic YAG phase is formed, and the temperature is required to increase further for the reaction to be completed. Such high temperature results in a microstructure consisting of a larger grain size that implies relatively poor mechanical strength compared to the ceramics with smaller grain sizes.

3.4.2 Powders from Solution

Powders prepared by a solid-state reaction at a high temperature end up in large agglomerates that require another comminution process such as ball milling or attrition milling to break the powders that may introduce contaminants from the milling media. Powders obtained by the liquid solutions, either by the evaporation of solution or by the precipitation of products from the solution by an addition of a chemical reagent, are found to be more suitable for transparent ceramics. Wet chemical methods such as sol-gel, precipitation, combustion and hydrothermal and emulsion synthesis methods have been developed over the past few decades to prepare very fine nano-scale powders with uniform morphology for these applications.

3.4.2.1 Co-Precipitation Process

Precipitation from solution occurs in two stages: the nuclei formation and their further growth driven by the supersaturation of chemicals. A control on these two processes decides the properties of the resultant product. Tiny crystals grow from precipitation by the transfer of solute from the solution, forming an embryo. Eventually the radius of the embryo, formed from the solution, grows above a critical radius given by:

$$r_c = \frac{2\gamma v_l}{kT \ln\left(\dfrac{p}{p_0}\right)} \qquad (3.4)$$

Where γ is the specific surface energy of the cluster, v_l is the volume of the molecule in the liquid drop, k is the Boltzmann constant, T is the absolute temperature and p_0 and p are the saturated and supersaturated vapor pressures, respectively. When the matter is transferred from liquid state with higher chemical potential (μ_l) to solid state with lower potential (μ_s), the change in molar Gibbs free energy ($\mu_l - \mu_s$) becomes negative, which is a condition for supersaturation to form the nucleus. In terms of vapor pressure and surface energy, nuclei formation requires the surmounting of an energy barrier during the transfer from the solution to solute, which is given by a change in Gibbs free energy:

$$\Delta G_n = 4\pi r^2 \gamma - \frac{4}{3}\pi r^2 \Delta G_v \tag{3.5}$$

Where ΔG_v is free energy changed due to the transfer from vapor to liquid given by $\frac{kT}{\upsilon_l}\ln\frac{p}{p_0}$,

$$\Delta G_n = 4\pi r^2 \gamma - \frac{4}{3}\pi r^2 \frac{kT}{\upsilon_l}\ln\frac{p}{p_0} \tag{3.6}$$

Where p/p_0 is the supersaturation ratio S. An increase in this supersaturation ratio (S>1) decreases the free energy activation barrier height given by $\Delta G_c = \frac{4}{3}\pi r_c^2 \gamma$ that results in the acceleration of the nucleation rate to form critical nuclei of liquid droplets. Concentration above the minimum supersaturation level results in the nuclei formation. To initiate this nuclei formation, a precursor solution can be supersaturated by various ways such as evaporation, cooling, heating, addition of an external component, applying pressure and induced by reaction.

However, the continuous growth of nuclei in a supersaturated solution is achieved by the transport of solute matter from the solution to the particles. Transport through the solution is followed by absorption on the crystal, surface diffusion and incorporation into the crystal surface. Controlling these processes can limit the growth rate of crystals. To get monodisperse powders from the solution with a narrow size distribution, one needs to refer to the LaMer diagram that depicts the concentration of solute with time during nucleation and growth of particles from the solution [25]. This diagram in Figure 3.6 defines an increase in solute concentration with time, which leads the solute to be at saturated concentration and eventually to a state of supersaturated concentration, which initiates the nuclei formation and growth.

Once the nuclei starts growing, solute concentration in the solution goes down and further growth occurs by the diffusion of solute from the solution. When the concentration goes further below a solubility limit, the crystal growth eventually stops. In order to achieve ultra-fine monodispersed powders, nucleation should be allowed to happen in a narrow time range when the concentration of solute is within supersaturated

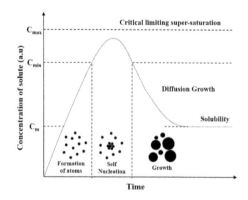

FIGURE 3.6
LaMer diagram depicting the nucleation and growth process of nanocrystals in solution.

condition. This can be practically achieved by using a solution with a low concentration of reactants and by releasing the solute slowly by allowing the solute to diffuse into the particles.

Nano-sized powders of Y_2O_3, Lu_2O_3 and YAG are prepared by the precipitation method to obtain transparent ceramics for laser applications. Such powders with good crystallinity, homogeneity and high purity with a single phase have been prepared at lower temperatures compared to the solid-state reaction method. Even though Ikesue et al. [6]. made the first laser ceramics using powders of sizes less than 2 µm produced by the solid-state reaction method, Konoshima Chemicals adopted a co-precipitation process to synthesize ultra-fine powders of Nd:YAG at a lower temperature. In chloride-based precursors, the chlorides of yttrium (YCl_3), aluminum chloride ($AlCl_3$) and neodymium chloride ($NdCl_3$) are dissolved in water and mixed well, added slowly and mixed with an aqueous solution of either ammonium sulfhate (($NH_4)_2SO_4$) or ammonium hydrogen carbonate (NH_4HCO_3) of a given concentration.

In the case of the modified urea precipitation method, an aqueous solution of urea is also added, maintaining a ratio of urea:Y^{3+} ions as 6.25 [26]. The wet chemical synthesis temperature was maintained at 95°C, and the obtained precipitate was separated and calcined in air above 1200°C to end up in sub-micron YAG oxide powders. The co-precipitation process of YAG powder and their later stage fabrication is given in the Figure 3.7.

Powders with a particle size below 100 nm have been synthesized by this technique by several researchers. Sarath et al. have studied the effect of different precursors and precipitant on the morphology and sinterability of YAG nanopowders produced by co-precipitation process [27]. Use of yttrium nitrate instead of yttria as the precursor chemicals and ammonium hydrogen carbonate instead of ammonia water yields soft agglomerates of near-spherical shaped particles that are more desirable for transparent ceramics because of their highly sinterable nature. In addition, powder can be produced by this process at a relatively low crystallization temperature, with greater yield and good reproducibility at a lower cost. Ceramics produced by co-precipitated powders are used to significantly increase the output power of lasers that made transparent ceramics eventually an alternative for laser crystals [26].

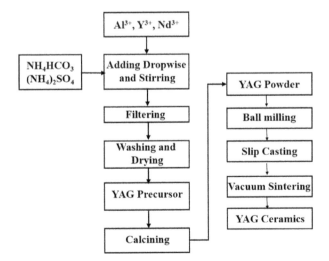

FIGURE 3.7
Schematic on the fabrication of nanocrystalline Nd:YAG ceramics [26].

3.4.2.2 *Other Solution-Based Processes*

In addition to the precipitation technique, other solution-based processes have been used to synthesize ultra-fine powders for transparent ceramics. The sol-gel process has the unique advantage of producing monodisperse free-flowing particles with spherical morphology at a cheaper production cost over other chemical processes such as hydrothermal or combustion synthesis methods. This process involves polymeric gel formation by hydrolysis followed by condensation and gelation. By drying and grinding the polymeric gels, which act as a steric barrier for nanoparticle formation, fine oxide powders such as Y_2O_3 of 50 nm particle size are reported. In another study, the yttria source chemicals ($Y(NO_3)_3.6H_2O$) were mixed with anhydrous citric acid along with ethylene glycol that formed polymeric complexes which were further dried and calcined to obtain ultra-fine powders of yttria [28]. Apart from yttria, this synthesis protocol is well established for several oxide particles as well.

In the case of the hydrothermal process, an aqueous solution of metal salts such as oxides or hydroxides is taken in an autoclave and subjected to both temperature and pressure in order to control the purity and morphologies of particles. Transparent yttria ceramics have been tried by spark plasma sintering of 50 nm Y_2O_3 particles prepared by hydrothermal process [29]. The combustion process is an exothermic reaction to produce oxide powders of high purity and crystallinity at a short reaction time between metal sources and fuel subjected to high temperature with fast heating rates. Nanopowders of spinel, YAG, LuAG and $Y_2Zr_2O_7$ have also been synthesized by the combustion process with the aim of developing transparent ceramics. However, in an industrial scale, co-precipitation and solid-state reaction techniques are preferred in comparison to other processes that produce powders for transparent ceramics.

3.5 Role of Additives in Transparent Ceramics

3.5.1 Dispersant

Most of the powder synthesis processes from the perspective of transparent ceramics use special additives, either organic or inorganic materials, to serve as dispersants, binders or sintering aids. The amount and choice of additive should be made in a way that it does not form a separate solid phase that eventually affects the optical quality of the end product. Each one serves different purposes and plays a vital role in their respective processes. Generally ceramic formation from wet methods is relatively easy because of uniform and dense packages of particles that form green bodies due to the relatively easy flow of material in a wet state (slurry) compared to the dry packing of powders. During the aging process of slurry in the colloidal method, the increased viscosity of slurry leads to the formation of aggregates by rapid sedimentation during collision. In order to stabilize the slurry, the critical content of dispersant deflocculates the particles and reduces the agglomeration tendency through the repulsion force between particles having same charge, i.e., by electrostatic stabilization. An ideal dispersant should help to achieve ceramic suspensions with a well-dispersed and highly concentrated slurry with a maximum solid content. In the case of colloidal processing of yttria ceramics, several dispersants such as polyethylenimine (PEI), polyacrylic ammonium acid (PAA), Dolapix CE64 and triammonium citrate (TAC) have been employed. A maximum solid loading of up to 35 vol% of ultra-fine yttria powders in the suspension has been achieved using TAC and tetramethyl ammonium hydroxide (TMAH) as dispersants. For YAG, dispersants such as PAA, [30] Duramax D3021 and Dolapix PC21 and 75 have been used, in which 50 vol% of solid loading was achieved using 0.40 wt% Dolapix PC21 [31]. The quantity of dispersant depends on the rheological properties of different materials, thus, the variation of viscosity with concentration of the material. Even a small increase in dispersant above the threshold decreases the zeta potential value of slurry particles leading to a poor package and densification [30]. A solid loading of 45 vol% of $MgAl_2O_4$ has been achieved using TMAH and Duramax D-3005 as dispersants [32]. For alumina, menhaden fish oil is also successfully used as dispersing agent.

3.5.2 Binder and Plasticizer

To further assist the powder compaction process, especially in casting, like slip casting or gel casting of slurry, a binder such as vinyls, acrylics and glycols is preferentially added, mostly as a solution to increase the compact strength. Binders are mostly organic materials such as polymers with long chains and higher molecular weight compared to dispersants, which have short-chain polymers and lower molecular weight. Such polymer-based binders form bridges between the particles to enhance the strength of the green body. For alumina particles, a mixture of polyethylene and paraffin wax was used as the binder. Plasticizers are sometimes added along with binders with a main purpose to soften the binder. Plasticizers have a lower molecular weight than the binder and are used to break the long chains of binders thereby softening them. For transparent YAG ceramics made by the tape casting method, polyethylene glycol (PEG-400) is used as a plasticizer whereas polyvinyl alcohol (PVA) was used as the binder. Different plasticizers at different quantities have been investigated to achieve suitable viscosity of YAG slurries [33]. In the case of transparent spinel ceramics such as MgAlON, polyvinyl butyral (PVB) was used as binder whereas PEG was used as plasticizer. The most important aspect is to remove these organic additives before the final stage of sintering. Thus, when used in low concentration, the binders and plasticizers can be easily burned out from the cast at low temperature and time, whereas at a higher concentration, binder removal causes microstructural inhomogeneities. Several processes such as wicking, solvent extraction and the most common binder debinding/decomposition can achieve the removal of binders by thermal treatment [23]. Binders are therefore chosen with good debinding properties along with substantial rheological and chemical properties.

3.5.3 Sintering Aids

Sintering aid is an important functional additive and is preferentially added with the precursors to enhance effective densification during the sintering process. The first laser quality Nd:YAG ceramics was made with ethyl silicate (TEOS) with 0.14 wt% of SiO_2 as sintering aid [6]. Different ceramics use different sintering aids to promote sintering: MgO for transparent alumina, TiO_2 for ZrO_2 ceramic, LiF for Y_2O_3 ceramics, B_2O_3 for spinel ceramics, Y_2O_3 for AlON ceramics and so on. Similar to the additives mentioned earlier, the choice and amount of sintering aids for different ceramics are important as they influence the final microstructure that affects the optical and thermo-mechanical quality of the product. For example, yttria ceramic fabrication is challenging without sintering aid due to the presence of oxygen vacancies, which lead to abnormal grain growth during sintering at a high temperature. In order to avoid such large grains which contain large number of pores, sintering aids such as ThO_2, ZrO_2 and HfO_2 are used in a small percentage to limit the grains smaller in size by a solute drag mechanism. Unfortunately, the addition of an ZrO_2 additive reduces the thermal conductivity of the Y_2O_3 ceramic and it also suffers from poor chemical stability due to the charge difference between Zr^{4+} and Y^{3+}. Recently, an optimal amount of ZrO_2 (0.8 wt%) and MgO (0.075 wt%) were found to end up in high-quality Y_2O_3 ceramics without affecting thermal conductivity [34]. In many cases, sintering additives end up along the grain boundaries that form a thick layer at a higher concentration, making the ceramics opaque. Si_3N_4 ceramics with 2 wt% of Y_2O_3 sintering aid reduced to 31% transparency, and a further decrement up to 0% transparency in the presence of 12 wt% Y_2O_3 content was noted [35]. LiF at 0.25 wt% was found to be an optimum level for Al_2O_3 rich spinel ceramic [36]. For garnet ceramics, MgO and SiO_2 are the most commonly used sintering aids. The difference in valence, for example MgO in LuAG, can be compensated by further air annealing of the sintered ceramics. For YAG, even though other sintering aids such as CaO, MgO, B_2O_3–SiO_2, TEOS–MgO have been attempted, TEOS as a SiO_2 source is widely used to produce laser quality YAG ceramics fabricated by the solid-state reaction method. Stevenson et al. intensively studied the effect of SiO_2 on the densification, microstructure and optical quality in YAG ceramics [37]. According to thermodynamic modeling, SiO_2 has been long believed to form a liquid phase in the YAG system that enhances the diffusion mechanism that results in a high densification rate. But as later results showed, SiO_2 forms a solid solution with YAG instead of settling as a separate liquid phase. Direct substitution of Si^{4+} in the Al^{3+} site in the YAG matrix increases grain-boundary diffusion and, hence, eventually densification. In addition, because of the smaller ionic radius of Si^{4+} (0.04 nm) compared to Al^{3+} (0.054 nm), lattice expansion can be avoided when dopant ion is added. The most accepted defect mechanism possible during this substitution is represented using the Kroger–Vink notation as [38]:

$$SiO_2 + \frac{5}{6}Al_{Al}^x + \frac{1}{2}Y_Y^x \rightleftharpoons Si_{Al}' + \frac{1}{3}V_Y'' + \frac{1}{6}Y_3Al_5O_{12} \qquad (3.7)$$

Where Si_{Al}' represents Si^{4+} replacing Al^{3+} preferably in its tetrahedral site bringing an excessive positive charge, V_Y'' means Y^{3+} vacancy. Substitution of Si^{4+} in Nd:YAG system increases vacancies in Y^{3+}/Nd^{3+} sublattice which in turn increases the diffusion coefficient and thereby the densification rate. It has been found that an increase in SiO_2 leads to full densification even at a relatively low temperature. Polycrystalline ceramics show 99.9% density at 1,600°C when 0.28 wt% SiO_2 is added whereas samples doped with 0.035 wt% of SiO_2 require temperatures above 1750°C. Interestingly, ceramics with high silica content result in larger grain growth by coarsening mechanisms. Coarsening occurs when the grain growth rate (dG/dt) given by D_{gb}/G^2 dominates over densification due to an increase in grain boundary diffusion (D_{gb}) with an increase in SiO_2 [37]. Therefore, small concentration (0.14 wt% or below) of SiO_2 is recommended to obtain fine grain YAG ceramics. In addition to its effects on grain size, TEOS also affects optically in some cases, such as Cr:YAG or Yb:YAG, where the conversion efficiency of Cr^{4+} and Yb^{3+} are severely affected by the SiO_2 addition, respectively. Recently divalent ion-based additives such as CaO and MgO have been found to improve the optical quality with moderate grain growth at Ca/Mg=1:4 molar ratio [39]. As can be seen, the influence of sintering aids on the microstructure and optical quality of ceramics as well as the choice of additive for different applications is still critical and an area of interest for several research groups who are working on transparent ceramics.

3.6 Conditions of Forming

In addition to the choice of precursors, additives, powder synthesis methods and conditions, compaction or packing of powders is another crucial step to make green bodies before the final sintering, whose efficiency mainly depends on the early stage of powder packing efficiency. Despite several powder consolidation techniques, the prime uniaxial pressing, cold-isostatic pressing, casting and *in situ* (spark plasma sintering and hot isostatic pressing) sintering are predominantly focused in the perspective of the development of high-density transparent ceramics. However, it is essential to remember that the compaction also depends on both the physical and chemical characteristics of the particles, not only fabrication protocols.

3.6.1 Pressing of Powders

Uniaxial compaction using die is widely used for green body formation that involves the steps such as filling die with free-flowing powders, compaction and ejection of the green body. During uniaxial compaction, as illustrated in Figure 3.8a, the dry powder is compressed and density increases through the closing of intergranular pores by rearrangement of granulated powders followed by granule deformation and densification. Granules are generally obtained by the spray drying of slurries along with additives as mentioned earlier. Spray-dried granules have uniform morphology and improved particle flow compared to aggregates or submicron powder, and they also undergo plastic deformation as the pressure is increased. If the granules are hard, rearrangement is easy but difficult to deform, whereas soft granules suffer from deformation without rearrangement that leads to poor packing. Granules with medium hardness (soft agglomerates) that can deform and rearrange easily in the applied pressure are preferred. When the applied pressure is released, stored elastic energy in the inorganic powders or trapped air result in a springback effect along the axial direction. This leads to a decrease in the resultant density over a critical pressure due to pressing defects such as bulging, delamination, end capping and cracks that cannot be removed by sintering. Organic additives such as binders and plasticizers help improve the green body strength and efficient compaction. In addition, lubricants are also added either internally with slurry or externally on the inside walls of the die to avoid die friction. Green bodies formed by such uniaxial pressing suffer from inhomogeneity of load distribution due to the low packing density near the die walls compared to the center. Radial stress is different from shear stress under uniaxial compaction which results in non-uniform packing in the green body.

The packing of powders by isostatic compaction, either wet bag or dry bag, avoids any stress difference along axial and radial directions because of uniform hydrostatic pressure applied from all directions. In the wet bag pressing method, a flexible rubber mold is filled with powders, and the mold is immersed in a pressurized liquid medium such as oil or water. The mold is taken out to remove the pellets, whereas in the case of dry bag pressing, the mold is fixed in the vessel. If the pressing is done at room temperature, it is known as cold isostatic pressing (CIP), whereas hot isostatic pressing (HIP) occurs at a high temperature using a pressurized gaseous medium such as nitrogen or argon. Compared to uniaxial compaction, CIP is used to form complex shapes and to achieve relatively high-density green compacts. A typical schematic of the CIP is demonstrated in the Figure 3.8b. Most YAG transparent ceramics employ a uniaxial pressure of 20 MPa followed by cold isostatic pressure of 200 MPa. Andreas Krell and Jens Klimke investigated the consolidation

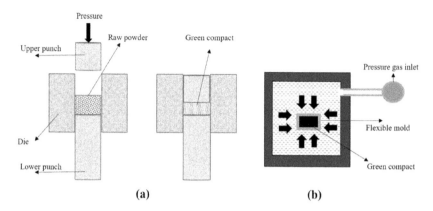

FIGURE 3.8
Schematic representation of (a) uniaxial pressing and (b) cold isostatic pressing process.

protocol of transparent alumina, where they demonstrated uniaxial pressure of 200 MPa with larger pores of 75–200 nm and further cold isostatic pressure of 700 MPa eliminated the pore effectively [40]. Apart from the pressing technique, casting techniques such as slip casting and gel casting of the same powders reduce the size of pores below 50 nm and exhibit better homogeneity and density.

3.6.2 Casting of Powders

Casting methods are based on colloidal systems in which particle consolidation is achieved by removing the liquid from the slurry through a porous medium, like gypsum. There are several casting methods such as slip casting, gel casting, tape casting and pressure casting being employed to achieve uniformly packed green bodies with high density even with complex shapes. As discussed earlier, slurry with high solid content and good rheological properties is desirable to achieve homogenous packing at a high casting rate because suspension with high concentration does not suffer from significant shrinkage during drying. In an industrial scale, slurries are made out of liquids having a high evaporation rate such as toluene, ketones, ethanol and trichloroethylene. In addition, binders are added to the slurry in order to control the rheological properties, such as an increase in the viscosity with the increase in concentration. This is where the plasticizers are helpful in softening the binders to increase the flexibility of the green body. In the slip casting method, liquid is removed by both the mold and consolidated layer. The flux of liquid removed by flowing through a porous medium is given by Darcy's law:

$$J = \frac{K(\mathrm{d}p / \mathrm{d}x)}{\eta_L} \tag{3.8}$$

Where K is the permeability of the porous medium, $\mathrm{d}p/\mathrm{d}x$ is the pressure gradient in the liquid and η_L is the viscosity of the liquid.

The capillary suction pressure of the mold leads to the pressure gradient that causes the liquid flow. Typical slip casting processing steps are shown in Figure 3.9. Slip casted transparent ceramics are limited to small thickness due to a reduction in consolidation rate over time. This consolidation rate can be increased if the permeability of the cast K_c is increased. For monodisperse particles with spherical morphology, K_c is given by:

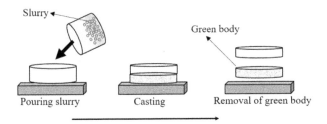

FIGURE 3.9
Schematic representation of slip casting processes.

$$K_c = \frac{d^2(1-V_C)^2}{180V_c^2} \tag{3.9}$$

Where d is the particle diameter and V_C is the volume fraction of solids in the cast. By varying solid content in the slurry, permeability can be tuned to control the consolidation rate. Other factors such as optimum pore size of the mold, particle size in the slurry and slurry temperature all play a major role in efficient consolidation by slip casting.

Tape casting, widely used in solid oxide fuel cells, is popularly known as the doctor-blade or knife coating method in which the slurry is spread over a surface using a controlled blade (doctor blade). As can be seen in Figure 3.10, the slurry is poured in the doctor blade assembly and the blade is moved over the surface to form a tape cast of the slurry. The surface is mostly covered with plastic or a removable sheet of paper and mainly used to make thin sheets of material with thickness between 10 μm to 1 mm. By consolidating several thin tape casted sheets, bulk ceramics can be made, especially with a unique advantage of fabricating multi-layered composite ceramics. A typical process of tape casting and the doctor blade assembly is shown in Figure 3.10.

Laser quality layered YAG ceramics with a wide doping profile have been made with the tape casting technique which has been found to be more cost effective than the conventional diffusion bonding technique for making such ceramics [41]. Similar to slip casting, the choice of solvent in the slurry, binder and solute concentration play a major role in achieving ceramic sheets of different thickness with full density. There are other casting methods, such as pressure casting to enhance the consolidation process by applying pressure and gel casting in which the slurry is dispersed with a monomer and then poured into a mold to form the cast. Monomer limits the movement of particles, thereby avoiding any particle segregation, and is helpful in fabricating thick ceramics. Transparent ceramics of YAG and AlON have been successfully fabricated using a gel casting technique.

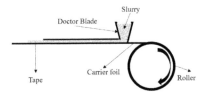

FIGURE 3.10
Tape casting of ceramics from slurry made of ceramic particles.

3.7 Sintering Protocols

Sintering is the final and crucial step in achieving complete densification by fusing particles together to attain a desirable microstructure. When the particles are joined together, pores are formed both inside the grains (intra-crystalline) and along the grain boundaries (inter-crystalline). Pores experience different optical properties as they have a different refractive index ($n_{air} = 1$, $n_{YAG} \sim 1.82$ @ 900 nm, $n_{YAP} \sim 1.89$ @ 900 nm, $n_{Al_2O_3} \sim 1.75$ @ 900 nm, $n_{MgAl_2O_4} \sim 1.70$ @ 900 nm) value compared to surrounding grains, which lead to the refraction of incoming light in different directions with respect to incident direction. Opacity of ceramics is due to a large number of pores. Average pore size is dependent on average particle size (pore radius ∝ particle radius) which in turn makes nanocrystalline powders more desirable for transparent ceramics. Green bodies made of such powders can also be quickly densified because of the higher mobility of nano-sized pores during sintering. A reduction in surface free energy of consolidated mass of particles is the driving force of sintering. Since a curved surface always wants to become flat, the curved surface of powders having different pressures is assumed as the primary driving force. The pressure difference between the inside and outside of a curved surface, ΔP, is related to surface free energy (γ) as $\Delta P = 2\gamma/r$, where "r" is the radius of curvature which is positive when the center of the radius is inside the material and vice versa [42]. For spherical particles of radius "a," decrease in surface free energy is given by:

$$E_s = \frac{3\gamma_{SV}V_m}{a} \tag{3.10}$$

Where V_m is molar volume and γ_{SV} is surface energy per unit area of the particles. It can be seen that the smaller the particles, the larger the driving force for sintering with a decrease in E_s. The driving force can be enhanced by applying pressure like in the case of hot pressing and hot isostatic pressing. Along with temperature, uniaxial pressure is applied to the sample in a die that is mostly made of graphite because of its low cost and creep resistance at high temperature. Removal of porosity can be achieved by several matter transport mechanisms as shown in Figure 3.11.

This implies the probable modes of matter transport from grain volume, grain boundaries and from outer surface. Among them, diffusion from outer surface dominate the matter transport to shrink the pores. The diffusion of atoms or ions is the predominant mechanism of matter transport due to the various structural defects like point defects, line defects and planar defects present in crystalline solids. The defects can be intrinsic such as the Schottky or Frenkel defects, which are related to lattice site occupancy of ions or extrinsically influenced by the sintering atmosphere (partial pressure of the gas) or dopant ions. Therefore, several variables such as temperature, gaseous atmosphere and dopant concentration plays a major role in the diffusion of species. In crystalline solids, Fick's first law defines the flux of diffusing species as:

$$J = -D\,dC/dx \tag{3.11}$$

Where dC/dx is the concentration gradient and D is diffusivity. For materials with time independent composition, D is a temperature-dependent material parameter to define the rate of diffusive mass transport. The negative sign represents the diffusion direction toward lower concentration. In most of the cases, the concentration is time dependent (dC/dt) and is given by Fick's second law:

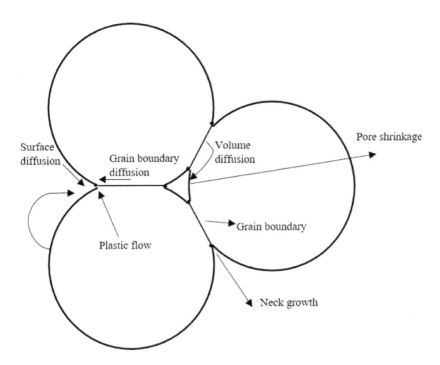

FIGURE 3.11
Different diffusion paths of matter. (GB is grain boundary, SD is surface diffusion, VD is volume diffusion, PF is plastic flow, E-C is evaporation–condensation, P is pore.)

$$\frac{dC}{dt} = D\frac{d^2C}{dx^2} \tag{3.12}$$

An atom has to overcome an energy barrier called activation energy to jump from one lattice site to another by the diffusion mechanism. The schematic representation of the diffusion and the activation energy is shown in Figure 3.12. To surmount this high-energy intermediate site between the species, external temperature or pressure is applied to the compacts. Diffusion can be a lattice diffusion, grain boundary diffusion and surface diffusion. Lattice diffusion occurs due to the movement of the point defect either by the vacancy site or interstitial site present in the bulk of the lattice. Diffusion increases with an increase in the number of vacancy sites in the lattice. Grain boundary (GB) diffusion is faster than lattice diffusion due to the disorder nature of grain boundary and needs less activation energy.

GB diffusion is dependent on grain size because smaller grain sizes result in more grain boundary volume. The defect nature of surfaces of crystalline solids causes surface diffusion. In addition to the concentration gradient, matter transport can also be defined as the flux of diffusing species over the gradient of chemical potential which is the Gibbs free energy per mole at the temperature and pressure in question. This results in a modified version of Fick's law in terms of chemical potential as:

$$J_x = -L_{ii}\frac{d\mu}{dx} \tag{3.13}$$

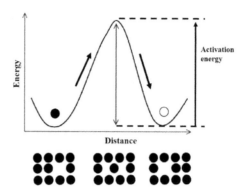

FIGURE 3.12
Illustration of the activation energy for vacancy diffusion.

Where L_{ii} is the transport coefficients and μ is the chemical potential. Flux of species can be defined separately for atoms, vacancies in terms of chemical potential. Detailed theory can be found elsewhere [23]. Herein, different types of sintering such as solid-state sintering, liquid phase sintering, spark plasma sintering, vacuum sintering and hot isostatic pressing are discussed that are preferentially used for the fabrication of transparent ceramics.

3.7.1 Solid-State Sintering

Solid-state sintering occurs by the bonding and growth of necks between adjacent particles due to mass transport from regions of higher chemical potential to regions of lower chemical potential. During sintering, GB diffusion results in densification, whereas vapor diffusion, lattice diffusion and vapor transport can lead to non-densifying mechanisms such as coarsening (grain growth) with neck growth without densification. Since most of the properties such as fracture strength are improved for smaller grain sizes, control of grain growth is important for potentially engineered or technical ceramics. Grain growth due to grain boundary diffusion was predicted by Burke and Turnbull as a parabolic grain growth law given by:

$$G^2 - G_0^2 = Kt \tag{3.14}$$

Where G is the grain size at time t, G_0 is the initial grain size and K is the temperature dependent growth factor given by $K = 2\alpha\gamma_{gb}M_b$, where α is boundary shape dependent geometrical constant, γ_{gb} is specific grain boundary energy and M_b is boundary mobility. Experimentally, grain growth does not follow this parabolic law which is then modified into the general grain growth equation given by:

$$G^m - G_0^m = Kt \tag{3.15}$$

Where, value of m is between 2 and 4. Most practical grain growth follows an exponent value of $m = 3$ that defines lattice diffusion. Exponent value of $m = 2$ and 4 defines vapor transport and surface diffusion, respectively. Densification without coarsening is desirable for high-density ceramics with controlled grain size which requires an increased densification rate over coarsening rate). Hot pressing that increases the driving force for densification with small grain size can suppress coarsening. Pinning of grain boundaries is also used to inhibit grain growth. This has been achieved by the addition of solid phase inclusions or dopants, for example, MgO additions in Al_2O_3 limit grain growth of alumina. The presence of inclusions along the grain boundary limit curvature and boundary mobility by solute drag effect. Limiting grain size d depends on the particle size of the inclusion d_i and their volume fraction fd_i [42].

$$d = d_i / fd_i \qquad (3.16)$$

However, an optimum amount of dopant is only beneficial to restricting the abnormal grain growth, including uniform microstructure, otherwise, excessive dopants facilitate the formation of the secondary phase along the grain boundaries and reduce the resultant optical and mechanical properties.

3.7.2 Liquid Phase Sintering

Sintering of powder in the presence of a liquid phase is coined as liquid phase sintering (LPS) that enables the easy rearrangement of particles to enhance densification rates along with an accelerated grain growth phenomenon known as the Oswald ripening process. At the sintering temperature, the liquid phase (~25–30 vol%) immediately rearranges particles due to surface tension forces of the liquid bridge applied to the particles. Because of enhanced rearrangement and matter transport through the liquidus media, liquid phase sintering is very effective for materials that are difficult to densify by solid-state sintering only. Figure 3.13 shows the plausible liquid phase sintering mechanisms.

Materials such as non-oxide Si_3N_4 with a high degree of covalent bonding are preferentially sintered with MgO as a sintering additive that enhances the formation of the liquid phase at sintering temperature. Reduction of the liquid–vapor interfacial area provides the driving force for densification of the system. The number of liquid phases influences the shape of grains and also grain size. During the sintering, liquid fills small pores first followed by large pores. Following the initial rearrangement of particles because of this, solution precipitation leads to further densification and coarsening. Coalescence of small grains with large grains is part of densification accompanied by Oswald ripening. For LPS, the additives are selected using phase diagrams in a way that it forms a liquid with the major component at a eutectic temperature well below the sintering temperature. For Si_3N_4 ceramics, MgO additive forms a liquid phase with SiO_2 present on the particle surface at a eutectic temperature of 1550 °C, whereas the sintering temperature is more than 100–300°C above this point.

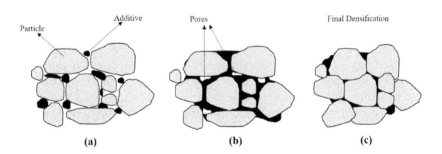

(a)   (b)   (c)

FIGURE 3.13
The role of liquid phase in sintering the solid material.

3.7.3 Vacuum Sintering

Vacuum sintering is a more popular technique for laser quality transparent ceramics including the first YAG ceramics in which the sintering process is carried out at a high vacuum level ($\sim$10^{-6} Torr). Vacuum sintering avoids the problem of gases being trapped which are otherwise hard to be removed if present inside the grains. Such gases can cause a scattering of light, reducing the transmittance of the ceramics. In the vacuum-based sintering processing of garnets like YAG, powders of Y_2O_3 and Al_2O_3 are mixed well with a small amount of TEOS using ball milling and compacted by both uniaxial and isostatic pressing at a respective pressure of 20 and 250 MPa. The compacts are then sintered at a temperature of $\sim$1750°C under a vacuum of 5×10^{-6} Torr. Because of the vacuum environment, the oxide specimens end up with a lack of oxygen that cause coloration of ceramics due to oxygen vacancies. This can be fixed by further annealing of the ceramics in the air at 1450°C for 24 hours. Grain size of the ceramics increases with sintering time and temperature. Rapid vacuum sintering can suppress this coarsening while maintaining the densification.

3.7.4 Spark Plasma Sintering

Spark plasma sintering (SPS) is relatively new field-activated sintering method to achieve rapid densification rates using rapid heating along with uniaxial pressing. Rapid heating is provided by passing an electrical current through the die that contains the compacts or the powders. With the ability to sinter the ceramic powder quickly, full density can be achieved with nanostructured grains at shorter sintering times and temperatures compared to conventional sintering techniques. Especially for ceramics made from nanopowders, rapid shrinkage occurs initially during SPS which is then followed by plastic deformation of nanoparticles due to the applied pressure along with high temperature. This helps maintaining nano or micro-structures consisting of nanopores located at the grain boundaries and it can be eliminated further without much coarsening of Yb^{3+} doped CaF_2 transparent ceramics. This class of material have also been made through the SPS technique at a relatively lower sintering temperature and time with an average grain size of 130 nm [43]. In recent times, several other transparent ceramics have been fabricated with the SPS technique. Nevertheless, this technique is not commercially popular because of size and processing limitations. Similar to this SPS technique, hot pressure sintering helps to suppress the grain growth due to the applied pressure along with relatively low sintering temperature. There are other sintering methods such as hot isostatic sintering, microwave sintering and electric discharge sintering that have also been used to fabricate transparent ceramics.

3.7.5 Hot Isostaic Pressing (HIP)

Another effective way of achieving transparency without much grain growth is hot isostatic press (HIP) sintering. In general, HIP is the final densification process of transparent ceramics in which the presintered/sintered samples are subjected to experience high isostatic gas pressure as well as elevated temperature in a metallic high-pressure containment vessel. An isostatic pressure of 200 MPa can increase the driving force by $\sim$50 times than that of normal pressureless sintering. Due to the driving force achieved by high pressure, the samples can be sintered at a relatively lower temperature that can help restrict the grain growth. Lee et al. fabricated transparent ceramics of Nd:YAG with smaller grain size

(2–3 μm) using HIP-assisted sintering [44]. Recently, MgAlON transparent ceramic prepared by the HIP process shows very fine microstructure and thereby superior mechanical properties such as better flexural strength and hardness. Flexural strength of presintered MgAlON ceramics was around 232 MPa whereas HIP treatment increased the flexural strength to 268 MPa [45]. HIP has been successfully used to fabricate transparent ceramics of YAG, TAG, $MgAl_2O_4$, Al_2O_3, sesquioxides, YSZ, MgO, $La_2Hf_2O_7$ and so on. The HIP process has its own limitations, such as high cost, and it demands highly dense (>98%) sintered samples before subjected them to HIP treatment. For instance, HIP treatment remains ineffective for alumina ceramics when sintered porosity was greater than 8% [46]. The transparency of the final HIP-treated ceramic depends on the pre-sintering conditions. Also, in many oxides, HIP treatment yields a dark coloration because of the high temperature treatment in inert atmosphere, which requires an additional high temperature annealing in air.

3.8 Transparent Ceramics—Applications

The first translucent alumina was developed in General Electric (GE) by Robert L. Coble and used as an envelope for sodium vapor lamps in the early 1950s. Over the last few decades, transparent ceramics gained popularity for several applications such as solid-state lasers, scintillators, optical elements, ceramic armors, windows and even medical implants, especially as a replacement of single crystals in many cases. The important applications of ceramics that made significant breakthroughs in the field of lasers, scintillators and armor windows are summarized.

3.8.1 Ceramic Lasers

In solid-state lasers, the gain media acts as an amplifying media that implies larger gain media always facilitate a high degree of amplification. Due to the limitations of growing large size single crystals and poor thermal conductivity of large size glass gain media, transparent polycrystalline ceramics find themselves as an alternative laser gain media that can be made in large dimensions at a relatively cheap cost, at a lower temperature and in a short time. After Ikesue et al [6]. first demonstrated ceramic lasers and found them to exhibit similar properties to that of single crystals in 1995, in 2002 Lu et al [26]. achieved kW power using Nd:YAG ceramic and then 100 kW power was reported by Northrop Grumman Corp. in 2009. Such results eventually ascertained the possibility of replacing single crystals with ceramics for high-power lasers, microchip lasers and even ultrashort lasers. J. Sanghera et al. have given brief review on ceramic laser materials and their evolution [47]. In addition to large dimensions, ceramic technology also makes it possible to develop composite ceramic lasers with multiple layers to improve thermal conductivity of the overall gain media during laser operation. In the case of single crystals, such composite media is generally made by costly diffusion bonding whereas in the case of ceramics, this can be done just by a layer-by-layer approach of stacking layers with different compositions. Thermal conductivity drops with the addition of dopant ions in YAG and the ceramics are designed in a way to handle high pump power for high-power laser applications. During pumping, the front face of the ceramic experiences a high heat load from the pump and it is desirable to use an undoped cap in the front face followed by doped regions. The doped regions can vary in dopant concentration to handle the thermal load of the pump beam. This can be realized by using either multi-slab ceramics with different concentration or by graded doped ceramic layers starting. The layers are arranged with an undoped YAG end cap followed by lower dopant concentration layers and then higher dopant concentration layers. In this manner, one can avoid severe thermal loading in the front face that actually cause damage to the gain medium. By stacking several layers made by tape casting, graded ceramics can be made as a monolith with good optical quality and laser performance [48]. Recently, high average laser power of more than 1000 W has been achieved in the HiLASE Research Center using multi-slabs cryogenic gas cooled Yb:YAG ceramics (100 mm square) with Cr:YAG cladding layer (10 mm wide), as demonstrated in Figure 3.14a. An average energy per pulse of over 100 J at 10 Hz was achieved using six square-shaped ceramic slabs (120 mm x 8.5 mm) with different Yb concentration. The slabs are arranged in a way with the lowest Yb concentration (0.4 at%) slab in the front face followed by 0.6 and 1 at% doped YAG slabs. Such arrangement helps to equalize the gain and heat load along the pump direction [22].

Unlike graded ceramics made by stacking layers as mentioned above, these slabs are separated by a small gap of 2 mm for efficient cooling. The main amplifier head containing six Yb:YAG ceramic slabs being used in HiLASE and the achieved pulse energy from these slabs are shown in Figure 3.14b.

3.8.2 Ceramic Scintillators

Scintillators are luminescent materials that emit visible light when irradiated with high-energy radiation. Solid-state scintillators with good transparency, high stopping power, more light yield with high resolution and fast response times are widely used for various applications such as medical imaging, high-energy physics, dosimetry and nuclear medicine. Different inorganic materials are studied for different radiation detection. For example, gadolinium-based garnets were found to be effective for gamma ray spectroscopy whereas Tl:CsI and $CdWO_4$ were used as x-ray detectors for CT scanners. Some of the most commonly used single crystalline scintillators such as $CdWO_4$, GdO_2S and $BaCl_2$ belong to non-cubic crystal structure. At the same time, garnet-based scintillators with cubic structure have also been found to be attractive for scintillator applications with an outstanding performance. A high-density compound like a cerium-doped $Gd_3(Al,Ga)_5O_{12}$ (GGAG:Ce) single crystal with good chemical and physical properties has been investigated as an

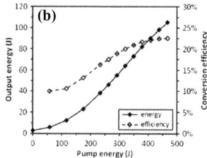

FIGURE 3.14

(a) Yb: YAG ceramic slabs used to achieve 100 J average energy per pulse in HiLASE; (b) the variation of conversion efficiency and output energy with respect to pump energy [22].

ideal material for Positron Emission Tomography (PET) and gamma ray detection applications. This class of material exhibit unique scintillation properties such as high light yield (55,000 photons/MeV) with an energy resolution of about 3.7% at 662 keV [49]. This cubic nature enables the material to be made in its polycrystalline form as well. Transparent composite ceramic of Ce:GGAG/Cr:YAG have been fabricated opening the possibility to tune the wavelength and density of the product for scintillation applications.

With the improved ceramic technology, even non-cubic scintillator ceramics have been made. Lutetium ortho silicate doped (Lu_2SiO_5; LSO) with cerium (Ce:LSO) belonging to a monoclinic structure are widely used in various industrial purposes, and their crystalline form has demand for PET and high-energy physics applications in research institutions. Ce:LSO have been made in transparent ceramic form by dispersing the ceramic particles and orienting by slip casting in a strong magnetic field (9 T) which is followed by hot-isostatic pressing (HIP), and the transmittance is shown in Figure 3.15 [50]. LSO:Ce ceramics are found to have almost the same quality as their single-crystalline counterpart. Similarly, $BaCl_2$ transparent ceramics have been achieved by converting the orthorhombic structure into a fine grained cubic structure by stress-induced displacive polymorphic phase transformation at a low sintering temperature of 450°C [51].

3.8.3 Ceramic Armor Windows

Glasses of more than five inches thickness are widely used as windows for aircrafts and armored vehicles, face shields and visors. Glasses have low wear resistance that lead to significant replacement cost. Single crystals of sapphire with high strength as a replacement to the glasses are limited to sizes due to crystal growth difficulties.

In order to overcome this situation, the polycrystalline $MgAl_2O_4$ spinel and aluminum oxynitride (AlON) with smaller grains have been proven to be better than sapphire single crystals for armor-related applications with no or minimal scratches during ballistic testing. Surmet Corporation is a pioneer in manufacturing large size (>18 × 35 inch) aluminum oxynitride transparent ceramic with high hardness and durability and has been proven as an efficient armor window and domes to be used in military application in extreme environment. A 1.3-inch-thick AlON transparent ceramic produced by Surmet is shown in

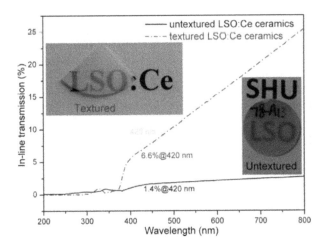

FIGURE 3.15
Transmittance of Ce:LSO textured and untextured transparent ceramics [50].

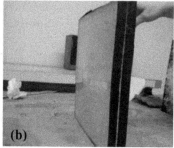

FIGURE 3.16
(a) The 1.3-inch-thick AlON transparent ceramic; (b) the exit side of 1.6 inch-thick AlON ceramic after being fired upon by 50 cal. M2AP rounds. The ceramic completely blocked the bullet from exiting [52 SPIE].

Figure 3.16a. The same transparent AlON restricts the motion of 50 cal. M2AP rounds after being fired, no damage is observed in the back side (Figure 3.16b). With their wide transparency from UV to mid-IR, AlON is useful in infrared-guided missiles with long-lasting performance compared to glasses. Transparent ceramics of lead lanthanum zirconate titanate (PLZT) and lead magnesium niobate titanate (PMN-PT) have been used for electro-optic applications because of their good electro-optic properties. Similarly, zirconia (ZrO_2) transparent ceramics can be used as lenses with high scratch and impact resistances.

3.9 Conclusions

Transparent ceramics have attracted tremendous attention among researchers and industries in a wide range of applications that require transparent materials with large dimensions along with properties similar to that of single crystals. With continuous progress in this technology, transparent ceramics have reached a level of being capable of replacing single crystals for several applications because of ease of fabrication at a fast production time in a cost-effective manner compared to time-consuming crystal growth processes. Relatively thin transparent ceramics are also potential candidates to replace thick glasses in several applications because of their superior thermo-mechanical properties over glasses. This chapter briefly discussed the scope and development processes of transparent ceramics. Several processing parameters and the possible mechanisms involved in obtaining good quality transparent ceramics have been discussed. Because of the importance of achieving the desirable microstructure that determines the property of the end product, the role of chemicals such as binders, dispersants, sintering aids and processing parameters such as temperature, pressure have been discussed. Different types of powder synthesis processes, green body-forming methods and sintering techniques have been briefly explained. Few applications of transparent ceramics in different fields were discussed. With the ability to make these ceramics in large dimensions, laser energy per pulse as high as 100 J @ 10 Hz has been achieved in HiLase using large aperture transparent ceramic laser gain media for the first time. Similarly, large-sized ceramic windows find unique applications as windows in armored vehicles and laser-guided missile domes. Research is being carried out to achieve non-cubic ceramics as well for applications such as scintillators and lasers. The brief information preludes how to obtain extreme dense components like transparent ceramics and this philosophy eventually can be employed to make a high-density compact of either traditional and advanced ceramics. Opaque ceramics result if the process condition fails to make transparent ceramics, and thus any intermediate-density opaque ceramics are categorized in between transparent ceramics and porous ceramics.

References

1. Coble, R. L. 1961. Sintering crystalline solids. II. Experimental test of diffusion models in powder compacts. *Journal of Applied Physics* 32(5):5793–799.
2. Goldstein, A., and Krell, A. 2016. Transparent ceramics at 50: Progress made and further prospects. *Journal of the American Ceramic Society* 99(10):3173–3197.
3. Geusic, J. E., Marcos, H. M., and Van Uitert, L. G. 1964. Laser oscillations in Nd-doped yttrium aluminum, yttrium gallium and gadolinium garnets. *Applied Physics Letters* 4(10):182–184.
4. Shawley, C. 2008. Optical and Defect Studies of Wide Band Gap Materials. PhD Dissertation, Washington State University. p. 124.
5. Hatch, S. E., Parsons, W. F., and Weagley, R. J. 1964. Hot—pressed polycrystalline $CaF_2:Dy^{2+}$ laser. *Applied Physics Letters* 5:153–154.
6. Ikesue, A., Kinoshita, T., Kamata, K., and Yoshida, K. 1995. Fabrication and optical properties of high-performance polycrystalline Nd:YAG ceramics for solid-state lasers. *Journal of the American Ceramic Society* 78:1033–1040.
7. Goldman, L. M., Balasubramanian, S., Kashalikar, U., Foti, R., and Sastri, S. 2013. Scale up of large ALON windows. In: *Proceedings of SPIE, Window and Dome Technologies and Materials XIII*, 8708 870804-1-15 (June 4, 2013), SPIE.
8. Bayya, S. S., et al. 2017. Recent developments in spinel at NRL. (Conference Presentation). In: *Proceedings of SPIE, Window and Dome Technologies and Materials XV*, 10179 101790H-1 (June 7, 2017), SPIE.
9. Stevenson, A. J., et al. 2011. Effect of SiO_2 on densification and microstructure development in Nd:YAG transparent ceramics. *Journal of the American Ceramic Society* 94(5):1380–1387.
10. Boulesteix, R., Maître, A., Baumard, J.-F., Rabinovitch, Y., and Reynaud, F. 2010. Light scattering by pores in transparent Nd:YAG ceramics for lasers: Correlations between microstructure and optical properties. *Optics Express* 18(14):14992–15002.
11. Fukabori, A., An, L., Ito, A., Chani, V., Kamada, K., Yoshikawa, A., Ikegami, T., and Goto, T. 2012. Correlation between crystal grain sizes of transparent ceramics and scintillation light yields. *Ceramics International* 38:2119–2123.
12. Apetz, R. and van Bruggen, M. P. B. 2003. Transparent alumina: A light-scattering model. *Journal of the American Ceramic Society* 86(3):480–486.
13. Rothman, A., Kalabukhov, S., Sverdlov, N., Dariel, M. P., and Frage, N. 2014. The effect of grain size on the mechanical and optical properties of spark plasma sintering-processed magnesium aluminate spinel $MgAl_2O_4$. *International Journal of Applied Ceramic Technology* 11:146–153.
14. Tokiwa, K., and Uchino, K. 1989. Grain size dependence of electro-optic effect in Plzt transparent ceramics. *Ferroelectrics* 94:87–92.
15. Zhou, L., Zhao, Z., Zimmermann, A., Aldinger, F. and Nygren, M. 2004. Preparation and properties of lead zirconate stannate titanate sintered by spark plasma sintering. *Journal of the American Ceramic Society* 87(4):606–611.
16. Sarkar, D. 2019. *Nanostructured Ceramics: Characterization and Analysis*. CRC Press: Boca Raton, FL.
17. Wollmershauser, J. A., Feigelson, B. N., Gorzkowski, E. P., Ellis, C. T., Goswami, R., Qadri, S. B., Tischler, J. G., Kub, F. J., and Everett, R. K. 2014. An extended hardness limit in bulk nanoceramics. *Acta Materialia* 69:9–16.

18. Sokol, M., Halabi, M., Mordekovitz, Y., Kalabukhov, S., Hayun, S., and Frage, N. 2017. An inverse Hall–Petch relation in nanocrystalline $MgAl_2O_4$ spinel consolidated by high pressure spark plasma sintering (HPSPS). *Scripta Materialia* 139:159–161.
19. Patel, A. P., Levy, M. R., Grimes, R. W., Gaume, R. M., Feigelson, R. S., McClellan, K. J., and Stanek, C. R. 2008. Mechanisms of nonstoichiometry in $Y_3Al_5O_{12}$. *Applied Physics Letters* 93:191902-1–191902-3.
20. Fabrichnaya, O., and Aldinger, F. 2004. Assessment of the thermodynamic parameters in the system ZrO2–Y2O3–Al2O3. *Zeitschrift für Metallkunde* 95:27–39.
21. Pandey, S. J., Martinez, M., Pelascini, F., Motto-Ros, V., Baudelet, M., and Gaume, R. M. 2017. Quantification of non-stoichiometry in YAG ceramics using laser-induced breakdown spectroscopy. *Optical Materials Express* 7(2):627–632.
22. Mason, P., Divoký, M., Ertel, K., Pilař, J., Butcher, T., Hanuš, M., Banerjee, S., Phillips, J., Smith, J., De Vido, M., Lucianetti, A., Hernandez-Gomez, C., Edwards, C., Mocek, T., and Collier, J. 2017. Kilowatt average power 100 J-level diode pumped solid state laser. *Optica* 4:438–439.
23. Rahaman, M. N. 2003. *Ceramic Processing and Sintering*, 2nd Edition. CRC Press, Marcel Dekker Inc: New York.
24. Carter, C. B., and Norton, M. G. 2007. *Ceramic Materials Science and Engineering*. Springer-Verlag: New York.
25. LaMer, V. K., and R. H. Dinegar. 1950. Theory, production and mechanism of formation of monodispersed hydrosols. *Journal of the American Chemical Society* 72(11):4847–4854.
26. Lu, J., Ueda, K. I., Yagi, H., Yanagitani, T., Akiyama, Y. and Kaminskii, A. A. 2002. Neodymium doped yttrium aluminum garnet $(Y_3Al_5O_{12})$ nanocrystalline ceramics—a new generation of solid state laser and optical materials. *Journal of Alloys and Compounds* 341:220–225.
27. Sarath Chandra, K., Monalisa, M., Ulahannan, G., Sarkar, D., and Maiti, H. S. 2017. Preparation of YAG nanopowder by different routes and evaluation of their characteristics including transparency after sintering. *Journal of the Australian Ceramic Society* 53:751–760.
28. Hajizadeh-Oghaz, M., Razavi, R. S., Barekat, M., Naderi, M., Malekzadeh, S., and Rezazadeh, M. 2016. Synthesis and characterization of Y2O3 nanoparticles by sol-gel process for transparent ceramics applications. *Journal of Sol-Gel Science and Technology* 78:682–691.
29. Ghaderi, M., Shoja Razavi, R., Loghman-Estarki, M. R., and Ghorbani, S. 2016. Spark plasma sintering of transparent Y2O3 ceramic using hydrothermal synthesized nanopowders. *Ceramics International* 42(13):14403–14410.
30. Appiagyei, K. A., Messing, G. L., and Dumm, J. Q. 2008. Aqueous slip casting of transparent yttrium aluminum garnet (YAG) ceramics. *Ceramics International* 34:1309–1313.
31. Esposito, L., and Piancastelli, A. 2009. Role of powder properties and shaping techniques on the formation of pore-free YAG materials. *Journal of the European Ceramic Society* 29:317–322.
32. Ganesh, I., Olhero, S. M., Torres, P. M. C., and Ferreira, J. M. F. 2009. Gel casting of magnesium aluminate spinel powder. *Journal of the American Ceramic Society* 92(2):350–357.
33. Messing, G. L., Kupp, E. R., Lee, S. H., Juwondo, G. Y., and Stevenson, A. J. 2009. Method for manufacture of transparent ceramics. US Patent No. 2 US 7,799,267 B2.
34. Ning, K., Wang, J., Luo, D., Li Dong, Z., Kong, L. B., and Tang, D. Y. 2017. Low-level sintering aids for highly transparent $Yb:Y_2O_3$ ceramics. *Journal of Alloys and Compounds* 695:1414–1419.
35. Yang, W., Hojo, J., Enomoto, N., Tanaka, Y., and Inada, M. 2013. Influence of sintering aid on the translucency of spark plasma-sintered silicon nitride ceramics. *Journal of the American Ceramic Society* 96(8):2556–2561.
36. Sutorik, A. C., Gilde, G., Cooper, C., Wright, J., and Hilton, C. 2012. The effect of varied amounts of LiF sintering aid on the transparency of alumina rich spinel ceramic with the composition $MgO \cdot 1.5\ Al_2O_3$. *Journal of the American Ceramic Society* 95(6):1807–1810.
37. Stevenson, A. J. 2010. The Effects of Sintering Aids on Defects, Densification, and Single Crystal Conversion of Transparent Nd:Yag Ceramics. PhD Dissertation, The Pennsylvania State University.
38. Kuklja, M. M. 2000. Defects in yttrium aluminium perovskite and garnet crystals: Atomistic study. *Journal of Physics: Condensed Matter* 12(13):2953–2967.

39. Zhou, T., Zhang, L., Li, Z., Wei, S., Wu, J., Wang, L., Yang, H., Fu, Z., Chen, H., Tang, D., Wong, C., and Zhang, Q. 2017. Toward vacuum sintering of YAG transparent ceramic using divalent dopant as sintering aids: Investigation of microstructural evolution and optical property. *Ceramics International* 43:3140–3146.

40. Krell, A., and Klimke, J. 2006. Effects of the homogeneity of particle coordination on solid-state sintering of transparent alumina. *Journal of the American Ceramic Society* 89(6):1985–1992.

41. Hostasa, J., Piancastelli, A., Toci, G., Vannini, M., and Biasini, V. 2017. Transparent layered YAG ceramics with structured Yb doping produced via tape casting. *Optical Materials* 65:21–27.

42. Carter, C. B., and Grant Norton, M. 2013. Sintering and grain growth. In: C. Barry Carter and M. Grant Norton (Eds.), *Ceramic Materials Science and Engineering*, pp. 439–456. Springer–Verlag: New York.

43. Li, W., Mei, B., Song, J., Zhu, W., and Yi, G. 2016. Yb^{3+} doped CaF_2 transparent ceramics by spark plasma sintering. *Journal of Alloys and Compounds* 660:370–374.

44. Lee, S.-H., Kupp, E. R., Stevenson, A. J., Anderson, J. M., Messing, G. L., Li, X., Dickey, E. C., Dumm, J. Q., Simonaitis-Castillo, V. K., and Quarles, G. J. 2009. Hot isostatic pressing of transparent Nd:YAG ceramics. *Journal of the American Ceramic Society* 92:1456–1463.

45. Ma, B., Zhang, W., Wang, Y., Xie, X., Song, H., Yao, C., Zhang, Z., and Xu, Q. 2018. Hot isostatic pressing of MgAlON transparent ceramic from carbothermal powder. *Ceramics International* 44(4):4512–4515.

46. Petit, J., Dethare, P., Sergent, A., Marino, R., Ritti, M. Landais, S., Lunel, J., and Trombert, S. 2011. Sintering of α-alumina for highly transparent ceramic applications. *Journal of the European Ceramic Society* 31(11):1957–1963.

47. Sanghera, J., Kim, W., Villalobos, G., Shaw, B., Baker, C., Frantz, J., Sadowski, B., and Aggarwal, I. 2013. Ceramic laser materials: Past and present. *Optical Materials* 35:693–699.

48. Hostaša, J., Biasini, V., Piancastelli, A., Vannini, M., and Toci, G. 2016. Layered Yb:YAG ceramics produced by two different methods: Processing, characterization, and comparison. *Optical Engineering* 55(8):087104-1-8.

49. Sibczynski, P., Iwanowska-Hanke, J., Moszyński, M., Swiderski, L., Szawłowski, M., Grodzicka, M., Szczęśniak, T., Kamada, K., and Yoshikawa, A. 2015. Characterization of GAGG:Ce scintillators with various Al-to-Ga ratio. *Nuclear Instruments and Methods in Physics Research Section A* 772:112–117.

50. Fan, L., Jiang, M., Lin, D., Shi, Y., Wu, Y., Pi, L., Fang, J., Xie, J., Lei, F., Zhang, L., Zhong, Y., and Zhang, J. 2017. Grain orientation control of cerium doped lutetium oxyorthosilicate ceramics in a strong magnetic field. *Materials Letters* 198:85–88.

51. Shoulders, W. T., and Gaume, R. M. 2017. Phase-change sintering of $BaCl_2$ transparent ceramics. *Journal of Alloys and Compounds* 705:517–523.

52. Goldman, L. M., Twedt, R., Balasubramanian, S. and Sastri, S. 2011. ALON optical ceramic transparencies for window, dome, and transparent armor applications. In: *Proceedings of SPIE 8016, Window and Dome Technologies and Materials XII*, 8016 801608-1-14, SPIE.

4

Porous Ceramics

Samuel Paul David and Debasish Sarkar

CONTENTS

4.1 Introduction

Despite metallic and polymeric dense structures, ceramics consisting of full or maximum density without any pores are highly preferred for structural and functional applications, in which pores have to be eliminated deliberately to attain better performance. Contrary to high-density ceramics, the lowest density ceramics comprising of certain porosity (fraction of pore volume to the total volume) are also expected to attain high permeability, desirable mechanical properties, chemical and thermal stability, corrosion resistance and wear resistance to serve many traditional and advanced technological applications. Thus, intense research and economic interest on porous ceramics have moved to the forefront over the last few decades. Such advanced porous ceramics envisage low relative density with porosity of over 30% (*Porosity* = 1-ρ_r where ρ_r is the relative density given by ρ_{bulk}/ρ where ρ_{bulk} is bulk density and ρ is theoretical density or solid density) and are preferentially used in thermal insulating material, catalytic supports, sound absorption, chemical sensors, filters and even as implantable bioceramics [1]. Usually, the porous ceramics are classified in consideration of chemical composition, porosity, physical state of products, inner structure, refractoriness and their application area. In terms of chemical composition, porous ceramics are of several types such as silicate, aluminosilicate, oxide, silicon carbide, corundum and cordierite. From the perspective of porosity characteristics, it can be divided into three types: moderate porosity (30–50%), high porosity (50–75%) and super-high porosity (>75%). They can also be further classified based on their inner hierarchy as cellular, granular and fibrous [2]. A well accepted IUPAC has classified porous ceramics with respect to pore diameter (d); preferably, there would be three types: macro-porous (d > 50 nm), meso-porous (50 nm > d > 2 nm) and microporous (d < 2 nm).

Extensive research has been made over recent decades to control the pore architecture (the size, shape and content) by various processing techniques to lead unique properties for different applications. For instance, cellular ceramics with closed porosity are used in insulators because of their light weight, high melting point and high specific strength whereas nano-filtration is possible using microporous ceramics to filter out or separate matters in fluids. The open porosity with accessibility from outside elements is preferred for filters because of their easy permeability whereas closed porosity is desirable for ceramic thermal insulators. Therefore, excellent characteristics of porosity make them more promising for their use in wide areas including chemical engineering, metallurgy, pharmaceutical, electronics, environmental protection and even in energy applications. Several critical issues are discussed, including fundamental concepts, classification, fabrication protocols, pore size phenomena, properties and applications.

4.2 Classification of Porous Ceramics

Porous ceramics can be divided into different types based on various factors as briefed in the previous section. The most common classification is shown in the flowchart in Figure 4.1. Herein, prime emphasis has been given to the pore architecture that eventually synchronize the most common mechanical reliability of porous ceramics. In this context, the cellular ceramics are well studied in the perspective of synchronization of pore content, structure and sizes (macro-, meso- and microporous ceramics).

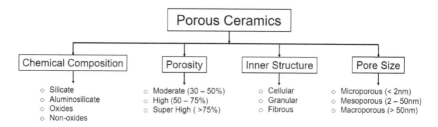

FIGURE 4.1
Major classification of porous ceramics.

With high porosity content (>70%), these ceramics exhibit high specific area and permeability as well as low thermal conductivity that make them attractive for heat exchangers and other thermal-based applications. They also exhibit high chemical resistance, low dielectric constant and low density. Cellular ceramics with open cells (interconnected cells) show different properties than that of closed cells (isolated cells inside the skeleton) due to their different structure and thus are useful in different applications. Different processing techniques and conditions play a major role in achieving desirable structures with cell sizes and distribution. Apart from the cell size and porosity, the property of struts is also crucial to tune the mechanical strength of the final product. Characteristics of struts, such as their thickness, density and shape, can be controlled by different processing methods. Figure 4.2a and b show the typical structure of porous ceramics and the difference between closed and open pore structure. Another class of porous ceramics, coined as ceramic foams, are made of hollow polygonal structures in three-dimensional arrays with randomly oriented cell walls. In the case of ceramic foams, if the surrounding faces are fully solid, then these pores or cells are closed cells, whereas the cells are called open cells if the faces are just voids (Figure 4.2b).

Ceramic foams consisting of open-cell cellular materials with a network of voids are called reticulated ceramics and are generally made by replication of a sacrificial foam template. Extreme porosity, interconnected void volume, adjacent pores and their lightweight make reticulated ceramics attractive for filters for molten metals and catalytic supports. These porous structures have hollow struts that lead to the reduction in their mass without change of their properties. However, if the fabrication parameter, such as heating, is not controlled properly, macroscopic triangular voids are formed in struts that reduce the mechanical strength of the ceramics. The mechanical strength of foams can be improved by several other ways: by increasing ceramic strut size, by filling the voids of hollow struts, by coating or electro-spraying the template using pre-ceramic polymer, by reaction bonding and so on. Strut size is increased by repeating the process of impregnating more slurry into the polymeric template and the recoating of struts in reticulated foam, whereas the filling of voids and micro-cracks are carried out by immersing sintered foam in a suspension of refractory oxides such as silica and alumina or others. In order to increase the compression strength of final ceramic foams, low viscous melts such as Si are generally infiltrated into the struts.

Honeycomb is another class of porous ceramics with a two-dimensional array of polygonal columnar pores. It is highly attractive for catalytic conversion and as a filter to trap diesel particulates. Honeycombs are primarily made by green body extrusion followed by drying and sintering, and they exhibit a high surface area and structural strength. They are also chemically inert with good refractory qualities and made in different cross-sectional

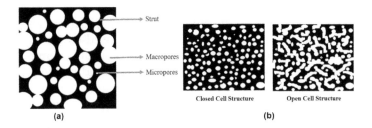

FIGURE 4.2
Schematic representation of (a) typical microstructure of porous ceramics, consisting of strut and size variation of extreme two ends, micro and macro porous (b) closed- and open-cell structure in porous ceramics.

shapes, such as a triangle, circle, square and hexagon, with cell size typically in the order of a millimeter. Large dimensions of honeycomb cellular ceramics can be made by the paste extrusion technique and are made in large sizes by stacking several lower dimensional parts. Glass-based honeycomb structures are mostly made by the extrusion method, and they find several unique applications such as capillary flow controllers and bio membrane reactors in addition to their applications as highly efficient filters. Beyond the most common cellular ceramics, such as ceramic foams and honeycomb structures, other structures such as connected fibers, hollow spheres and bio template structures are demanded for several applications.

Ceramic fiber mats are made by stacking layers of fiber deposits, achieved by collecting randomly oriented fibers along the length and width direction. They are made into rigid boards by casting a slurry of loose fibers containing organic or inorganic binders followed by pressing and sintering processes. These ceramic-connected fibers are used as filters and in thermal management. Cellular ceramics with hollow spheres of different sizes (1–10 mm coarse, 1–100 µm fine) are generally made by a sacrificial core process and sol-gel techniques. The size and shape of these hollow sphered cellular ceramics can be effectively controlled for different applications.

Converting biological organic structures into ceramics offers unique advantages such as anisotropic nature in mechanical, thermal and electrical properties. Such biologically derived preforms or biomorphous ceramics are made by a biotemplating technique in which templates are based on biological materials, such as wood cellulose or biopolymers and so on. For example, biomorphous silicon carbide (SiC) cellular ceramics have been made by the pyrolysis of wood-based organic material. Wood undergoes a pyrolysis reaction above 700°C to form the carbon template. The carbon template is then infiltrated by nano-sized silica sol, and the infiltrated structures undergo a carbothermal reduction reaction at a temperature of 1575°C to yield SiC biomorphous ceramics with unique structural design of the wood template [3]. These ceramics with micro- and mesoporous structure are interesting in the field of biotechnology and medicine as bio-inert and corrosion-resistant immobilization supports for living cells or enzymes and also as pseudomorphous to biological tissues.

4.3 Fabrication of Porous Ceramics

The fabrication protocols and processing parameters play a major role in obtaining porous ceramics with desirable cell structure, cell size, distribution, thickness of struts and thereby they tailor the overall porosity of the end product. There are several ways to produce porous ceramics starting from the traditional techniques, such as replica, sacrificial template and direct foaming, to the modern techniques, such as freeze drying, wood ceramics and so on. Fabrication techniques are chosen based on the targeted pore morphology in porous ceramics. For example, pores in translationally symmetric bodies such as two-dimensional honeycombs are made by extrusion molding, whereas the common three-dimensional porous structures are made by partial sintering, pyrolysis of the pore-forming agent, polymer replica method, impregnating organic foam, biotemplating and recently developed rapid prototyping techniques (details in Section 9.2). Apart from conventional ways to produce ceramics with uncontrolled microstructure, more techniques were developed to have more control on the microstructure. Herein, some of the common fabrication methods as well as the most recent techniques are discussed.

4.3.1 Particles Coalescence

The simplest way to achieve relatively low porosity of below 60% is traditionally carried out by partial sintering (reverse to obtain transparent ceramics) of low density porous powder compacts. When powder mixtures are subjected to partial sintering that is mostly done at a lower temperature and shorter time compared to fully dense material, the process results in interconnected pores during their solid-state reaction. The size of the starting powders determine the pore size whereas the amount of porosity is preferentially controlled during the sintering process. Particle stacking sintering is a complimentary process to form porous ceramics by stacking and sintering the aggregate ceramic particles. Adapting this process, one can control the final pore structure. In some cases, other solid-state additives that have suitable wettability are added during higher temperature processing. As an example, lead silicate or lead borosilicate glasses are used as glass additives for quartz-based porous ceramics because of their good wettability between grains and glass. Added advantages of using such glass additives are the lower sintering temperature and shorter sintering time. Ceramics with large pore sizes can be achieved by using large aggregate particles whereas a wide pore distribution is achievable with smaller aggregate particles. If the starting aggregate particles have a narrow size distribution, ceramics with higher porosity and uniform pore sizes are obtained at a relatively low sintering temperature. Such structures have been achieved in porous Al_2O_3 ceramics made by fine particles of α-Al_2O_3 with a small amount of sintering additive SiO_2-Al_2O_3-RO-R_2O (RO-CaO, SrO, BaO, MgO; R_2O – K_2O, Na_2O, Li_2O) [4]. The starting materials are mixed in a wet condition at a proper ratio, dried, molded and sintered to make the targeted ceramics. Incorporation of a small amount of additives helps to improve the mechanical properties of the end product. A small amount of Y_2O_3 acts as the binder for fabricating porous Si_3N_4 and SiC ceramics in which this addition of binder eventually facilitates the strength enhancement of struts or grain boundary strength in such an incomplete sintered body. Flexural strength of SiC ceramics was increased by almost three times when 1.5 wt% of Y_2O_3 was added [5]. This technique offers the advantage of achieving ceramics with smaller grain sizes due to the lower sintering temperature.

4.3.2 Pyrolysis of Pore-Formers

A selective amount of different organic or combustible pore-forming agent facilitates the formation of different degrees of pores and their morphology in complex-shaped ceramics after the final stage of high temperature processing. Both organic materials, such as paraffin, flour, starch and naphthalene, and dissolvable inorganic salts act as pore-forming agents. Addition of these chemical agents enhances the open-cell porosity. Organic materials are removed during sintering whereas inorganic materials are dissolved through high temperature solid solution or secondary phase. Apart from these, carbon powders, wood scraps and fibers are also used as pore-forming agents. The strength of the ceramics is projected to improve by increasing the sintering temperature and thereby increasing the density. Another way to increase the strength is by adding other additives such as feldspar or sodium fluoride. A mixture of polyvinyl chloride and ZnO was used to make porous calcium magnesium zirconium phosphate porous ceramic, whereas organic compounds such as naphthalene, olefin and PMMA are commonly used as pore-forming agents for porous hydroxyapatite (HA) [6]. One of the drawbacks of adding organic material with HA is that this class of organic-based pore-forming agents undergoes significant thermal expansion during thermal treatment, as it has nearly ten times the high thermal expansion coefficient (CET) of HA. During heating, this difference in CET results in the formation of cracks that eventually reduce the mechanical strength of the material. Carbon powders along with biological glass and organic phosphate are found to increase the strength of porous hydroxyapatite due to their relatively similar thermal expansion coefficient.

Instead of adding pore-forming agents with the starting powders, their preferential addition in slurry provides controlled properties of porous components. For example, mullite porous ceramics have been made using alumina aggregate powder with ethyl silicate as a binder and carbon powders (~20%) as the pore-forming agents [7]. During sintering, ethyl silicate decomposes into ethanol and water, which leaves a large number of pores during evaporation (50–56% porosity). Organic materials such as starch, corn or potato powders have also been used as the pore-forming agents. Owing to their densification, binding and membrane-formation capabilities, starch powders have been found to be attractive for porous alumina and Si_3N_4 [8]. Starch is stable in water below 50°C, whereas it becomes dissolvable at 55–80°C. When added with the slurry of ceramic powders, starch expands with the absorption of water at 60–80°C. This forms a solid green body with much strength by binding the ceramics particles. In sintering, the starch particles undergo combustion leaving a large number of pores analogous to the amount of starch mixed with the slurry in the initial process.

4.3.3 Replica Technique

The replica technique was introduced in 1960s by Schwartzwalder and Somers, and it is one of the widely employed traditional methods of fabricating porous ceramics in which a ceramic suspension infiltrates a cellular structure such as a synthetic or natural template [9]. After the organic template burnout, porous ceramics are made with a cellular structure similar (replica) to that of the template with pores slightly smaller than that of the template. Ceramic foams with open-cell cellular structures with interconnected voids are fabricated by this technique and are considered well-accepted industrial fabrication protocol. Such ceramics are also called reticulated ceramics. Early organic templates were made of polymers such as polyurethane, PVC, polystyrene and cellulose in order to achieve the targeted dimension and properties. These polymeric sponges are impregnated with ceramics slurry such as oxides or non-oxides along with binders and dispersants. The excess slurry in the sponge is then squeezed out, wiped and dried. The dried sponge is then subjected to slow heating burnout of the polymer template, which is followed by high temperature sintering [10]. Uniform green body formation and removal of the excess slurry is critical to avoiding closed cells, which can degrade the permeability and mechanical properties of the porous ceramics. This technique helps to achieve hollow struts, thereby reducing the overall mass without reducing the properties. A wide range of porosity (40–95%) is the result of such a process. The steps involved in the replica technique are shown in Figure 4.3.

By stacking templates that have different morphological features, graded pore structures can also be achieved by this technique. In order to increase the strength of ceramic foams made by the replica technique, the template impregnation and drying processes are repeatedly done to increase the size and strength of the strut and foam, respectively. The strength is also improved by immersing the sintered foam in a suspension of refractory oxides such as alumina or even colloidal silica. These fine suspended particles fill-up the tiny voids, fine cracks and hollow struts, incrementally strengthening the structure after the second stage of sintering. To obtain different cell sizes and distribution, several factors starting from the choice of organic template, ceramic powders, slurry preparation, drying and sintering protocol play a major role.

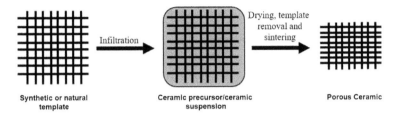

FIGURE 4.3
Schematic of replica technique to fabricate porous ceramics.

The organic template should have an open-cell structure to allow and absorb the ceramic slurry, and also to squeeze out the excessive slurry. The template volatilization temperature should be less than the sintering temperature of ceramic particles. Commercial-grade ceramic slurry is made with a high solid content (50–70 wt%) and consists of a narrow range of particle sizes (between 45 and 175 μm) mixed with liquid solvents. Additives such as binders, plasticizers, dispersants and defoamers are preferentially added to control the slurry viscosity and improve the strength. The slurry with less solid content results in shrinkage and cracking, whereas the slurry with high solid content will find it difficult to infiltrate the template structure. Therefore, a well-dispersed slurry with optimum solid content is desirable. To increase the ceramic strength, sometimes an inorganic sol is also added with the slurry. After impregnation of the slurry, the excess material should be removed to avoid any pore blockage which is critical for applications. This green body is then dried to remove excess water. The dried compact is then subjected to slow heating to remove the organic foam (template) followed by high-temperature sintering to increase the strength of the ceramics. Recently, the porous SiC ceramics were prepared using polyurethane sponge as the organic template, carboxyl methyl cellulose as the stabilizer to improve the rheological property of the slurry and silica sol as the binder. The green body prepared by this slurry is dried at room temperature followed by drying at 110°C for 24 hours. The dried compact is then subjected to sintering at 1350–1450°C in a controlled atmosphere. The strength of the SiC ceramics can be enhanced by coating slurry followed by coating silicon powders. After sintering at 1600°C, the compressive strength is drastically improved due to the siliconizing process [11].

4.3.4 Sacrificial Template Technique

This is analogously opposite to the replica technique because it produces the negative replica of the original sacrificial template. Porous ceramics with high open porosity are made through a similar protocol as the replica technique by mixing ceramic powders with sacrificial materials or sacrificial fugitives that can be organic (natural and synthetic), inorganic, metallic or even liquid. Organic materials such as polymer beads, potato starch and organic fibers are more commonly used as pore formers; however, non-organic pore formers such as nickel and carbon are also popularly used to fabricate porous ceramics. Apart from solid-based pore-forming agents, liquids such as water, gel and emulsions are also used. The schematic process involved in the sacrificial template technique is shown in Figure 4.4. Microbeads of polymethylmethacrylate (PMMA) act as successful sacrificial fugitives for porous ceramic foam made of SiOC and β-tricalcium phosphate (β-TCP) with 70–80% porosity.

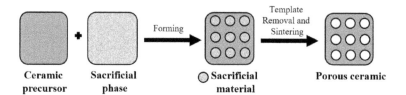

FIGURE 4.4
Sacrificial template technique to fabricate porous ceramics.

Sacrificial expandable micro-spheres are also used to achieve porosity of up to 86% with more advantages such as low-cost and low-gaseous byproducts during pyrolysis. Duplex structure of SiC ceramics with large pores were made using such expandable organic micro-spheres (461DU40, Expancel, Sundsvall, Sweden) that exhibit high gas permeability coefficient for corrosive gases filtering and other industrial applications [12]. Corn starch has been used as sacrificial material for porous ceramics of barium strontium titanate $((Ba, Sr)TiO_3)$ whose porosity, grain size and electrical properties were dependent on the addition of corn starch [13]. Porous alumina with unidirectionally aligned pores of continuous channels are made using long fibers of cotton thread or natural tropical fiber as sacrificial fugitives. To overcome the difficulties of handling long narrow fibers, short fibers or whiskers are also added. Porous Si_3N_4 made using organic whiskers with 45% porosity showed high gas permeability because of the randomly oriented rod-shaped pores [14]. Liquid sacrificial fugitives are also desirable in some cases because of their ability to form as a solid and sublime without release of any harmful burn out materials. In this case, slurry made with water is poured into a mold and then the bottom of the slurry is frozen which leads to the formation of ice along the vertical direction. Once the ice eventually sublimes under reduced pressure, the pores are formed vertically, yielding unidirectionally aligned porous ceramics with the structure replica of the original ice structure. Pores are formed with anisotropic morphology due to the different rates of sublimation along the vertical direction. Using such a freeze-casting/freeze-drying technique, the amount of porosity and their dimensions can be decided by controlling the amount of initial compositions of slurry and the freezing conditions. The shrinkage of the ceramic depends on the amount of water content in the slurry, and a high-level porosity of up to 90% can be obtained with uniform open-cell pores. Porous alumina and SiC ceramics have been made by this technique using ethanol as refrigerant liquid at –50°C. After freezing at this temperature, the mold is then dried under low pressure for 24 hours followed by sintering in air at high temperatures (1400–1500°C) for 2 hours [15]. Ceramics produced by this technique find an appreciable market in several classic applications including chemical sensors, adsorbent carriers, biological reactors and so on.

4.3.5 Foaming Process

The direct foaming process was invented in 1970s. It is an easy and low-cost production method to achieve ceramics with high porosity (>95%) and good thermo-mechanical and thermochemical stability with impressive strength by sintering ceramic suspensions foamed by gas or air. During the direct foaming process, the gaseous phase is intentionally introduced into the ceramic suspension by mechanical frothing. In addition to mechanical incorporation, generation of foam is also achieved by gas injection, where gas is released by several factors including exothermic reaction, foaming agent and evaporation of low melting point solvent. Ceramic suspensions usually use polymers such as silicone resin and are therefore known as pre-ceramic polymers. Despite introducing a physical foaming agent such as inert gas or low boiling-point liquids (pentane, hexane, methylene dichloride), several chemical foaming agents such as calcium carbonate, calcium hydroxide, aluminum sulfate and hydrogen peroxide are also used. Organic materials such as polyurethane, sulfonyl hydrazine, azo and nitro compounds are also an effective choice because of their higher foaming efficiency. These materials show a stable release of gas at their pyrolysis temperature.

A prerequisite foaming agent is usually added with the ceramic suspension, which are then placed in a mold and heated at a high temperature (900–1000°C) under oxidizing atmospheric conditions. The foaming/blowing agents can also be volatile liquids or gases produced *in situ* or by gas injection by bubbling. The pyrolysis of these foaming agents form pores through instant liberation of gaseous products. These pores are formed by different stages: nucleation, expansion and shaping of bubbles/pores. By controlling the pressure and temperature of the heat treatment, both open and closed-cell porous ceramics can be obtained. This results in a large number of gas–liquid interfaces and a large free energy. Such high free energy results in thermodynamic instability and hence reduces the total Gibbs free energy thereby collapsing the foam. To avoid this, surfactants are generally added to achieve a stable system by the reduction of gas–liquid boundary interfacial energy. A typical schematic of the direct foaming process is shown in Figure 4.5.

These surfactants facilitate the increase of the surface viscosity and create electrostatic forces to prevent foam from coalescing. Porous SiC ceramics have been fabricated by blowing a polymeric precursor such as polycarbosilane (PCS) or polysiloxane along with foaming agents such as azodicarbonamide (59–85% porosity) or n-octylamine (70–90% porosity) [16]. The amount of gas present in the suspension eventually controls the amount of porosity. To increase the mechanical strength of such porous ceramics, one can increase the solid content in the slurry, although the pore size depends on the size and distribution of the starting precursor aggregates. Even though this method helps to achieve better strength with high dense struts, it is difficult to make ceramics with narrow pore size distribution.

4.3.6 Sol-Gel Method

The sol-gel method prefers a low processing temperature and thus is often used to fabricate porous ceramics with tailored properties. In the sol-gel process, inorganic or organometallic precursors are hydrolyzed either by water or organic solvents such as alcohol mixed with water [17]. After condensation, the amorphous mixture forms inorganic polymer gel with M-O-M or M-OH-M bonds (M-metal). The dried gel undergoes decomposition at a higher temperature that results in porous metal oxide ceramics. Porosity can be monitored through starting powder size, packing, drying and the selection of templates. To achieve ceramic membranes with smaller pores for

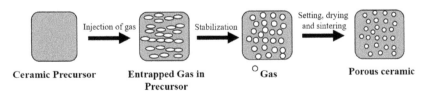

FIGURE 4.5
Schematic representation of direct foaming process to fabricate porous ceramics.

applications such as gas filters, smaller particles are desirable. To disperse such fine particles without agglomeration, different approaches have been adopted either by forming an electrostatic double layer by controlling pH or by forming core-shell composites. Even though aggregation is undesirable in particulate sol deposition, they can still be exploited for the porosity of films. In the case of porous ceramic films made by polymeric sols, the volume fraction of porosity increases with an increase in the average size of the polymers comprising the sol. In such cases, the desirable aggregation develops by varying the aging process, resulting in a controlled microstructure. With an increase in aging time, one can increase the volume fraction porosity, surface area and pore size [17]. The stiffness of the network without pores collapsing during drying is dependent on the rate of condensation as well as adjusting the pH and the aging time. At a slower condensation rate, polymer interpenetration occurs, ending up with high-density porous films and small pore sizes. In addition, a reduction in the condensation rate can collapse the structure of porous films during drying. Therefore, to obtain desirable pore sizes or commercial-grade large-sized products, it is desirable to create pores within a dense matrix by using colloidal crystalline grains as the template. For example, PMMA-based close-packed polymer colloidal crystals act as a template to make macroporous ceramics of TiO_2 and ZrO_2 with uniformly distributed pore architecture. Such a pore template is made by filtering of colloidal particles through a filter membrane with the desirable shape and size. Thus, the tailored porous ceramics is expected by pouring sol into the voids. Porous ceramics of titania, alumina and zirconia have been made using metal chlorides as precursors and olephylic and hydrophilic polyoxyalkylene triblock copolymer as the guiding agent. Compressed organic foam is another choice of a template to make porous ceramics consist of porosity as high as 97% achieved by dipping or coating the organic foam into a colloidal solution prepared by the sol-gel method. After the removal of excess solution, the foam is then dried, aged and then finally sintered. To increase the strength, the sintered body can be further immersed in sol-gel solution.

Aluminum sol is attempted by the hydrolysis of aluminum powder in $AlCl_3$ solution. Similarly, porous silica ceramic membranes have been made with silica synthesized by the sol-gel process. Silica powders prepared by controlled hydrolysis and condensation polymerization of tetraethylorthosilicate ($Si(OC_2H_5)_4$) shows superior properties such as high surface area and microporosity. This process can be seen as per the following chemical reactions:

$$\text{Hydrolysis}: Si(OC_2H_5)_4 + H_2O \rightarrow Si(OH)(OC_2H_5)_3 + C_2H_5OH \tag{4.1}$$

$$\text{Condensation polymerization}: Si(OH)(OC_2H_5)_3 + Si(OH)(OC_2H_5)_3$$
$$\rightarrow (OC_2H_5)_3 Si\text{-}O\text{-}Si(OC_2H_5)_3 + H_2O \tag{4.2}$$

After heat treating the gel, fine particles of silica are obtained [18]. The pore hierarchy preferentially synchronizes by the hydrolysis and condensation parameters. In a similar process, synthetic zeolite achieves a desirable porous structure by agitating silica and alumina source solutions at a low temperature (100°C) with a high pH and find it useful for molecular sieving applications. The precursors of alumina and silica are typically derived from aluminides and double silicates, respectively.

4.3.7 Other Techniques

The gel casting method is similar to the foaming process and is commonly used to fabricate 3D reticulated ceramics. In this technique, the gel is formed from the slurry containing ceramic powder, water, dispersers and monomers. Polymerization takes place by further adding surfactants, initiators and catalysts. The gel is then dried and sintered at a higher temperature to form dense ceramics by burning out the polymers. The gel casting method has been widely adopted to fabricate ceramics with complex structures, strong pore struts and high solid content, achieving porosity of 40–50%. For example, porous alumina with uniform pore distribution is achieved using methacrylamide as the gel monomer, ammonium persulfate as the initiator, tetramethylethylenediamine as the catalyst and dispersant Duramax D3019 and graphite powder as the pore-forming agent [19].

Porous ceramics analogous to wood hierarchy, coined, as *wood ceramics* is a relatively new kind of porous materials studied using wood as a biological template. The anisotropic nature of wood due to the cell morphology and cell wall alignment acts as a hierarchically structured template that can enable the development of novel ceramics with anisotropic cellular structure. This provides the potential to fabricate microcellular ceramics with the analogous structural features of the wood. Wood is made of several polymers, cellulose, hemicellulose and lignin that are easy to modify chemically and physically to get carbon, porous carbonates and oxides. Several porous ceramics of alumina, silica, titania and zirconia have been made from pine wood. By controlled thermal decomposition, wood tends to transform into a carbon template without any change in anatomical features of the wood. Wood ceramics obtained by high temperature carbonization of wood-based materials are also known as carbon-based ceramics. Porous ceramics produced by this process show good mechanical properties due to the presence of amorphous carbon with the anisotropic cellular architecture of the wood. Another type of wood ceramic obtained by high-temperature pyrolysis of natural wood in an inert atmosphere followed by infiltration into liquid Si is known as SiC-based ceramics. Matovic et al have fabricated porous silica ceramics using tetra-ethyl orthosilicate (TEOS) as the precursor and linden, a deciduous wood as the biological template [20]. The carbon template is made from this wood either by calcining at 1000°C in Ar atmosphere or by soaking the dried wood in 1 M HCl solution to leach out the lignin. This wood is then soaked in an ethanol solution of TEOS (molar ratio of TEOS: ethanol is 1:4) at pH of 4. The silica sol obtained is then dried at 100°C followed by a heat treatment at 800°C and further annealing at 1300°C in inert atmosphere (N_2) to yield porous SiO_2 ceramics with wood-like microstructure [20]. Figure 4.6 shows the microscopic images of SiO_2 ceramics with microstructure similar to the wood sample with a wide range of pore sizes.

Self-propagating high-temperature synthesis (SHS) or combustion synthesis is a fast one-step porous ceramic fabrication process which is initiated by point-heating a small part of the sample (reactant mixture). With minimum energy, this process results in chemical exothermic reactions and can be self-propagating due to the heat released by the reaction. This technique requires less energy and cost compared to conventional sintering; however, it is troublesome to control the porosity and microstructure of ceramics. Porous and reticulated Ti-Si-Al-C ceramics have been made using TiAl, Ti_5Si_3 and TiC as the precursors with ammonium chloride and calcium peroxide as the agents for the combustion process by gasification [21]. Despite several processes, the microwave-assisted sintering of green bodies made of hollow spheres is another protocol used to fabricate closed-cell porous ceramics.

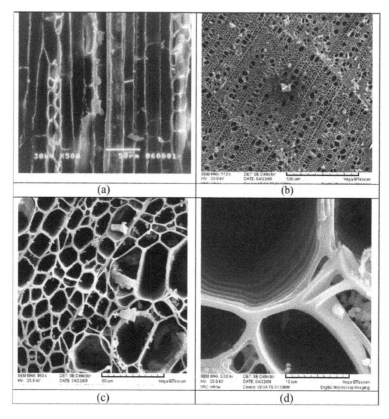

FIGURE 4.6
SEM images porous SiO_2 ceramics obtained at 1300°C for 4 h, (a) cross-section parallel to axial direction, (b–d) cross sections perpendicular to axial direction [20].

4.4 Characterization of Pore Size, Shape and Distribution Phenomena

Pore morphology, including size, distribution and pore connectivity, plays a major role in their properties and applications of porous ceramics. An optimum porosity and pore size distribution is obvious to enhance the functional properties of such ceramics. Pore size distribution is represented by the derivatives $(dA_p / dr_p$ or $dV_p / dr_p)$ as a function of pore radius (r_p), where A_p is the wall area and V_p is the wall volume. In order to tailor and engineer the ceramic materials for specific applications, it is essential to quantitatively measure and determine the pore characteristics such as pore size, volume, distribution and other porosity related properties of the materials. Commonly, such measurement processes are known as porosimetry. Practically, it is difficult to quantify these parameters just by a single technique. Different techniques are used depending on the size of porosity and on the nature of the porous material. The most common and convenient method to determine pore size, distribution and volume porosity is by processing the images obtained by several imaging methods such as optical microscopy, scanning electron microscopy (SEM), field emission SEM (FESEM) and transmission electron microscopy (TEM), which provides a 2D image of the sample. By using several image processing software, like Image J and Image Pro, one can estimate the distribution and size of the pores. Apart from a 2D view, techniques such as atomic force microscopy (AFM) and X-ray tomography are useful to visually analyze the void space inside the material.

Both transmitted and reflected light microscopes are commonly used depending on the opacity of the samples. Scanning electron microscopy helps to understand the surface topography by analysis of the signals from the secondary electrons. Higher magnification range, higher resolution and a greater field depth of SEM help to determine the pore information. Similarly, the transmission electron microscope (TEM) is an excellent tool to analyze the pore information when the size is below 1 nm pore diameter in size. By using image processing software, one can quantify the spatial characteristics of pores ranging from macropores to micropores. Unlike these techniques, X-ray Computed Tomography (XRCT) provides a three-dimensional image of the specimen microstructure. Such 3D images are constructed by combining several tomographs taken along the z-direction at different angles during the rotation of the sample. The use of x-rays provide high-resolution information but makes this technique an expensive one. In addition, it is hard to make a 3D map for a larger sample. In this context, gas adsorption protocol is widely accepted and less expensive to illustrate the details pore phenomena. Thus, it is important to have a detailed discussion on gas adsorption and mercury porosimetry techniques that are common in practice to determine pore size and distribution along with specific pore volume and specific surface area. Porosimetry techniques are based on physisorption (physical adsorption) and are effective for a wide range of pore sizes ranging from 0.35–300 nm, thereby covering the complete range of pore types. The gas adsorption technique is mainly used to determine the surface area of the pores based on the principle that a non-reactive gas or a non-wetting liquid undergoes adsorption or desorption onto a solid surface when subjected to pressure. Compared to the use of a solution such as in immersion calorimetry, the gas adsorption technique is easy to interpret. This was proposed in the early 1930s based on the Langmuir theory on adsorption. As per Langmuir, if one can find out the average area occupied by the absorbed molecule on a monolayer, it is easy to calculate the surface area of the adsorbent. Extending this idea to porous solids, Brunauer–Emmett–Teller came up with the theory (BET theory) to determine the surface area of iron synthetic alumina using low-temperature gas adsorption. The most common gas used in this BET technique is nitrogen (adsorption at 77 K), in addition, helium and carbon dioxide also are used frequently. By passing the gas adsorbate with known density through the sample, the pores are filled slowly and eventually saturated with adsorbate at equilibrium pressure. By measuring the volume of the adsorbate absorbed into the pores, the volume of the pores using the adsorbate density is determined. This method helps to determine the surface area and size of the pores based on the adsorption isotherm, which is a plot between the amount of adsorbate (gas) absorbed on the surface of adsorbent vs relative saturation pressure (p/p_0 where p_0 is the saturation pressure) at constant temperature.

The BET equation is used to determine the surface area using physisorption isotherm data, and is given by:

$$\frac{p}{n^a(p^0 - p)} = \frac{1}{n_m^a C} + \frac{(C-1)}{n_m^a C} \cdot \frac{p}{p^0} \qquad (4.3)$$

Where n^a is the amount of gas adsorbed at p/p^0 (the relative pressure), n_m^a is the monolayer capacity of gas and C is constant which is dependent on the type of isotherm. Six types of isotherms are hypothetically proposed by the BET theory depending on the different porous structures of the solids. Figure 4.7 describes six types of adsorption isotherms for gas–solid equilibria, as classified by IUPAC [22]. For a monolayer, if the average

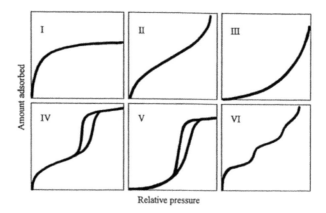

FIGURE 4.7
Six types of adsorption isotherms classified by IUPAC for gas–solid equilibria. Type I for adsorption on micro-porous adsorbants, Type II and III for macroporous adsorbents with strong and weak adsorbate–adsorbent interactions, Type IV and V for mono-and-multilayer adsorption plus capillary condensation, and Type VI adsorption isotherms for more steps [22].

cross-sectional area (a_m) of the molecules are known, surface area (A) or so-called BET area can be obtained by the following relation:

$$A\,(BET) = n_m^a \cdot L \cdot a_m \qquad (4.4)$$

The value of a_m of nitrogen monolayer at 77 K is 0.162 nm², which may vary depending on the level of adsorption. Pore sizes and their distribution are then calculated based on computational methods such as the classical Barrett–Joyner–Halendra (BJH) method for nitrogen-based isotherms. In the case of carbon dioxide, the most common analytical models based on Dubini-Astakhov (D-A), Dubinin–Radushkevich (D-R) and density functional theory (DFT) are in practice to determine accurate micropore size, volume and area. The degree of adsorption depends on the type of gases and temperatures. In a recent article, the effect of different adsorbates like CO_2, N_2, CH_4 and Ar are studied at both 293 K and 77 K on amorphous activated carbon AX21, which has a high specific surface area of 3,080 m²/gm [23]. This classic study revealed that the different pore size (H) distribution is significantly affected by the presence of adsorbates and characterization temperature, and such analysis is plotted in Figure 4.8. The observed pore size distribution characteristics are different when measured using N_2 or Ar at 77 K and CO_2 at room temperature (293 K). It is found that carbon dioxide is an appropriate adsorbate for analyzing micropore structure at 293 K compared to nitrogen at 77 K. Because of the simplified assumption of multilayer adsorption by BET theory, the measurement an empirical approach can be ascertained by comparing the measured data with a standard adsorption isotherm data of a fully dense reference material [24].

Mercury (Hg) porosimetry is similar to the technique mentioned above in which mercury is chosen instead of gases like CO_2, N_2, Ar and CH_4. This technique is suitable to characterize materials with pore sizes varying between 500 μm to 3.5 nm. Mercury is a non-wetting liquid and does not go through pores until a sufficient external pressure is applied. The Washburn equation governs the relation between the applied pressure (p) and pore radius (r), which is given by:

$$p.r = -2.\gamma.\cos\theta \qquad (4.5)$$

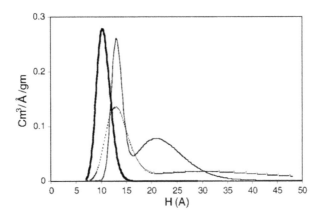

FIGURE 4.8
Bold line shows pore size distribution from fit to CO_2 isotherm at 293 K; solid line and dashed line show PSD from isotherms based on N_2 and Ar at 77 K, respectively [23].

Where γ is the surface tension of mercury (480 dyne/cm @ 25°C) and θ is the contact angle between mercury and the pore wall (~140°). The minus sign indicates that the friction force is opposite to the fluid flow. As per the equation, the pressure required to equilibrate the porous body is inversely proportional to the pore size. When the pressure is increased, mercury intrudes into the pores and the total intruded mercury volume is measured which is equal to the total pore volume. The sample is initially dried, then weighed and kept in a chamber along with mercury. Progressive pressure is applied step by step and the intrusion volume is monitored.

Mercury porosimetry measurements are made with several assumptions that make the data sometimes different from other measurement techniques like imaging methods. One of the main assumptions in this method is the hypothetical consideration of the cylindrical pore geometry that is open to the outer surface with direct contact with the surrounding mercury. This assumption is not applicable to all materials, such as hydrated systems like cement. Assuming a reversible intrusion process, surface area can be calculated using the pore volume data using the Rootare and Prenzlow equation given by:

$$A = -\frac{1}{\gamma_{Hg}\cos\theta}\int_0^V P\,dV \tag{4.6}$$

An error in the pressure value measurement may provide completely wrong information relating to the surface area by several orders. Therefore, care should be taken to accurately measure pressure during porosimetry analysis. Using the surface area value, mercury porosimetry also helps to calculate the particle size (r) of powder material using the following formula:

$$r = \frac{3m}{\rho A} \tag{4.7}$$

Where ρ is specific density of the material and m is the mass. Such estimation is also restricted to spherical particles and for a particular powder packing structure [25]. For materials that retain mercury even after the extrusion cycle and for materials that have a pore network, this technique can give different values which need to be verified by other techniques or by measuring the reference samples. Although mercury porosimetry remains a powerful tool for the estimation of pore characteristics for mesoporous materials as well.

4.5 Properties of Porous Ceramics

4.5.1 Mechanical Properties

The foremost mechanical reliability assessment is an essential parameter in order to develop and successfully handle the transportation of porous ceramics. In this context, fracture strength, Young's modulus, tensile strength, compressive strength and Poisson's ratio are critically discussed in relation to porosity content, size and distribution analogous to grain size phenomena. In actuality, the amount and nature of porosity along with the material properties play a major role in the mechanical properties of the product. Accordingly, the ceramics are engineered using theoretical models by adjusting the density, porosity and cell dimensions to improve the mechanical strength from the perspective of different applications. Among the earlier models, the Gibson and Ashby model [26] predicted fracture toughness (T) of open-cell brittle foams, based on the cell dimension (L) and the relative density (ρ/ρ_s) (ρ is the bulk material density, ρ_s is the solid density).

$$T = C\sigma_{fs}\sqrt{L}\left(\frac{\rho}{\rho_s}\right)^{3/2} \tag{4.8}$$

Where σ_{fs} is mean flexural strength of struts and C is geometric constant [27]. This model predicted the increase in fracture toughness with the increase in cell size whereas in reality, the mechanical response of the strut increases with the decreases in cell size as confirmed for vitreous carbon foams by Brezny and Green [28]. Thus, mechanical strength is not only influenced by the cell size, but mainly on its structural changes and on the strength of the strut. Accurate theoretical predictions are somewhat difficult because of the precise measurement of the strut strength. The strength of the strut also depends on the fabrication conditions and methods. Minimum solid area (MSA) model covers a wide range of porosity and different types of pores, such as tubular, spherical, cubic and polyhedral shaped pores.

The minimum solid area is the area of the neck formed with adjacent particles during sintering, whereas for the spherical pore, it is the area of surrounding solid material in the equatorial plane of the cell normal to the stress direction. This model is based on an assumption of how the mechanical property increases in relation with the cell structure. The mechanical property is assumed to increase with the ratio of MSA to the cross-sectional area. It is to be noted that MSA is considered normal to the stress, whereas the cross-sectional area of the cell is in the same plane of the MSA. This ratio (MSA value divided by cross-sectional area of the cell in the same plane), known as relative MSA, gives the correlation of mechanical properties and is generally determined by plotting various MSA values with different porosity values (P) with semi-log vertical axis. In such semi-log plots as shown in Figure 4.9, the value of MSA decreases linearly and then at a steeper rate as the value of porosity (P) increases [29]. The linear region of the plot can be easily approximated and related with Young's modulus as follows:

$$E / E_0 = e^{-bP} \qquad (4.9)$$

Where E is the Young's modulus, E_0 is the Young's modulus at P = 0, i.e., for the samples with theoretical density, b is the slope of the linear plot on this semi-log plot. The MSA value changes with pore types and has been found to be consistent with the experimental data. Extending this MSA model, computer-based models improved the strength dependence on porosity with an equation:

$$E / E_0 = (1 - P/P_c)^n \qquad (4.10)$$

Where n is the exponent number dependent on the type of pores, P is porosity and P_c is the critical value of volume-fraction porosity P at which the solid body behaves like liquid at which the elastic moduli value becomes zero. For a honeycomb structure with tubular pores aligned along the stress direction, this equation becomes:

$$E / E_0 = 1 - P \qquad (4.11)$$

With n = 1 for tubular pores. For the stress applied normal to the direction of tubular pores, the value of P_c is less which leads to anisotropic nature with lower mechanical

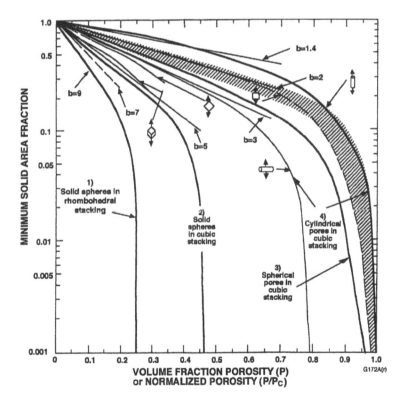

FIGURE 4.9
Semi-log plot of key MSA models versus P and plotting MSA versus P/P_c (normalized). Such normalization brings the wide diversity of MSA models shown into a modest band (cross-hatched area) at the upper limit of porosity dependent behavior, i.e., for tubular pores aligned with the stress axis [29].

strength compared to the axial strength. Pores having a circular cross-section are isotropic in nature irrespective of their direction with the aligned pores. Similarly, the properties vary from isotropic to anisotropic characteristics with pores of different geometrical cross-sectional shapes. Other anisotropic natures of honeycomb structures such as tensile strength, crushing strengths and thermal stress are studied using this model, which are essential for different applications.

Similar analyses can be made on ceramic foams and, in a similar way, the mechanical strength can be related as:

$$E / E_0 \sim (1-P)^2 \tag{4.12}$$

This equation can be rewritten in terms of shear modulus (G) as

$$G / G_0 \sim 3/8 (1-P)^2 \tag{4.13}$$

Where G_0 is the value of the shear modulus at $P = 0$. Zwissler and Adams devised qualitative models to study Young's modulus and fracture toughness of glass foams and ceramic foams with uncertainities [30]. However, such uncertainties were then fixed by the Green and Hoagland model [31]. Practically, ceramics with improved mechanical properties can be developed by using bimodal powder with different sizes. Because of various parameters involved in porous ceramics both in the macroscopic level such as porosity, pore size, and shape, and in microscopic level such as atomic level coordination, neck growth, it is always difficult to build a single model to understand the mechanical properties of porous ceramics completely. In general, the presence of both coarse and very fine starting particles can enhance particle packing, resulting in uniform microstructures with enhanced surface and grain boundary diffusion at intermediate temperature, followed by retarded densification at a higher temperature.

The other essential concept to describe the mechanical property of porous ceramics is effective elastic moduli (M_e), which can be expressed in terms of elastic moduli (M_0) of fully dense medium and the amount of porosity (P) of the material. Due to the complex nature of this property, theoretical predictions are always difficult to match accurately with the experimental values for porous ceramics. With the application of micromechanics, several techniques such as the composite sphere method, self-consistent method, differential method and Mori-Tanaka method have been developed based on the microstructure and the porosity of porous ceramics to closely predict the values of M_e [32]. The most common empirical equations based on these methods that relate porosity and elastic moduli are given by:

$$\text{Differential method} \quad M_e = M_0 e^{-dP} \tag{4.14}$$

$$\text{Self-consistent method} \quad M_e = M_0 (1-cP) \tag{4.15}$$

$$\text{Composite sphere method} \quad M_e = M_0 \frac{(1-P)}{(l+aP)} \tag{4.16}$$

Where a, d and c are data-fitting constants or adjustable constants. For example, constant *a* is related to the pore morphology and Poisson ratio (*V*) of the medium. The expression of '*a*' for spherical pores is given by *(13 - 15V)(1 - V)/(14 - 10V)*. These models have been found to be closely consistent with the experimental values of porous ceramics such as 3Y-TZP ceramics [33].

With an increase in computational abilities, the finite element method (FEM) makes it possible to analyze three-dimensional models of the material with more accurate predictions consistent with experimental values, even for materials with different microstructures. Figure 4.10 shows the predictions of FEM with the experimental data of spinel ceramics of different porosity with overlapping solid spheres [34]. As can be seen, the effective elastic modulus (E/E_s) decreases with an increase in porosity (E_s is the solid elastic modulus). Similar to elastic moduli, fracture strength is another important mechanical response that decreases exponentially with an increase in porosity. In terms of porosity, fracture strength is mathematically given by $\sigma = \sigma_0 \exp(-bP)$, where σ_0 is the flexural strength of material with zero porosity and b is the pore structure dependent parameter [35]. Apart from the amount of porosity, size of the pores also influences the fracture strength.

Yoshida et al. studied the influence of pore size on fracture strength of alumina-based porous ceramics having the same amount of porosity. The four-point bending method was used to evaluate the fracture strength and Figure 4.11 shows the change in strength behavior with respect to pore size. As can be seen, the fracture strength decreases with the

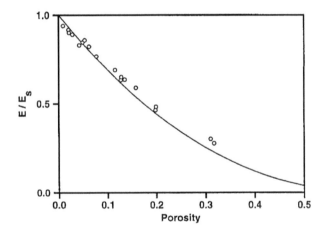

FIGURE 4.10
Effect of porosity on the elastic moduli of spinel ceramics. The solid line shows theoretical fit of FEM theory of overlapping solid spheres. E_s is chosen as 284 GPa (41.2×10^6 psi) [34].

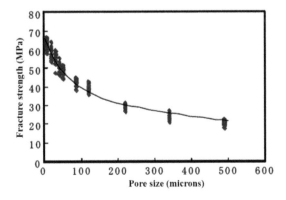

FIGURE 4.11
Effect of pore size on fracture strength of alumina-based porous ceramics. The pore content was maintained near to identical for all the specimens [36].

increase in pore size [36]. Fracture in general happens along the grain boundary because of the segregation of inclusions along the grain boundary and due to the presence of defects along with pores.

4.5.2 Permeability

Permeability is defined as the measure of the ease at which liquid or gas flows through the voids of a porous medium under a pressure gradient. It is a macroscopic property and not a property of the individual fluid or the porous medium but defines the level of effective interaction between the two. The nature of permeability in porous ceramics helps to remove soluble and suspended contaminants from liquid or air and thus it is an essential property to be investigated for several applications in geoscience, oil refining, concrete research and air and liquid filtration. Even though permeability can be enhanced by increasing the number of pores and their sizes in porous ceramics, however, the mechanical strength and stability of the ceramics can be reduced during this process. Therefore, it is worth developing porous ceramics that balance both the pore architecture and mechanical reliability. By controlling the microstructure in porous ceramics by different preparation methods, permeability improvement research is at the forefront. For instance, unidirectionally aligned cylindrical open pores with uniform size were made in alumina porous ceramics by the extrusion method. These pores made parallel to the flow direction are found to have a high gas permeability with high strength and controlled microstructure [37]. The fluid permeability in porous media is represented by one of the fundamental equations, known as Darcy's equation, and is given by:

$$\Delta P = \frac{\mu}{k} v_s \qquad (4.17)$$

Where ΔP is the pressure gradient along the flow direction, μ is the absolute viscosity of the fluid and V_s is the superficial fluid velocity or filtration velocity given by $V_s = Q/A$, where Q is volumetric flow rate and A is the exposed surface area, k is permeability coefficient or Darcian permeability. This equation holds well even for gases or mixtures but matches well with the experimental data in the presence of low flow velocities or less turbulence. At low velocity, the pressure difference and the fluid velocity are linearly dependent; however, high flow velocity encompasses turbulence and parabolic pressure gradient with fluid velocity. Such a turbulent flow can be governed by Forchheimer's equation, given by:

$$\Delta P = \frac{\mu}{k} v_s + \beta \left(\rho v_s^2 \right) \qquad (4.18)$$

Where β is the Forchheimer or non-Darcy coefficient. The permeabilities k and β depend only on the structural features of the porous medium, independent of the type of the fluid or the flow conditions. Practically, the limit of Darcy's law is defined by the Reynolds number, which is given by:

$$R_e = \frac{\rho v_s d}{\mu} \qquad (4.19)$$

Where μ is absolute viscosity and d is the hydraulic diameter. The value of d is simply the cell diameter for circular channels and cell width for square channels. For non-circular channels, the hydraulic diameter is the ratio of four times the cross-sectional area of flow through the channel to the perimeter of the channel wetted with fluid contact. In general, the liquid flow obeys Darcy's law when their Reynolds number is between 1 and 10 (R_e =1 – 10). The flow generally follows the linear regime between 2 to 5 of Reynolds number [38]. The integration of the pressure gradient changes these respective equations as following:

$$\frac{P_i^2 - P_o^2}{2PL} = \frac{\mu}{k} v_s \tag{4.20}$$

$$\frac{P_i^2 - P_o^2}{2PL} = \beta \rho v_s^2 \tag{4.21}$$

Where $\Delta P = P_i - P_o$, L is the thickness of the medium along the direction of flow, P_i is the inlet pressure of the medium, P_o is the outlet pressure of the liquid and P is the absolute pressure of the fluid at which V_s, μ and ρ are measured. Permeability also depends on the porous ceramics' fabrication protocols as they comprise pore geometry and their distribution. Cellular ceramics with honeycomb or foam-like structures are highly desirable for fluid separation, because of their ability for easy fluid flow due to their interconnected pore structure. Cellular ceramics made by the replica technique exhibit better Darcian permeability compared to the cellular ceramics prepared by gel-casting foams [39]. This is due to larger pore sizes and the web-like structure in replica-made ceramics compared to gel-cast foams that have the same amount of porosity. When the flow velocity was increased, i.e., under non-Darcian conditions, the permeability was the same for both of the ceramics. Figure 4.12 shows the value of Darcian and non-Darcian permeability for the cellular ceramics made by replica and gel-casting techniques.

Unlike the liquids, the fluid velocity of gases and vapors is higher at the exit than at the entrance due to their expansion along the flow path. The compressibility of gases at different pressures also needs to be considered for gas flow. For such a case, the mass/volumetric flow rate (Q_{mass}) can be related with pressure using the Darcy equation and Hagen-Poiseuille law as follows:

$$\frac{Q_{mass}}{A} = \frac{p_0 r^2}{16\mu} \frac{p_i^2 - p_f^2}{RTt} \tag{4.22}$$

Where Q_{mass} is volumetric flow through the porous ceramics, r is the pore radius, p_i and p_f are the initial and final pressures when gas travels through a porous body, μ is the viscosity, T is the temperature, R is the gas constant and t is the tube length. Even

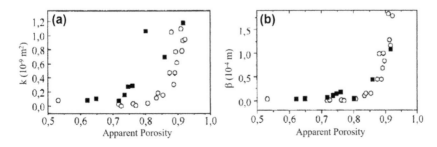

FIGURE 4.12
Effect of apparent porosity on (a) Darcian permeability, k and (b) non-Darcian permeability, β of cellular ceramics prepared by replica (■) and gelcasting (○) techniques [39].

the method of preparation decides the ability to permeate gas. For example, more than 34% porosity is needed for permeation in conventionally made ceramics, whereas effective permeation occurred even with a porosity of 18% in ceramics made by extrusion method. Gas permeability (q) in such ceramics can be theoretically predicted based on porosity (P) and pore size (d) by a capillary permeability model that gives the formula for permeability as [37]:

$$q = \frac{d^2}{32} P \qquad (4.23)$$

In most cases, the observed permeability values are always lower than that of the calculated ones due to the imperfect microstructure unlike the ideal ones. Thus, the structure also plays a major role in the amount of permeability. More detailed formulations for each structure, such as granular media, fibrous media, honeycomb structure and ceramic foam cellular ceramic media, are extensively discussed by Innocentini [40].

4.5.3 Thermal Properties

Porous ceramics find out several thermal management applications such as insulators, heat exchangers, solar receivers, high-temperature membrane reactors, gas sensors and porous burners. Therefore, in understanding various thermal properties, heat capacity and thermal shock information provide better insight into designing equipment for thermal applications. Heat transfer through a material is directly proportional to thermal conductivity in the steady state and is controlled by the combination of thermal conductivity and specific heat capacity in transient situations. In pure crystalline solids, thermal conductivity is inversely proportional to the temperature, whereas it is directly proportional to the temperature for glasses. For porous ceramic materials, the value varies with different materials based on their microstructure. In addition to temperature, thermal conductivity also depends on the size of grains [41]. D. S. Smith et al. demonstrates the effect of grain size on thermal conductivity for various materials [42]. For insulating applications, including high mechanical strength in ceramics consisting of low thermal conductivity and low density is desirable. In ceramics, heat conduction occurs by lattice vibrations in the solid phase, collision of gaseous molecules in the pores and radiation across large pores or through the semitransparent solid phase. It is obvious that the presence of pores reduces the heat transfer and hence decreases the resultant thermal conductivity. In addition to porosity, the intrinsic thermal conductivity and thermal resistance along grain boundaries influence the overall thermal conductivity of the material. Most of the insulators have thermal conductivity value below 2 W/m/K. To achieve extremely low thermal conductivity as low as <0.1 W/m/K, materials with very low intrinsic thermal conductivity such as kaolin-based foams, calcium aluminate foams and silica aerogels are desirable. To understand such mechanisms, a brief discussion on the theoretical aspects is important. For a homogeneous isotropic medium, Fourier's law describes the heat transport (Φ) which is proportional to thermal conductivity and temperature gradient given by:

$$\Phi = -\lambda \nabla T \qquad (4.24)$$

Where λ is the thermal conductivity and T is the absolute temperature along three dimensions. For a one-dimensional stationary case when the temperature through the medium is constant, this equation becomes $\Phi = -\lambda \partial T/\partial x$. Considering thermal conductivity as a phonon gas in non-metallic crystalline solids, we can apply the kinetic theory of gases to get the expression

$$\lambda = \frac{1}{3}\int_{0}^{\omega_D} c(\omega)v(\omega)l(\omega)d\omega \tag{4.25}$$

Where $c(\omega)d\omega$ is the contribution to the heat capacity of vibrational modes with frequencies from ω and $\omega+d\omega$, $V(\omega)$ is the group velocity, $l(\omega)$ is the phonon mean free path and ω_D is the Debye frequency. For transient cases this equation becomes:

$$\frac{\partial T}{\partial t} = \kappa\nabla^2 T \tag{4.26}$$

Where K is thermal diffusivity and T is the absolute temperature. Thermal diffusivity (K) is given by $\lambda/\rho C_p$ where ρ is the bulk density and C_p is the specific heat capacity. In addition to thermal resistance by solid structure of the grains, grain boundaries also influence thermal resistance. Considering both, thermal conductivity (λ_s) can be written as:

$$\frac{1}{\lambda_s} = \frac{1}{\lambda_{grain}} + nR^* \tag{4.27}$$

Where λ_{grain} is thermal conductivity of the grains, n is the number of grain boundaries per unit length and R^* is the thermal resistance for a grain boundary of unit area. The number of grain boundaries can be estimated by the linear intercept method and is given by:

$$n = \frac{N}{(L-P)} \tag{4.28}$$

Where N is the number of grains, L is the heat path length and P is the pore length crossed by the line. For porous ceramics with a heterogeneous nature of solid and pores, the heat transfer and temperature have to be mentioned as volumetrically averaged quantities, with an effective thermal conductivity. The overall/effective thermal conductivity (λ_{eff}) considering all the heat transfer mechanisms such as conduction, convection and radiation, can be written as:

$$\lambda_{eff} = \lambda_R + \lambda_S + \lambda_C \tag{4.29}$$

Where λ_R is radiative heat transport, λ_S is solid heat transport and λ_C is convective heat transport. Convection transfer can be neglected when the Grashof number (G_r) is less than 1000. The Grashof number is given by:

$$G_r = \frac{g\beta\Delta T D^3 \rho^2}{\eta^2} \tag{4.30}$$

Where $g = 9.81$ m/s², β is the volumetric thermal expansion coefficient of the pore gas, ΔT is the temperature difference across one pore, D is the pore diameter, ρ is the density of the gas and η is the gas viscosity. Similarly, the heat transfer by radiation in porous materials can be neglected based on the Stefan–Boltzmann radiation law when the dimensionless ratio (j*) given by the following equation is much smaller than unity (<<1).

$$j^* = \frac{4\gamma\varepsilon\sigma D\vartheta^3}{k_0} \tag{4.31}$$

where γ is a geometric factor (2/3 for spherical pores), ε is pore surface material emissivity, ϑ is the average Kelvin temperature, σ is Stefan–Boltzmann radiation constant and k_0 is the thermal conductivity of the solid material. Thermal conductivity of the pore phase containing air is negligible compared to ($k_{air} \sim 0.025$ W/m.K) that of solid material ($k_{Al_2O_3} \sim 33$ W/m.K). Therefore, different theoretical methods are appreciable depending on the pore geometries and their connectivity.

4.5.4 Electrical Properties

Even non-predominant, porous ceramics also find their places in electrical applications such as electric heaters and electrical conducting layers. For instance, SiC foam materials can be directly heated by applying electrical power due to their low specific electrical resistance (10^{-4}–10^4 Ωm) compared to other ceramics such as alumina (10^{10}–10^{13} Ωm) or mullite (10^7 Ωm). Also, for materials used in electric insulation, it is necessary to understand the electrical property, such as electrical conductivity. Pores have different electrical properties compared to the solid phase and are mostly passive at lower temperature, until high voltage is applied. Even though one can expect the electrical properties to be the same for a material in its solid phase, , they differ drastically depending on their internal architecture. In this context, it is worth mentioning that a bulk of ceramics exhibit different electrical properties from their foam ceramic counterparts because of their different microstructure. Therefore, the electrical properties will vary for the same material due to different geometry, solid network and their interactions with pores. This also helps to tailor the electrical property, such as electrical resistance, by designing different cellular structure and microstructure. Apart from electrical conductivity, other electrical properties, such as dielectric constant, can also be influenced by pore size and shape.

Specific electrical resistance is an important electrical parameter of a material. This is a constant for a material and varies with temperature, similar to electrical conductivity, because of the change in the number of free electrical carriers and mobility with temperature. Specific electrical resistance (ρ_{el}) is given by:

$$\rho_{el} = R \cdot \frac{A}{l} \tag{4.32}$$

Where R is the electrical resistance and A and l are the material cross-sectional area and length, respectively. Some ceramic materials like ZrO_2, which behave as insulators ($\rho_{el} = 10^6$ Ωm @ 20°C) at room temperature, become electrically conducting material ($\rho_{el} = 10^{-2}$ Ωm @ 1000°C) at high temperatures due to the generation or activation of ions. Some materials behave as semiconductors whereas others remain insulators that find applications such as ferroelectrics and piezoelectrics. As can be seen in the above equation, the electrical resistance is predominately modified by controlling the cross-sectional area, grains, formation of specific grain boundaries and the interaction of ceramic grains with pores. The types of porosity play a major role in the resulting electrical properties, which leads to different electrical nature for different types of cellular structures such as honeycombs, foams or fibers.

Porous piezo ceramics such as lead zirconate titanate (PZT) find specific applications such as hydrophones, ultrasonic imaging and underwater sonar because of the ease of tailoring their electric and acoustic properties by controlling porosity and pore morphology. Zhang et al. studied the effect of porosity on the ferroelectric, piezoelectric and acoustic properties of porous PZT ceramics fabricated by a pore-forming agent technique using stearic acid (SA) and polymethylmethacrylate (PMMA) [43]. It has been found that the dielectric constant, remnant polarization, piezoelectric constant and acoustic impedance decrease almost linearly with an increase in porosity indicating a decrease in ferroelectric domains compared to the fully dense PZT. The *modified* cubes model was used to predict the relative dielectric constant (ε_r) and piezeoelectric constant (d_{33}) of porous ceramics in terms of their porosity (P) and pore morphology [44]. These equations are written as:

$$\varepsilon_r = \varepsilon_r^0 \left\{ 1 + \frac{1}{P^{1/3}(\varepsilon_r^0 - 1)K_s^{2/3} + 1}\left(\frac{P}{K_s}\right)^{2/3} - \left(\frac{P}{K_s}\right)^{2/3} \right\} \tag{4.33}$$

and

$$d_{33} = d_{33}^0 \left\{ 1 - \frac{P^{1/3}}{K_s^{1/3}} + \frac{\left(\frac{P}{K_s}\right)^{1/3}\left(1 - \left(\frac{P}{K_s}\right)^{1/3}\right)}{1 - P^{1/3}K_s^{1/3}} \right\} \tag{4.34}$$

Where ε_r^0 and d_{33}^0 are relative dielectric and piezoelectric constant of fully dense material, respectively, K_s is the shape factor of the pore (K_s is 1 for spherical pore). Figure 4.13 shows the effect of bulk porosity on the piezoelectric constant and acoustic impedance in porous PZT ceramics. The difference in value rises from the different pore morphology formed by different pore-forming agents. Stearic acid forms pores of irregular shapes that are easily interconnected, whereas PMMA-derived pores are spherical in shape. This ease of interconnection in the case of SA-derived pores results in the mismatch of theoretical dielectric constant to experimental values at higher porosity values as shown in Figure 4.13. Acoustic impedance is not dependent on the pore morphology and hence both the ceramic samples show the same trend.

With the advantage of controlling acoustic impedance by varying the porosity, porous PZT ceramics with graded acoustic impedance are potential candidates for use in medical imaging and in underwater sonar because of the good match of acoustic impedances at their interface. A similar trend of a decrease in the dielectric constant with an increase in porosity has been recently seen in other porous ceramics also [45].

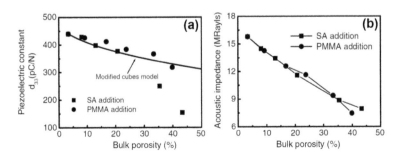

FIGURE 4.13
(a) Piezoelectric constant and (b) acoustic impedance as a function of bulk porosity in porous PZT ceramics [43].

4.6 Applications of Porous Ceramics

Porous ceramics experience low thermal conductivity, low density and high temperature resistance, abrasion and wear resistance with good chemical inertness due to their wide range of porosities and microstructures. Owing to these properties, compared to other porous materials such as polymers or metals, porous ceramics are attractive in a wide range of applications in harsh environments such as high temperature and corrosive and abrasive scenarios. By tailoring the microstructure and chemical properties of pore surfaces, it can be used as catalytic carriers, high temperature gas purification filters, bio-scaffolds, separators, sensors, chemical fillers and absorbers. Herein, a few classic applications of porous ceramics are discussed.

4.6.1 Heat Insulation and Exchange

Despite protecting the surrounding environment from the heat generated near to the processing area, thermal insulating ceramics also helps to reduce power consumption during the thermal treatment of materials at an industrial scale. An ideal thermal insulator should resist all forms of heat transport such as conduction, convection and radiation. Porous ceramic foams are widely used as thermal insulation material because of their low thermal conductivity and thermal capacity, low density and high temperature stability and good shock resistance capability. These lightweight foams can be made in large sizes and in different shapes with a wide range of porosity. Even though higher porosity can lead to a good thermal insulation, it affects the mechanical strength of the ceramics. Similarly, closed porosity is desirable for thermal insulation by blocking convective heat transfer, but it suffers from high thermal conductivity values due to high solid content. Compared to other fabrication techniques such as replica or sacrificial templates, porous ceramics prepared by the freeze-casting method offer unique properties with high porosity and good mechanical strength. This technique helps to align the pores in ceramics by directional a freezing (ice-templating) process as described in Figure 4.14 that eventually helps to avoid convective heat transfer.

Despite insulation, large-area ceramic foams are also being used as heat exchangers. For example, chimney inlets are covered with ceramics foams and when heated to an extremely high temperature, the foams radiate back the heat toward the heat source thereby avoiding the heat loss through the chimney and saving the energy of the heat source.

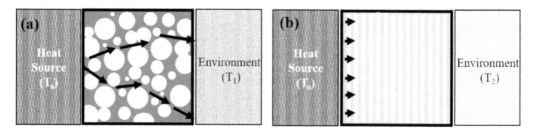

FIGURE 4.14
(a) Random porous microstructure and (b) aligned porous microstructure with closed channels for thermal insulation applications. (The temperatures are $T_0 > T_1 > T_2$.)

4.6.2 Filters and Separators

Together with consistent thermal and mechanical properties, porous ceramics for filter applications should also have a large filtration area and high permeability along with good fluid flow properties that finally maintain identical flow pressure without a drop of filtration efficiency. Ceramic foams with unidirectionally aligned channels made by the freeze-casting technique are desirable for filter applications because of their permeability and enhanced flow rate. Water filters are also fabricated by cheap porous ceramics made from clay along with pore-forming agents such as starch or other organic materials. With the ability to handle high temperatures, they are also able to filter hot gases, contaminants from molten metals and remove metallic or intermetallic inclusions from the molten metals during casting. Hot gas filters are used in several industries such as energy production, cement, steel, glass and metal production. The hot gas filters should be competent to handle temperatures between 260 and 900°C with consistent, good thermo-mechanical properties along with chemical stability that is capable of withstanding the high pressure and thermal shock experienced during the gas flow. Porous ceramics of alumina, mullite and SiC are commonly used to handle such hot gases from processes like coal gasification or the discharge of hydrocarbons and catalysts from power plants. To block the hot smoke particles from the diesel engine exhaust, 3D reticulated ceramics with porosity of 80–90% are preferred due to their good exhaust resistance and efficiency.

To obtain high pure metals for applications, such as aerospace, missile and electronics, molten metals are filtered through multiple layers of porous materials to remove inclusions (dissolved gases, non-metallic and intermetallic) that envisage a detrimental effect to the final metal cast, like the presence of FeS inclusion reduces the wear resistance of cast iron. The porous materials are chosen in a way that they do not undergo any chemical reaction with the metals. Open-cell porous ceramics of silica, mullite, SiC or ZrO_2, extruded honeycomb ceramics and large pore-sized reticulated ceramics with good flow properties are more preferable in this incidence. For cast-aluminum alloys, multilayers of ceramic foams based on cordierite with channeled pores are used to remove non-metallic inclusions [46]. Foam filters and extruded honeycomb structures operate in-depth filtration by passing the fluid through the walls of the filter utilizing larger surface area for filtration. There are also other filtration modes such as surface filtration and cake filtration mode (a kind of surface filtration) based mostly on the surfaces. The pore size manipulation is dependent on the type of material to be filtered and also on the applications of the end products. Ceramics with unidirectional pores aligned along the flow directions are more effective for filters and are made by the extrusion method. The freeze-casting technique adds another advantage of achieving hierarchical pores to filter out a wide range of particles due to the change in channel diameter along the channel direction. Porous ceramics with pores of sizes >100 nm are also used in microfiltration to filter and capture biological cells and macromolecules in the food industry and biological and pharmacological fields.

Separators based on inorganic ceramic membranes are used in several industries and research laboratories. They are generally made by the sol-gel process with a smaller and narrower range of pores along with supporting materials having larger pores (macroporous ceramics). By depositing the gel inside the porous channels, a porous film can be formed. When gas flows through the channels, the materials diffuse through the porous film and are collected on the membrane. These ceramics are also used as desiccants due to their good water absorbability and high water diffusivity with good chemical and physical stability. Mixed gases can also be separated by permeation using microporous ceramics. Similarly, non-mixed fluids and fluid with microparticles can also be separated by being passed through porous ceramics. Particles that are larger than the pore size prefer to trap on the ceramics surface, whereas the smaller particles deposit along the porous channels inside.

4.6.3 Biological Media

Synthetic biological materials have attracted more attention for the last few decades to treat or replace damaged human body parts such as bone or tissue. Damaged parts are usually treated by the transplantation of parts from one's own body or those of other human beings in limited and complex procedures. Bone replacement has been carried out using biological porous scaffolds as temporary support materials in the location of the damaged bone. Eventually, the growth of new tissues into the scaffold results in new bone formation and the simultaneous degradation of the scaffolds. For such applications, it is important that the porous materials are biocompatible, bioactive or biodegradable, and that the materials (bioceramics) consist of a competitive microstructure with interconnected pores to allow the flow of nutrients and oxygen for tissue development. The properties of these porous materials depend on pore size, pore volume and interconnected channel size. Pores have to be large (>100 μm) for the nutrients transport for inward bone growth and also small (2–50 nm) for cell adhesion and absorption of metabolism. The most common bioceramics for bone replacement are made of calcium phosphate, such as hydroxyapatite and tri-calcium phosphate, due to their similar composition to human bone and their biocompatible nature. Because of the different compressive strength of different human bones (cortical bone being 100 to 150 MPa, trabecular bone being 2 to 12 MPa), porous hydroxyapatite bioceramics with appropriate mechanical strength are fabricated with different microstructures by choosing and controlling the fabrication parameters. Hydroxyapatite ceramics with porosity of over 60% and interconnected pore size of 430 μm have been successfully implanted in a living body and are found to have good biocompatibility, chemical stability and good wear resistance [47]. Even though smaller pores are essential for cell adhesion, regeneration and absorption of protein and auxin for regeneration, interconnected pore sizes above 100 μm facilitate the tissue growth and vascularity of the implantation. Therefore, the fabrication of porous ceramics with a wide range of pore size starting from submicron to macro size, being analogous to natural bone structures, is a challenging task. Compared to conventional fabrication techniques, such as replica, sacrificial template and freeze-casting methods, additive manufacturing (AM) techniques offer better control over the pore characteristics such as interconnectivity, size and geometry. One of the most successful direct AM techniques is the use of a 3D printer with a micro-syringe system, and this has been successfully used to fabricate calcium polyphosphate ceramic components by layer-by-layer deposition. The fabrication was also done along with a sacrificial photopolymer network on specific layers using a micro-syringe [48]. Similarly, other direct AM techniques such as stereolithography [49] and selective laser sintering [50] are used to engineer scaffolds with desirable pore structure. Figure 4.15 shows 3D ceramic scaffold of hydroxyapatite fabricated based on the additive manufacturing technique. Such scaffolds show uneven surfaces that envisage the effective cell attachment *in vitro* [49].

By using a mold made of polymer or wax, these ceramic scaffolds can be indirectly fabricated by the AM method. By casting hydroxyapatite slurry into the wax mold having a negative replica of the structure, the 26 vol% porous scaffold is achieved by mold removal followed by sintering.

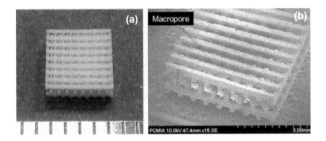

FIGURE 4.15
(a) Macroscopic and (b) SEM image of 3D cellular ceramic of hydroxyapatite [49].

4.6.4 Miscellaneous Applications of Porous Ceramics

3D reticulated ceramic foams with smaller pore sizes, high porosity volume and good mechanical strength are a fascinating choice in cinema halls to absorb the sound by converting the sound waves into thermal energy as the sound waves pass through the pores. Further, utilizing a ceramic ($Ca_3Co_4O_9$, $CaMnO_3$, $SrTiO_3$, In_2O_3)-based thermoelectric device, one can convert this thermal energy into electrical energy in order to operate the cinema hall with a low operating cost! Compared to organic and inorganic porous materials, porous oxide composites with good thickness and high porosity show a higher sound absorption coefficient. Due to a higher damping coefficient, metal-ceramic composites find applications as the supporting parts in mechanical tools to dampen the vibration during machine operation. As a catalytic carrier, porous ceramics of alumina or titania with interconnected pores sized between 6 nm and 500 μm are useful for carrying the tiny metal particles used in catalysts and play a major role in the promotion of the reaction. Ceramic foams are widely used as carriers for exhaust purifiers to convert toxic gases like CO, HC and NO_x into non-toxic gases similar to CO_2, H_2O and N_2, and they are useful in diesel engines to purify carbon granules. Porous ceramic electrodes are useful in chemical sensors for combustion process and exhaust control. Ceramic sensors such as ZrO_2 or zeolite with microporous structure are used to detect the composition of gas or liquid by absorbing part of the composition in the medium thereby measuring the change in electric potential. Porous ceramics are also used as combustors for the liquid fuel combustion that facilitates a high combustion rate and results in high energy output due to the effective radiation through the pore channels. Such combustion avoids the emission of toxic chemicals such as CO or NO_x and are useful in the incineration of dangerous wastes in liquid. An extensive list of applications based on porous ceramics are discussed elsewhere [51].

4.7 Conclusions

The type of voids in porous ceramics simulate different properties. Such porous ceramics find applications with extreme thermal, chemical or mechanical conditions that cannot be handled by metallic or polymeric counterparts. Starting from a typical thermal insulation, porous ceramics have extended their potential for other promising applications such as filters and bioceramics. Based on the type of applications, ceramics need to be fabricated to satisfy the desirable properties. Apart from the amount of porosity, the size of pores, the morphology of pores and the interconnected structure play a major role in tuning several characteristics, such as permeability and the thermal, electrical and mechanical properties. The fabrication techniques and the processing parameters trigger to control the desired microstructures for a particular application. With a wide range of fabrication techniques, porous ceramics are made by the conventional sintering process and advanced 3D printing techniques, and they are extensively demanded for socio-economic benefits.

References

1. Liu, R., Xu, T., and Wang, C. 2016. A review of fabrication strategies and applications of porous ceramics prepared by freeze-casting method. *Ceramics International* 42:2907–2925.
2. Guzman, I. Y. 2003. Certain principles of formation of porous ceramic structure, properties and applications. *Glass and Ceramics* 60:280–283.
3. Herzog, A., Klingner, R., Vogt, U., and Graule, T. 2004. Wood-derived porous SiC ceramics by sol infiltration and carbothermal reduction. *Journal of the American Ceramic Society* 87:784–793.
4. Gao, Z. Y., and Shi, X. L. 1999. Preparation techniques of porous ceramics. *Foshan Cera* 4:19–20.
5. Ding, S., Zhu, S., Zeng, Y., and Jiang, D. 2006. Effect of Y_2O_3 addition on the properties of reaction-bonded porous SiC ceramics. *Ceramics International* 32(4):461–466.
6. Hirschfeld, D. A., Li, T. K., and Liu, D. M. 1996. Processing of porous oxide ceramics. *Key Engineering Materials* 115:65–80.
7. Wang, L. X., and Dang, G. B. 1997. Research of porous ceramics with a series of pore sizes. *Journal of Functional Materials* 28(2):186–191.
8. Redington, W., Lonergan, J., Le Gonidec, G., Diaz, A., Bilgen, H., and Hampshire, S. 2002. Controlled porosity in silicon nitride. *Materials Science Forum* 383:19–24.
9. Schwartzwalder, K., Somers, H., and Somers, A. V. Method of making porous ceramic articles. U.S. Patent 3090094, 1963-05-21.
10. Swain, S. K., Bhattacharyya, S., and Sarkar, D. 2011. Preparation of porous scaffold from hydroxyapatite powders. *Materials Science and Engineering: C* 31(6):1240–1244.
11. Zhu, X. W., Jiang, D. L., and Tan, S. H. 2001. Impregnating process of reticulated porous ceramics using polymeric sponges as the templates. *Journal of Inorganic Materials* 16(6):1144–1150.
12. Song, I., Kwon, I., Kim, H., and Kim, Y. 2010. Processing of microcellular silicon carbide ceramics with a duplex pore structure. *Journal of the European Ceramic Society* 30:2671–2676.
13. Kim, J. G., Sim, J. H., and Cho, W. S. 2002. Preparation of porous (Ba,Sr)TiO$_3$ by adding cornstarch. *Journal of Physics and Chemistry of Solids* 11:2079–2084.
14. Yang, J. F., Zhang, G. J., Kondo, N., Ohji, T., and Kanzaki, S. 2005. Synthesis of porous Si_3N_4 ceramics with rod-shaped pore structure. *Journal of the American Ceramic Society* 88(4):1030–1032.
15. Fukasawa, T., Ando, M., Ohji, T., and Kanzaki, S. 2001. Synthesis of porous ceramics with complex pore structure by freeze-dry processing. *Journal of the American Ceramic Society* 84(1):230–232.
16. Fitzgerald, T. J., Michaud, V. J., and Mortensen, A. 1995. Processing of microcellular SiC foams. *Journal of Materials Science* 30:1037–1045.
17. Brinker, C. J., Sehgal, R., Hietala, S. L., Deshpande, R., Smith, D. M., Lop, D., and Ashley, C. S. 1994. Sol-gel strategies for controlled porosity inorganic materials. *Journal of Membrane Science* 94:85–102.
18. Liu, P. S., and Chen, G. F. 2014. *Porous Materials: Processing and Applications*, Elsevier: Oxford, England; Waltham, MA.
19. Cao, X. G., and Tian, J. M. 2001. Gel-casting of porous alumina ceramics. *Journal of Functional Materials* 32(5):523–524.

20. Matovic, B., Babic, B., Egelja, A., Radosavljevic-Mihajlovic, A., Logar, V., Saponjic, A., and Boskovic, S. 2009. Preparation of porous silica ceramics using the wood template. *Materials and Manufacturing Processes* 24:1109–1113.
21. Vadchenko, S. G., Ponomarev, V. I., and Sychev, A. E. 2006. Self-propagating high-temperature synthesis of porous Ti–Si–Al–C based materials. *Combustion, Explosion, and Shock Waves* 42(2):170–176.
22. Donohue, M. D., and Aranovich, G. L. 1998. Classification of Gibbs adsorption isotherms. *Advances in Colloid and Interface Science* 76–77:137–152.
23. Scaife, S., Kluson, P., and Quirke, N. 2000. Characterization of porous materials by gas adsorption: Do different molecular probes give different pore structures? *Journal of Physical Chemistry B* 104:313–318.
24. Rouquerol, J., Avnir, D., Fairbridge, C. W., Everett, D. H., Haynes, J. H., Pernicone, N., Ramsay, J. D. F., et al. 1994. Recommendations for the characterization of porous solids *Pure and Applied Chemistry* 66(8):1739–1758.
25. Giesche, H. 2006. Mercury porosimetry: A general (practical) overview. *Particle & Particle Systems Characterization* 23:9–19.
26. Gibson, L. J., and Ashby, M. F. 1999. *Cellular Solids, Structure & Properties.* Cambridge University Press: Cambridge, England.
27. Green, D., and Colombo, P. 2003. Cellular ceramics: Intriguing structures, novel properties, and innovative applications. *MRS Bulletin* 28(4):296–300.
28. Brezny, R., and Green, D. J. 1994. In: M. Swain (Ed.), *Materials Science and Technology*, VCH: Weinheim, 11:463.
29. Rice, R. W. 2005. Use of normalized porosity in models for the porosity dependence of mechanical properties. *Journal of Materials Science* 40:983–989.
30. Zwissler, J. G., and Adams, M. A. 1983. Fracture mechanics of cellular glass. In: R. C. Bradt, A. G. Evans, D. P. H. Hasselman, and F. F. Lange (Eds.), *Fracture Mechanics of Ceramics*, pp. 211–242. Plenum Press: New York.
31. Green, D. J., and Hoagland, R. G. 1985. Mechanical behavior of lightweight ceramics based on sintered hollow spheres. *Journal of the American Ceramic Society* 68(7):395–398.
32. Ramakrishnan, N., and Arunachalam, V. S. 1993. Effective elastic moduli of porous ceramic materials. *Journal of the American Ceramic Society* 76(11):2745–2752.
33. Luo, J., and Stevens, R. 1999. Porosity-dependence of elastic moduli and hardness of 3Y-TZP ceramics. *Ceramics International* 25:281–286.
34. Roberts, A. P., and Garboczi, E. J. 2000. Elastic properties of model porous ceramics. *Journal of the American Ceramic Society* 83(12):3041–3048.
35. Nyongesa, F. W., and Aduda, B. O. 2004. Fracture strength of porous ceramics: Stress concentration vs minimum solid area models. *African Journal of Science and Technology* 5(2):19–27.
36. Yoshida, K., Tsukidate, H., Murakami, A., and Miyata, H. 2008. Influence of pore size on fracture strength of porous ceramics. *Journal of Solid Mechanics and Materials Engineering* 2(8):1060–1069.
37. Isobe, T., Kameshima, Y., Nakajima, A., Okada, K., and Hotta, Y. 2007. Gas permeability and mechanical properties of porous alumina ceramics with unidirectionally aligned pores. *Journal of the European Ceramic Society* 27:53–59.
38. Sobieski, W., and Trykozko, A. 2014. Darcy's and Forchheimer's laws in practice. Part 1. The experiment. *Technical Science* 17(4):321–335.
39. Innocentini, M. D. M., Sepulveda, P., Salvini, V. R., and Pandolfelli, V. C. 1998. Permeability and structure of cellular ceramics: A comparison between two preparation techniques. *Journal of the American Ceramic Society* 81(12):3349–3352.
40. Innocentini, M. D. M., Sepulveda, P., and Ortega, F. S. 2005. Permeability. In: M. Scheffler, and P. Colombo (Eds.), *Cellular Ceramics: Structure, Manufacturing, Properties and Applications*. Wiley-VCH Verlag GmbH &; Co. KGaA, Weinheim pp. 313–341.
41. Zivcová, Z., Gregorová, E., Pabst, W., Smith, D. S., Michot, A., and Poulier, C. 2009. Thermal conductivity of porous alumina ceramics prepared using starch as a pore-forming agent. *Journal of European Ceramic Society* 29:347–353.

42. Smith, D., Alzina, A., Bourret, J., Nait-Ali, B., Pennec, F., Tessier-Doyen, N., Otsu, K., Matsubara, H., Elser, P., and Gonzenbach, U. T. 2013. Thermal conductivity of porous materials. *Journal of Materials Research* 28(17):2260–2272.
43. Zhang, H. L., Li, J. F., and Zhang, B. P. 2007. Microstructure and electrical properties of porous PZT ceramics derived from different pore-forming agents. *Acta Materialia* 55(1):171–181.
44. Banno, H., and Saito, S. 1983. Piezoelectric and dielectric properties of composites of synthetic rubber and PbTiO3 or PZT. *Japanese Journal of Applied Physics* 22:67–69.
45. Tan, J., and Li, Z. 2017. Effects of pore sizes on the electrical properties for porous 0.36BS–0.64PT ceramics. *Journal of Materials Science: Materials in Electronics* 28:9309–9315.
46. Liu, P. S., and Chen, G. F. 2014. *Porous Materials: Processing and Applications*, Elsevier: Oxford, England; Waltham, MA.
47. Mendes, S. C., Sleijster, M., Muysenberg, A. V. D., De Bruijn, J. D., and Blitterswijk, C. A. 2002. Cultured living bone equivalents enhance bone formation when compared to a cell seeding approach. *Journal of Materials Science* 13:575–581.
48. Vlasea, M., Shanjani, Y., Bothe, A., Kandel, R., and Toyserkani, E. 2013. A combined additive manufacturing and micro-syringe deposition technique for realization of bio-ceramic structures with micro-scale channels. *International Journal of Advanced Manufacturing Technology* 68:2261–2269.
49. Seol, Y. J., Park, D. Y., Park, J. Y., Kim, S. W., Park, S. J., and Cho, D. W. 2013. A new method of fabricating robust free form 3D ceramic scaffolds for bone tissue regeneration. *Biotechnology and Bioengineering* 110:1444–1455.
50. Shuai, C., Gao, C., Nie, Y., Hu, H., Zhou, Y., and Peng, S. 2011. Structure and properties of nano-hydroxypatite scaffolds for bone tissue engineering with a selective laser sintering system. *Nanotechnology* 22(28):285703-1-9.
51. Hammel, E. C., Ighodaro, O. L. R., and Okolin, O. I. 2014. Processing and properties of advanced porous ceramics: An application based review. *Ceramics International* 40:15351–15370.

5

Ceramic Coatings

K. Sarath Chandra and Debasish Sarkar

CONTENTS

5.1 Introduction

Ceramic coatings are the two-dimensional layered structures deposited on the surface of the substrate or object by several deposition procedures (e.g., plasma spraying, vapor deposition and sol-gel methods) to improve the performance and sustainability of the engineering materials against various disciplines of disparate wear processes acting under distinct service environments. In recent times, ceramic coatings have received a great deal of industrial and technological importance attributed to their unusual properties, which were recommended as potential candidates for their extensive usage in a wide variety of several miscellaneous engineering applications [1]. For instance, the extreme hardness coupled with the good toughening character of ceramic coating materials is the key property in endurable abrasive-resistant ceramic coatings primarily used for tool materials in machining and casting to provide excellent mechanical protection against abrasion and erosion. In a similar manner, the insulation property of electric insulation ceramic coatings is vital for applications in microelectronic circuits and heating elements, superior chemical inertness of corrosion-resistant ceramic coatings is a prerequisite property against hostile corrosive environments and high temperature resistance properties are indispensable for thermal barrier ceramic coatings employed in distinct zones of gas turbine engines, nuclear reactors and so on [2]. The important features of the several classes of ceramic coatings categorized according to their application area are listed in Table 5.1.

Failure of a ceramic coating often occurs in most aggressive service environments by the combined attack of disparate wear processes, including foreign object damage, abrasion and erosion at room and elevated temperatures, high temperature oxidation, hot corrosion, thermo-mechanical fatigue and creep [3]. However, the degree of action of these in-service active degradation mechanisms is typically dependent on the features of the structural component and severity of the service conditions. Further, these degradation processes may often lead to premature failure of the substrate material in the event of coating failure, challenges the ceramic engineers to develop durable ceramic coatings with extended performance to meet the growing demands. For instance, it is a widely considerable practice that the fabrication of thermal barrier ceramic coatings with enhanced performance should have the following important tailored properties: they are dense columnar grain morphology, good adhesion and cohesion to the substrate, and a low coefficient of thermal expansion with a close match to the substrate. These improved properties markedly limit the aggravating corrosive and oxidative wear processes under service, thereby reducing the origin of residual stresses and minimizing the thermal shock cracking and failure as a

TABLE 5.1

Features of the Various Classes of Ceramic Coatings

Ceramic Coatings	Property	Examples
Wear/Abrasion-resistant coatings	Resistance against abrasion, sliding wear, friction, fretting and erosion.	Al_2O_3, ZrO_2, TiO_2 and Al_2O_3–TiO_2.
Thermal barrier coatings	Resistance to oxidation, corrosion and thermal insulation.	ZrO_2, Al_2O_3 and YSZ.
Electrical insulation coatings	Good dielectric strength for effective insulation.	SiO_2–Al_2O_3, Al_2O_3, electric porcelain, Al_2O_3 and Al_2O_3–TiO_2.
Biomedical coatings	Biocompatibility in terms of tissuegrowth propensity.	Hydroxyapatite, TiO_2 and porcelain.
Aesthetics	Aesthetic qualities in terms of color, smoothness and glossy appearance together with resistance against disparate wear processes.	Porcelain, glass–ceramic, Al_2O_3 and ZrO_2.
Corrosion resistant coatings	Resistance against corrosion.	Al_2O_3, ZrO_2, MgO and Cr_2O_3.
Refractory emissive coatings	Compactable, highly chemical inert and radiative-coated structure promotes excellent resistance against in-service wear attack, improves thermal efficiency and service life.	Cr_2O_3, CoO, Fe_2O_3 and NiO.

result of the consequent spallation mechanism [4]. Therefore, it has been strongly insisted that the processing of sustainable ceramic coatings or coated structures with such afore-mentioned desirable properties for enhanced performance in distinct applications is typically dependent upon the choice of coating system that comprises the coating process and coating material selected on the basis of engineering considerations that certainly varies with the substrate component and the kind of process environment [4, 5]. However, such studies on ceramic coatings are comprehensively limited and utilized for the broad range of applications.

This text is interpreted in such a way to progress the exploitation of the fundamental understanding on principles of processing science and technology of ceramic coatings. These are:

- How to select a coating system (viz., a compatible coating material and coating process) depending upon the engineering considerations and kind of application?
- Processing methodology of miscellaneous ceramic coatings used for versatile engineering applications.
- Concise idea on decisive role of intrinsic properties of a ceramic coating on its failure behavior under service environment. Greater emphasis has been paid to the thermal barrier coatings.

5.2 Properties and Testing Protocols

The intrinsic properties of a ceramic coating or coated structure, which most commonly include chemical composition, morphology, wettability, adhesion, thickness, roughness, residual stress and hardness, as listed in Table 5.2 are the characteristic features that unambiguously reflect the degree of performance against failure caused by the combined attack of disparate wear processes acting under diverse process environments. However, these properties are typically controlled by the coating system that has been selected in accordance to the engineering considerations depending upon the kind of application [3, 6, 7]. This module discusses the significance of the intrinsic properties of a ceramic-coated structure and the associated characterization tools used to ensure the coating selection criterion.

5.3 Engineering Considerations

Engineering considerations are the decisive factors for selecting a particular coating system by ascertaining the compatibility of the coating process and coating material with respect to the components' base materials in order to develop a coated structure of requisite intrinsic properties. In this module, primary engineering considerations needed for the evaluation of a specific coating system, which typically include chemical or metallurgical compatibility, process compatibility, mechanical compatibility, and component coating efficacy are explained in detail as follows [4, 8].

TABLE 5.2

Significance and Characterization of Intrinsic Properties of the Ceramic Coatings

Property	Description	Test Method
1. Composition	Chemical composition of ceramic coatings has marked importance in determining the distinct properties for miscellaneous applications.	• Scanning electron microscopy • Transmission electron microscopy • X-ray fluorescence • Auger electron spectroscopy • X-ray photoelectron spectroscopy
2. Morphology	Morphological features of the coated structure vary depending upon type of deposition process and has considerable significance besides chemical composition attributed to the crucial role in control and manipulation of the engineering properties such as elastic modulus, hardness, and so on. For example, the molten or semi-molten state deposition processes produce the coated structures with splat-like morphological features whose boundaries tend to be parallel instead of normal to the surface of the substrate are the prerequisite for abrasion-resistant applications. Moreover, the Thornton model has depicted that the vapor deposition processes develop the ceramic coatings of dense columnar structure with equiaxed grain morphology at low homologous temperatures is advantageous for high-temperature applications. The homologous temperature (T_h) can be defined as: $$T_h = \frac{T_s}{T_c}$$ Where, T_s [K] is the temperature of the substrate and T_c [K] the melting temperature of the coating material.	• Scanning electron microscopy • Transmission electron microscopy
3. Wettability	Wettability or wetting behavior of the coated surfaces is an important factor that influences the key properties in miscellaneous applications majorly including chemical reactivity, thermal transfer properties in heat exchanger systems, *in vivo* properties of prosthetic devices in biomedical applications, corrosion properties in die-casting operations and so on. The degree of wettability of a surface can be estimated by measuring the wetting or contact angle (θ) between the air and the liquid droplet on a surface using Young's equation, can be expressed as: $$\cos\theta = \frac{\gamma_{sv} - \gamma_{sl}}{\gamma_{lv}}$$ Where, γ_{sv} [J/m^2], γ_{sl} [J/m^2] and γ_{lv} [J/m^2] are the specific energies of solid–vapor, solid–liquid and liquid–vapor interfaces, respectively.	• Sessile or pendant drop method • Wilhelmy plate method

(Continued)

TABLE 5.2 (CONTINUED)

Significance and Characterization of Intrinsic Properties of the Ceramic Coatings

Property	Description	Test Method
4. Adhesion	Adhesion or bond strength is a fundamental property which defines the ability of a coating to remain attached to the substrate material by interfacial bonding forces that majorly include valence forces and interlocking forces. Adhesion properties of the coated structure with different morphologies have considerable significance in the evaluation of interfacial debonding nature and consequent failure behavior. For example, a highly columnar and relatively porous PVD coating relatively experiences a low level of interfacial stresses, exhibits improved adherence with the substrate compared to a more cohesive and dense coating under tensile loading. It is noteworthy to mention that the terms *adhesive friction* and *adhesive wear* depicts the adhesive forces between the top surface of the coated structure and the sliding counterface.	• Direct pull-off adhesion test • Indentation test • Scratch test or Stylus method
5. Thickness	It is apparent to state that the thickness of the coated structure has an important role in the evaluation of the complex interplay between the effects primarily including adhesive forces or interfacial bond strength between the coated structure and substrate, hardness of the both coating and substrate and origin of intrinsic stresses within the coated structure under mechanical loading. However, the earlier-stated effects are significantly influenced by the change in density or porosity for a constant thickness of the coated structure.	• Scanning electron microscopy • X-ray fluorescence • Eddy current • Ultrasonic
6. Roughness	Roughness is the one important topography feature of a coated structure and it is typically dependent upon the type of deposition method employed. Surfaces of the ceramic coatings produced by electroplate deposition methods are considerably smoother than the thermally sprayed coatings. Thus the rough surfaces reflect the hydrophobic characteristics, exhibit large wetting angles with the corrosive melts and improve the corrosion resistance properties.	• Stylus profilometry • Atomic force microscopy
7. Residual Stress	Ceramic coatings often comprise residual stresses that may be either advantageous or disadvantageous, depending upon the level of stress and kind of application. An in-plane high compressive stress is beneficial to extending the fatigue life of a coated structure. However, the high stress can limit the thickness of the PVD coatings to be used due to coating detachment.	• Deflection method • X-ray diffraction method
8. Hardness and Toughness	Hardness and toughness are the ranking properties of ceramic coatings, particularly for protective and tribological applications. High hardness coupled with good toughness property is prerequisite for the fabrication of endurable abrasion-resistant ceramic coatings (TiC, WC–Co and SiC).	• Microindentation test

a. **Chemical or metallurgical compatibility:** This engineering consideration typically specifies that the deposited ceramic structure must be relatively stable with respect to the substrate component in order to control interdiffusion kinetics and limit chemical reactions under service. The interdiffusion process is indispensable for many coating systems to ensure improved adhesion characteristics with the substrate material. However, the excessive component interdiffusion is determinantal, leading to formation of the deleterious brittle reaction zones that precipitate brittle intermetallic compounds and other low-melting eutectic phases of poor hot strength near the chemical interface between the coated structure and the substrate material. For the case of thermal barrier coatings, these newly formed chemical interfaces primarily deteriorate the softening and creep resistance properties and often lead to premature failure of the both coated structure and substrate material by cracking and spallation.

b. **Process compatibility:** The material to be deposited or coated must be completely compatible with respect to the substrate component, but the coating or deposition process may be incompatible. This incompatibility usually arises due to the required high deposition process temperatures, differential coating thickness, special pre-coating surface treatments and necessary surface finishes for the complex geometries. For example, the process incompatibility for the physical vapor and chemical vapor deposition processes generally occur due to the requirement of higher deposition temperatures that are essential for the growth of the coated structure with improved adhesion properties and desirable microstructure features. However, these high deposition temperatures may impair the properties of the structural component due to warping of the substrate material. Annealing and mechanical straightening of the coated structures are the most common practices employed to restore the properties and dimensions. Furthermore, grit blasting is a surface treatment process performed prior to the thermal spray deposition processes in order to roughen the surface at an order to improve the adhesion properties but this pre-surface treatment induces residual strains in the coated structure and can be responsible for the consequent degradation.

c. **Mechanical compatibility:** Mechanical compatibility is an essential engineering consideration which dictates that the mechanical properties of the coating material at room temperature and elevated temperature must be designed to closely match those of the substrate components in order to maintain the protective features of the coated structure. The key mechanical considerations for the ceramic coatings include coating adhesion, coating cohesion, coating surface roughness, coefficient of thermal expansion (CTE) of the coating and substrate and the residual strains originated in the coated structure. For the case of thermal barrier coatings, the important mechanical considerations recommended for the excellent thermal shock resistance properties are stated as follows:

  i. Thermal expansion coefficient (CTE) of the coated material should be closely matched to the substrate component in order to limit the origin of residual stresses during processing and frequent thermal cycling.

  ii. *Strain tolerances* should be provided for the accommodation of induced inherent strains within the coated structure to control cracking and consequent spallation.

  iii. Coating surfaces with controlled roughness characteristics are pre-requisite to improve the adhesion characteristics and scale up the heat transfer performance.

   d. **Component coating efficacy:** This engineering consideration primarily details the ability of a coating or deposition process to deposit a coating on the targeted surface and is typically dependent upon complexity in the geometry and size of the substrate material. The uniformity of coating thickness and the amount of coat down (i.e., deposition of coating on inside of the cooling hole in structural components like turbine blades and nozzle guide veins) are two important factors under this consideration to evaluate the durability of the coated structure.

5.4 Principles of Coating Technology

Coating technology elucidates the variety of deposition procedures used to deposit the ceramic coating on diverse substrate components operating under various process environments. The general classification of surface coating procedures and their derivatives provided by Rickerby and Matthews, [9] are gaseous state, solution state, molten and semi-molten state, as shown in Figure 5.1.

A series of studies have claimed that the coating processes and their control play an integral role in determining the earlier-stated intrinsic properties of a coated structure and consequent performance. This section details the fundamental principles of various coating procedures in terms of deposition technology and growth mechanisms followed by a concise idea on the change in intrinsic properties of a coated structure with respect to distinct coating processes.

5.4.1 Gaseous State

Gaseous state methods or processes are the typical surface coating techniques which commonly involve conversion of the coating species into a gaseous or vapor state prior to the deposition onto the surface of the substrate material. These processes have a considerable importance attributed to opportunities yielded for the deposition of uniform, dense and well-adhered pure ceramic films. The process features of the primary generic coating methods in this category (Fig. 5.1), including chemical vapor deposition (CVD), physical vapor deposition (PVD) and ion beam-assisted depositions, are discussed below.

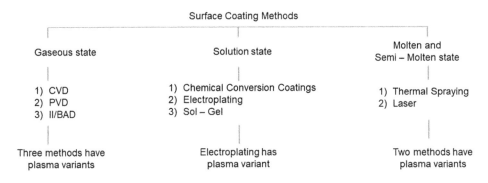

FIGURE 5.1
General classification of surface coating procedures. Definitions: CVD: Chemical vapor deposition; PVD: Physical vapor deposition; and II/BAD: Ion beam assisted deposition process, respectively.

5.4.1.1 Chemical Vapor Deposition

Chemical vapor deposition (CVD) is the process (Figure 5.2a) in which the volatile species containing the substance to be deposited is vaporized in a reaction chamber, and this vapor is thermally decomposed to atoms or molecules which further react with the other vapors or gaseous matter to yield non-volatile products that can be condensed onto the surface region of the substrate to form coating.

The normal deposition temperature of this process is in the range of 800°C–1200°C. The various characteristics and properties of the prepared coated structure via basic or thermally activated CVD process are controlled by the five reaction zones as schematically presented in Figure 5.2b. The characteristics, primarily including thickness of the coated structure is manipulated by managing the deposition rate through altering the transportation kinetics of diffusive species across zone I (boundary layer), microstructure features are refined by the heterogeneous reactions occurred at zone II and the effective adhesion nature is typically controlled by the formation of intermediate phases at zone IV due to diverse solid-state reactions, like phase transformation, precipitation and recrystallization, which can occur at elevated temperatures in zone III and zone V. There are many derivatives of the basic chemical vapor deposition technique, such as plasma-assisted, laser-assisted and electron beam-assisted CVD processes. These practices are commonly used to develop the coated structures with specific characteristics (e.g., epitaxial growth), to fabricate special types of coatings (e.g., diamond coating) and to control process parameters (e.g., deposition rate and temperature). This deposition practice (CVD) finds extensive applications in the field of high-abrasion-resistant coatings primarily including oxides, borides, carbides, carbo-nitrides and oxy-nitrides of the transition metals [3, 10].

5.4.1.2 Physical Vapor Deposition

Physical vapor deposition (PVD) processes utilize atomization or vaporization of the deposition material from a solid source through a variety of techniques (as shown in

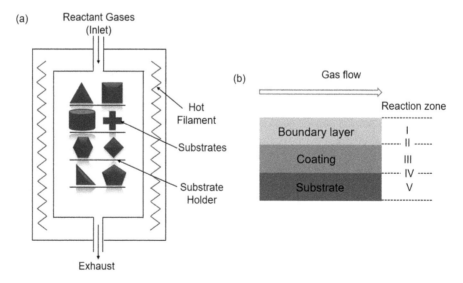

FIGURE 5.2
Schematics of the (a) typical hot-wall CVD process layout and (b) distinct reaction zones in the thermally activated chemical vapor deposition process.

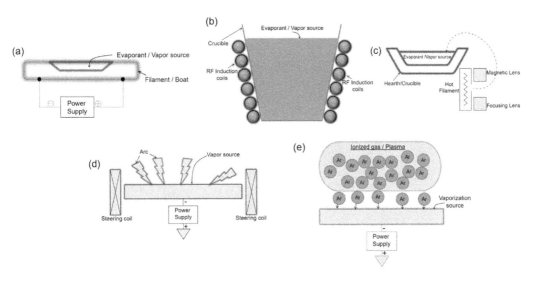

FIGURE 5.3
Schematic illustration of various atomization processes. (a) resistance evaporation method, (b) induction heating method, (c) electron beam evaporation method, (d) cathodic arc evaporation method and (e) cathode sputtering method.

Figure 5.3), including resistance evaporation, electron beam evaporation, arc evaporation, induction heating and cathode sputtering, to form a coating by the deposition of that material on to the surface of the substrate. Ceramic deposition can be obtained either by the use of a ceramic source or by using a composition mixture comprising a metal source combined with a reactive gas (e.g., O_2, N_2, CH_4) to produce coatings of oxides, carbides, nitrides and so on [3, 5].

PVD methods are often classified into evaporation and sputtering types, depending on the way of atomizing the source material, where evaporation-type PVD processes utilize the input of thermal energy for the atomization process and sputtering-type PVD processes use the transfer of kinetic energy because the sputtering mechanism is a momentum transfer process. Recently, there has been significant developments for the plasma-assisted physical vapor deposition (PAPVD) processes, where the conventional evaporation and sputtering methods are modified by introducing plasma within the deposition chamber to promote the control film nucleation and growth kinetics in order to obtain coated structures with the improved properties.

Figure 5.4 shows the process essentials of a typical PAPAVD process known as *ion plating* and key features of the sputtering mechanism on properties of the coated structures [11]. Ion plating-type deposition process (Figure 5.4a) is carried out under vacuum conditions and the coating material is treated as the negative electrode or cathode in a glow discharge of argon or plasma. Subsequently, the deposition material is vaporized into the plasma by a sputtering mechanism, which involves ejection of the coating atoms as a result of the transfer of momentum in the form of kinetic energy due to ionic bombardment between ions of inert gas and deposition material.

The vaporized material is further being ionized itself to a certain extent and arrived toward the surface region of the substrate with increased energy to form coating. Figure 5.4b displays salient features of the sputtering mechanism that involve bombarding

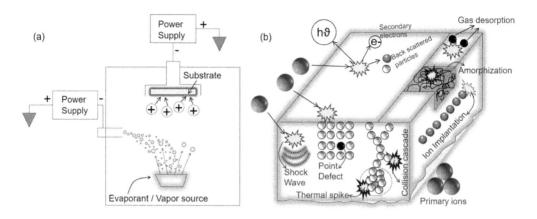

FIGURE 5.4
(a) Lay out of ion plating-type PVD process assisted with the cathode sputtering source and (b) salient features of the sputtering mechanism showing the surface effects occurring during ion bombardment.

the substrate with the accelerated ions and neutral ions. Typical properties of the coated structure have been detailed as follows [3, 12]:

- Improved adhesion properties of the coated structure attributed to the effects of sputter cleaning which involves cleaning and pre-heating the surface of the substrate
- Advancement in densification behavior of the coated structure is promoted by the continuous redistribution of surface coating atoms during film growth

The benefits of the plasma-assisted physical vapor deposition process (PAPVD) include lower deposition temperatures, controllable deposition rates, fabrication of highly pure-coated structures with improved adherence and uniform thickness, deposition of the multi-layer compositions and development of functionally graded coatings.

5.4.1.3 Ion Beam-Assisted Deposition

Ion beam-assisted deposition (IBAD) processes align more closely with the earlier-discussed plasma-assisted PVD process (PAPVD) and utilize the simultaneous combined action of ceramic deposition and ion-irradiation to develop coated structures with improved properties.

It has been widely reported that the ceramic coatings applied by IBAD processes significantly improved resistance properties against wear, friction, fatigue and corrosion of the diverse engineering materials, and it is noteworthy to mention that these improvements are controlled by the properties of the coated structure, which typically dependent upon the effects of an irradiating ion beam on the surface region of the substrate as earlier discussed (Figure 5.4b). There are four primary IBAD methods, presented in Figure 5.5, representing the process features as a function of the variable important effects of irradiating ionic beam on surface [5, 13].

Another special variety of IBAD processes are the reactive ion beam-assisted deposition (RIBAD) methods carried out with the use of irradiating reactive ion beams (e.g., N_2, C, O, etc.)

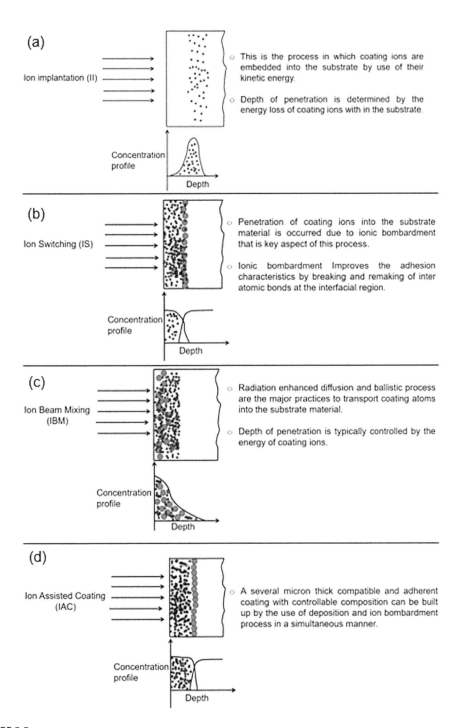

FIGURE 5.5
Four primary kinds of ion beam-based coating processes [13].

5.4.2 Solution State

The solution state processes generally utilize aqueous solutions to produce deposits of primitively hard ceramic compositions on the surfaces of metallic and non-metallic substrates by means of chemical or electrochemical reactions. The process essentials for the important classes of solution state processes, which include chemical conversion coatings, electroplating, and sol-gel deposition processes, are detailed below [3, 14].

5.4.2.1 Chemical Conversion Coatings

Chemical conversion coatings are the typical solution deposition methods in which a reagent in the solution reacts with the surface of the substrate material and consequently form an oxide compound that acts as a coating. For instance, chromated conversion coatings have been used to obtain the improved corrosion resistance properties for the metals like aluminum, magnesium, zinc and so on. These chromate coatings are usually applied by immersing or spraying onto the respective metal surface with the aqueous solution of chromic salts or chromic acids, leading to a chemical reaction that forms metallic chromate film which seals the surface of metals against corrosive wear.

5.4.2.2 Electroplating

Electroplating is a typical electrochemical deposition method that comprises deposition of the coated material on an electrode (substrate) by a process of electrolysis in which chemical changes are caused by the flow of current. Electroplated ceramic coatings are most commonly used to improve the wear resistance, oxidation resistance and high temperature resistance of the internal combustion engines.

5.4.2.3 Sol-Gel Deposition Procedure

The term *sol* refers to a colloidal dispersion of ceramic powder particles in a liquid which can be aqueous or an organosol. Sol-gel deposition processing can be performed in both batch- as well as continuous-mode operations and involves application of the sol onto the surface of the substrate by either one of the methods which majorly include dipping, spinning and spraying.

The applied sol undergoes hydrolysis and condensation reactions upon drying, leading to the formation of an agglomerated gel. At the final stages of drying, the formed agglomerated gel undergoes further condensation and develop a dense surface film on the substrate material. A schematic illustration has been presented to show the various stages of the batch-model dipping type sol-gel deposition process [14] in Figure 5.6. Sol-gel deposition techniques are useful for the preparation of composite coatings with controlled structural characteristics and most commonly applied to improve the electrical insulation property, corrosion resistance, abrasion resistance and scratch resistance of metals and polymer substrates.

5.4.3 Molten and Semi-Molten State

Thermal spraying and laser surface treatments are primitive deposition methods categorized under molten and semi-molten state processes [3, 5]. However, thermal spraying process has gained predominant importance due to its feasibility to produce high-performance ceramic coatings of various classes including electric insulation coatings, biomedical coatings, corrosion resistant coatings, thermal barrier coatings and many others, as listed earlier in Table 5.1 [15, 16].

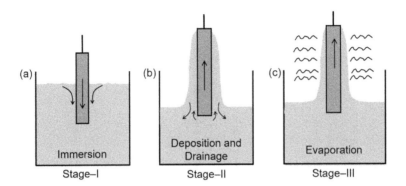

FIGURE 5.6
Stages involved in a batch-model dip-type sol-gel deposition process.

5.4.3.1 Thermal Spraying

Thermal spraying is the one important type of layer deposition process where the ceramic source material is in the form of either powder or wire, rapidly heated in a hot gaseous medium to a molten or semi-molten liquid and simultaneously projected at a high velocity using a spray gun onto the surface of the substrate component in order to form coating.

Figure 5.7 shows the schematic illustration of the thermal spraying process. Further, microstructural features of the thermally sprayed ceramic coating are usually evolved in such a manner that exhibits a continuous layered structure formed by an interlocking network that typically constitutes the build-up of one zone of overlapped splats onto another zone of overlapped splats via successive impacts of solid particles or molten droplets. Much more information on these microstructural details can be found elsewhere [2]. The process features of the important types of thermal spraying processes [17–20] are listed in Table 5.3.

5.4.3.2 Laser Surface Melting

Laser surface melting or alloying is an advanced ceramic coating practice that utilizes laser radiation as the heat source. This method can be exploited in a dual manner as a coating process similar to earlier technologies like thermal spraying or can be used to irradiate the formerly deposited coated structure on the metal/alloy substrate for further improvement in the microstructure development and consequent tribological behavior. More information on laser surface melting can be found elsewhere [3].

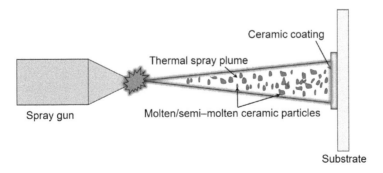

FIGURE 5.7
Schematic illustration of the thermal spray process.

TABLE 5.3

Features of the Various Thermal Spraying Processes

Thermal Spraying Process		Description	Advantages	Disadvantages
1. Combustion	Low–velocity oxy-fuel spraying (LVOF)	It is also known as flame spraying. The ceramic source material to be deposited is in the form of wire or powder. The heating source is the oxy-fuel gas mixture that provides high temperature (~3,000 °C) under combustion to melt the source material into a molten or semi-molten state which is being carried by a compressed carrier gas to reach the coating surface.	• Reproducibility of coated structure characteristics and quality • Ease of operation and economical	• Limited flame temperature (~3,000 °C) is insufficient to completely melt the coarse fractions of many ceramic powders • Possible distortion of work piece or substrate due to long flame
	High-velocity oxy-fuel spraying (HVOF)	Powdered form of ceramic deposition material is injected in the carrier gas (Ar, N₂, He) and process principle is similar to the earlier discussed one. High flame temperatures (~3,000 °C) with high flame velocities (1,500–2,000 m/s) in the HVOF process have been obtained by the combustion of fuel gases (propane, propylene, hydrogen) with oxygen	Coated structures of the following characteristics can be prepared. • Well adherence • High density • Fine grained size	• Limited flame temperature (~3,000 °C) is insufficient to completely melt the coarse fractions of many ceramic powders • Development of residual stresses in the coating
	D-gun spraying	This method is also known as the detonation spray technique. In this process, the sprayed material is in powder form and the flammable material is a oxy–acetylene gaseous mixture. A detonation wave with velocity (~2,600 – 3,500 m/s) and temperature (~3,500 °C) is formed instantaneously after ignition, causing rapid expansion of the reacted gaseous products that accelerate the molten/semi-molten ceramic particles with a velocity (~700–1,000 m/s) and impinged against the substrate being deposited.	Coated structures of the following characteristics can be prepared: • Extreme density with negligible porosity • Fine grained size	Brittle coatings are difficult to synthesize.

(Continued)

TABLE 5.3 (CONTINUED)

Features of the Various Thermal Spraying Processes

Thermal Spraying Process	Description	Advantages	Disadvantages
2. Electric Arc Spraying	Electric arc spraying process is performed in a similar fashion to that of combustion spraying method. The feed material is in the powder or rod form which is being atomized under high temperatures (~5,000°C) obtained from the intense electric arc heat source. Consequently, the atomized material is impinged on the surface being coated by a blast of compressed air.	• Larger deposition rates up to 40 kg/h can be achieved • In situ alloying • Attainment of the extremely high temperatures (>5,000 °C)	• Brittle materials can't be sprayed • Formation of the residual stresses in the coated structures
3. Cold Spray	Cold spraying is the process which utilizes a contracting-expanding nozzle (de-Lava-type) to accelerate the powder particles to supersonic speeds at the temperatures much lower than the melting temperature of the feed material and allow the deposition of particles in the solid state with the aid of carrier gas (He, N₂).	Coated structures of the following characteristics can be prepared: • High density • Fine grained size • High hardness • Thickness < 25 μm	
4. Plasma Spraying	This process utilizes a plasma-spraying gun to form the plasma (e.g., ionized Ar gas) which can attain temperatures in excess of 12,000 °C. Ceramic deposition powder particles are fed into a stream of plasma through a carrier gas that accelerates the powder particles to a velocity (~100–500 m/s). The coated structure on the surface of the substrate can be obtained by the successive impacts of semi-molten or molten ceramic particles.	Coated structures of the following characteristics can be prepared: • Well adherence • High density • Columnar submicrostructure with equiaxed finer grain sizes • Composition uniformity along the thickness gradient for the case of functional graded structures	• Reproducibility of coated structure characteristics and quality is difficult due to complexities in processing

5.4.4 Growth Mechanisms of the Ceramic Coatings

The growth mechanisms of deposited or coated films prepared via earlier-discussed various deposition processes have been demonstrated in this context as follows. There are two kinds of film growth processes depending on the relationship between the deposited film and the substrate component. They are homogeneous and heterogeneous film growth [14, 21]. Homogeneous film growth corresponds to the growth of deposited film on a single crystalline substrate component of the same materials and crystal orientations, but heterogeneous film growth refers to the growth of a deposited film on the substrate component which is composed of different materials. It is noteworthy to mention that the heterogeneous film growth processes have relatively predominant importance in the coating technology ascribed to the production of durable ceramic coatings of high performance.

Heterogeneous growth of deposited films follows the conventional laws of heterogeneous nucleation and growth mechanisms as similar to the phase transformations detailed elsewhere [22, 23]. Irrespective of the deposition process, the kind of interaction between the coated structure or film and substrate has a decisive role in determining the nucleation process and consequent crystalline and microstructural features of the resultant coatings. The critical nuclei size (r^*) and the corresponding Gibbs free energy barrier (ΔG^*) for heterogeneous nucleation can be expressed as:

$$r^* = \left(\frac{2\pi\gamma_{vl}}{\Delta G_v} \right) \left(\frac{Sin^2\theta\,Cos\theta + 2Cos\theta - 2}{2 - 3Cos\theta + Cos^3\theta} \right) \tag{5.1}$$

$$\Delta G^* = \left(\frac{16\pi\gamma_{vl}}{3\Delta G_v^2} \right) \left(\frac{2 - 3Cos\theta + Cos^3\theta}{4} \right) \tag{5.2}$$

Where, θ [°] is the wetting or contact angle, γ_{vl} [J m^{-2}] is the interfacial energy between vapor/liquid interfaces and ΔG_v [J mol^{-1}] is the change in Gibbs free energy per unit volume.

A series of studies have reported that the growth of a critical sized nuclei follows one of the three mechanisms (Figure 5.8) such as island growth, layer growth, and island – layer growth that typically depend on the combined effects of interfacial energies between the surface of the substrate component and interface of the coated structure governed by Young's law [14, 24]. They are:

1. Volmer – Weber growth or island growth

    Island growth of a deposited structure usually occurs when the contact angle must be greater than zero ($\theta > 0$) as expressed by the Equation 5.3 in accordance with Young's Law:

$$\gamma_{sv} < \gamma_{ls} + \gamma_{vl} \tag{5.3}$$

    The aforementioned statement asserts that the growth species are relatively more strongly bonded to one another than to the substrate component. Subsequent growth of these species results in islands that further coalesce to form a continuous film (Figure 5.8a).

2. Frank – van der Merwe growth or layer growth

    Layer growth of a deposited structure generally occurs when the coating material completely wets the substrate component by maintaining a contact angle equal to zero ($\theta = 0$). Therefore, the corresponding Young's equation can be expressed as:

$$\gamma_{sv} = \gamma_{ls} + \gamma_{vl} \tag{5.4}$$

The statement in reference to Equation 5.4 claims that the growth species are relatively strongly bound to the substrate component more than to each other. In this process, the species primarily form a complete monolayer before the deposition of second layer occurs (Figure 5.8b).

3. Stranski – Krastonov growth or island – layer growth

The island–layer growth of a coated structure commonly occurs when the specific surface energy of the substrate component (γ_{sv}) is greater than the combination of both the specific surface energy of the coating (γ_{vl}) and the specific interfacial energy (γ_{ls}) between the substrate component and coating. Therefore, the modified Young's equations correspond to the present situation and can be expressed as:

$$\gamma_{sv} > \gamma_{ls} + \gamma_{vl} \tag{5.5}$$

The above statement, in relation to Equation 5.5, emphasizes a new situation which states that a continuous coating film is formed through the island–layer type of growth process by an intermediate combination of layered growth followed by island growth (Figure 5.8c). However, coating films that have been grown by this method usually comprise residual stresses primarily due to lattice mismatch between the coated structure and the substrate component caused by differences in the chemical composition.

In this module, the various deposition procedures and conventional growth mechanisms of a coated structure have been discussed in a detailed way. However, the discussions in the earlier sections (5.1, 5.2 and 5.3) have strongly insisted that the choice of a coating process and process control variables in the selection of a coating system have an integral role in determining the intrinsic properties of a coated structure and consequent performance against the disparate wear attack depending upon the kind of service environment. In regard to these claims, a concise note representing the comparison of various process control variables of the important coating processes has been presented in Table 5.4.

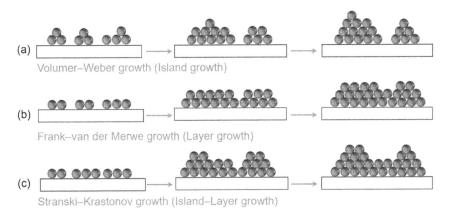

(a) Volumer–Weber growth (Island growth)

(b) Frank–van der Merwe growth (Layer growth)

(c) Stranski–Krastonov growth (Island–Layer growth)

FIGURE 5.8
Three basic heterogeneous growth mechanisms of the ceramic coating. (a) Volmer–Weber growth (island growth), (b)Frank–van der Merwe growth (layer growth) and (c) Stranski–Krastonov growth (island–layer growth).

TABLE 5.4

Comparative Study on Various Process Control Variables of Miscellaneous Coating Procedures

Coating Process		Coating Process Parameters				Coated Structure Characteristics	
		Deposition Rate (kg/h)	Deposition Temperature (°C)	Coating Thickness (µm)	Substrate Material	Coating Uniformity	Bonding Mechanism
Gaseous state process	PVD	Up to 0.5 as per source	50–500	0.1–1,000	Metals, ceramics and polymers	Good	Atomic
	PAPVD	Up to 0.2 as per source	25–500	0.1–100	Metals and ceramics	Good	Atomic + diffusion
	CVD	Up to 1.0 as per source	150–12,000	0.5–2,000	Metals and ceramics	Very good	Atomic
	PACVD	Up to 0.5 as per source	150–700	1–20	Metals and ceramics	Good	Atomic + diffusion
	Ion implantation	–	50–200	0.01–0.5	Metals, ceramics, and polymers	Line of sight	Integral
Solution state process	Sol – gel	0.1–0.5	25–1,000	1–10	Metals, ceramics, and polymers	Fair/good	Surface processes
	Electro – plating	0.1–0.5	25–100	10–500	Metals, ceramics, and polymers	Fair/good	Surface processes
Molten/Semi-molten state process	Thermal spraying	0.1–10	100–3,500	50–1,000	Ceramics and super alloys.	Variable	Mechanical/chemical/ metallurgical

* PVD = Physical vapor deposition; PAPVD = Plasma-assisted physical vapor deposition; CVD = Chemical vapor deposition; PACVD = Plasma-assisted chemical vapor deposition.

5.5 Processing and Failure Analysis

5.5.1 Abrasion Resistant Coatings

Abrasion-resistant coatings are an important class of ceramic coatings primarily used to reduce or eliminate the abrasive wear occurring between any of the two rough contacting surfaces, and to extend the durability of engineering materials by improving the intrinsic properties of the coated structure which consist of adherence, surface roughness, hardness and, importantly, the toughness [2]. A series of studies have reported that the plasma-sprayed Al_2O_3–13 wt.% TiO_2 abrasion-resistant ceramic coatings can be considered as the feature materials since they possess good adherence nature, high hardness and relatively improved toughness. These properties can be further tailored by introducing nanoparticles as coating media instead of conventional micron-scale approaches [25–27]. This section demonstrates the processing and failure analysis of conventional and nanostructured plasma-sprayed Al_2O_3–13 wt.% TiO_2 abrasion-resistant ceramic coatings.

Hot plasma-spray process technology is used to prepare the conventional and nano-structured Al_2O_3–13 wt.% TiO_2 ceramic coatings and the corresponding procedures are detailed as follows. Conventional ceramic coatings are deposited on the steel substrate in accordance with the methodology demonstrated in Section 5.4.3.1 with the use of a feed stock powder comprising typical angular powder particles having a size range of about 30–50 μm and chemical composition of ground-fused alumina (87 wt.%) clad with a titania layer (13 wt.%). Whereas a novel two-step process approach has been used for the preparation of nanostructured ceramic coatings in order to control the microstructure features and consequent intrinsic properties of the resultant coated structure. The primary step involves mixing the nanosized feed stock powders comprising alumina (~30 nm), titania (~30 nm) and additives in suitable proportions and spray drying this mixture to obtain agglomerate particles of size range about 30–50 μm.

The second step is known as plasma processing, in which the obtained agglomerates are injected into the plasma torch, sprayed and quenched in water to form bimodal microstructure features comprising the both fully molten and partially molten interfacial regions of ceramic droplets in the coated structure. The microstructural characteristics of the hot plasma-sprayed conventional and nanostructured Al_2O_3–13 wt.% TiO_2 abrasive wear-resistant ceramic coatings have been presented in Figure 5.9. The conventional coatings prepared using feed stock powders have displayed a typical plasma-spray *layered microstructure* which is comprised of fully melted regions (FM) together with the solidified splats of varying contrast due to chemical inhomogeneity (Figure 5.9a). However,

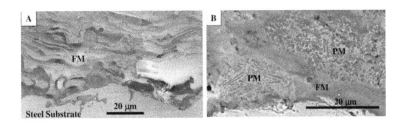

FIGURE 5.9
Scanning electron micrographs showing the cross-sectional views of interfacial region between ceramic and steel in the (a) conventional and (b) nanostructured, Al_2O_3–13 wt.% TiO_2 abrasive wear-resistant ceramic coating deposited steel substrate [27].

novel bimodal microstructural features have been exhibited by the nanostructured coating that constitutes a darker reinforcement region of partially melted substructure (PM) surrounded by a lighter matrix phase comprising both fully molten and solidified splats (Figure 5.9b). The evidenced partially melted substructure is about 25–50 μm in diameter, which constitutes nearly ~1 μm sized α-alumina grains (α) surrounded by a titania-rich amorphous phase. Further, the existence of a partially melted reinforcement is believed to be the beneficial feature for the significant improvement in interfacial toughness of nanostructured high-abrasion-resistant ceramic coatings compared to conventional type [27] and can be understood with the following discussion.

The estimates of interfacial fracture toughness property have gained predominant importance because of the key role in the development of abrasion-resistant coated structures with greater performance. The mathematical expression is provided to estimate the interfacial fracture toughness (G_R) for the case, emphasizing the delamination of coating leaving behind a thin annular plate of coating without exhibiting coating buckling is:

$$G_R = \frac{Et}{2(1-v)(1+v)}(\varepsilon_r(R)v\varepsilon_\theta(R))^2 \qquad (5.6)$$

Where G_R [J m^{-2}] is the interfacial toughness, E [GPa] is the elastic modulus, t [μm] is the thickness, v [-] is the Poisson's ratio, $\varepsilon_r(R)$ [-] is the radial strain and $\varepsilon_\theta(R)$ [-] is the circumferential strain, respectively.

The method proposed by Drory and Hucthinson is a widely practiced tool to measure the interfacial fracture toughness, where the Rockwell indentation load in the range of 20–50 N is applied normally to the ceramic/metal interfaces of ~40 μm-thick polished conventional and nanostructured coatings. Figure 5.10 shows the estimates of interfacial fracture toughness determined using Equation 5.6 for conventional and nanostructured coatings against the indentation loading (Figure 5.10a) and their corresponding failure behavior (Figure 5.10b). The mean interfacial fracture toughness (G_R) values for conventional and nanostructured coatings are noted to be independent of the indentation load and obtained as 22 J m^{-2} and 45 J m^{-2}, respectively. Phenomenological correlations derived from the

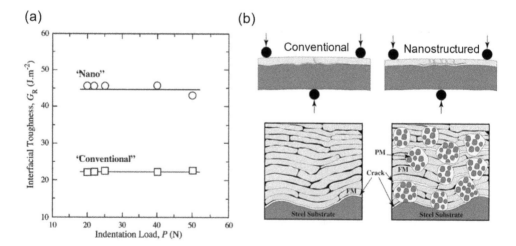

FIGURE 5.10
(a) Interfacial toughness measurements and (b) schematic illustration of failure behavior in the conventional and nanostructured coatings under Rockwell indentation loading [27].

earlier studies have strongly insisted that the two-fold increase in interfacial toughness of the nanostructured coating over the conventional type will promote a two-fold increase in adhesive strength characteristics that is believed to play a key role in their failure behavior. The vast deviation in the failure behavior of conventional and nanostructured coatings caused by the important effects of morphological features are summarized schematically in Figure 5.10b. The weaker chemical interfaces between the FM splat and steel substrate are being spontaneously cracked in both conventional and nanostructured coatings, but the tougher chemical interfaces between the PM splat and steel substrate are observed to be adherent; this has been corroborated as the primary responsible factor for this considerable improvement in interfacial toughness of the nanostructured coatings and much more information can be detailed elsewhere [28].

In accordance with the above discussion, selecting a coating system comprising nano-sized coating materials together with a modified spray process approach is found to be considerably important to introduce new chemical interfaces of tough morphological structures for the development of high-performance abrasive-resistant coatings with the tailored adherence, hardness and toughness properties against disparate abrasive and erosive wear attack in distinct process environments.

5.5.2 Corrosion Resistant Coatings

Corrosion-resistant coatings are a typical class of ceramic coatings that are primarily used to protect common metal and alloy engineering materials from corrosive media and should have the requisite properties of excellent chemical stability and high abrasion resistance in ambient and high temperature conditions [29–32]. Engineering properties of the plasma-sprayed Al_2O_3 corrosion-resistant ceramic coatings have strongly recommended as potential candidates to serve the purpose in such hostile environments. This module discusses the processing and failure analysis of plasma-sprayed Al_2O_3 corrosion-resistant ceramic coatings in a highly corrosive hydrochloric acid (HCl) environment, which is one of the most common severe process atmospheres for many metals and alloys.

The plasma-spray technology that has been used to prepare the bi-layer corrosion resistant ceramic coatings (typically a top ceramic coat together with a bond coat) can be described as follows. The cold-rolled Q235 plain carbon steel sheet specimens (30 mm × 25 mm × 3 mm) have been ground to obtain a rough surface, which is coated primarily with a bond coat (1Cr18Ni9Ti; 18-8 stainless steel) followed by a secondary coating with a ceramic top coat (platelet-shaped Al_2O_3 particles, d≈40–60 μm) using plasma-spraying technology (c.f. section 5.4.3.1). An intermediate Ni-based bond layer is employed to improve the adhesive strength characteristics of the ceramic coating to the substrate thereby enhancing the corrosion resistance properties which is subsequently confirmed with the microstructure analysis.

The SEM studies of the plasma-sprayed, two-layered, corrosion-resistant Al_2O_3 ceramic coating (Figure 5.11a) displays a microstructure which reveals that the bond layer (1Cr18Ni9Ti; 18-8 stainless steel) possesses good adherence to the Al_2O_3 ceramic coating and is indicative of good corrosion-resistant properties. Figure 5.11b shows the corrosion rate of Q235 plain carbon steel without Al_2O_3 ceramic coating (Sample 1, substrate), 18-8 stainless steel without Al_2O_3 ceramic coating (Sample 2, bond coat), Q235 plain carbon steel with Al_2O_3 ceramic coating (Sample 3), Q235 plain carbon steel with Al_2O_3 ceramic coating treated with porous sealing (Sample 4) and Al_2O_3 ceramic coating (Sample 5) in a boiling 5% HCl solution environment. It is observed that the corrosion rate of steel substrate with alumina coating treated with porous sealing (Sample 4) is greatly reduced and noted as

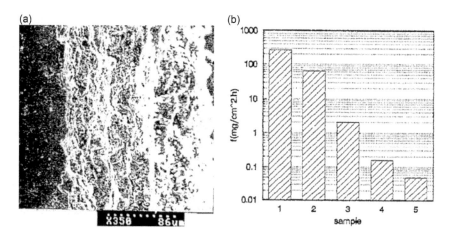

FIGURE 5.11
(a) Scanning electron micrograph of the sprayed Al_2O_3 coating deposited on steel substrate and (b) corrosive wear rate of the 18-8 stainless steel and Q235 plain carbon steel with and without Al_2O_3 spray coating in a boiling 5% HCl solution at a temperature of 150 °C [30].

1/130 parts of that of the sample having without coating (Sample 1). The improved corrosion resistance in this instance is attributed to the Al_2O_3 ceramic coating that isolates the substrate material away from the corrosive solution and protects the Q235 steel (substrate) from being corroded. These experimental results validate the good adherence characteristics of Al_2O_3 ceramic coating as demonstrated in Figure 5.11a.

Failure analysis of the spray-processed Al_2O_3 ceramic coating-deposited steel substrate in the aforementioned corrosive environment has been sequentially presented in reference to the Figure 5.12, as follows:

- Pore corrosion is the initial stage of damage, which involves the diffusion of generated corrosion products through the pore channels that have a large length to diameter ratio formed in the Al_2O_3 ceramic-coated structure during spray processing. The rate of corrosion is very low at this early stage and is controlled by the diffusion rate of corrosive media.

- Electrochemical corrosion is the next stage of corrosion that is initiated when the corrosive media penetrates the bond coat (1Cr18Ni9Ti; 18-8 stainless steel) and interacts with the steel substrate due to the occurrence of an electrochemical reaction between the bond coat and substrate. The rate of the corrosion process is relatively increased and governed by the difference between polar potentials of these bond coat and steel substrate.

- Electrochemical corrosion in the interlocking pores is an aggravated situation of the second stage of corrosion that involves rapid corrosion rates offered by the diffusion of corrosive media into the coalesced pores which allows corrosion products to accumulate near the interface between coating and the substrate, leading to the formation of bulges, consequent microcracking and delamination that even results in failure of the coating.

Therefore, the experimental results have evidenced that the spray-processed alumina ceramic coating deposited on the steel substrate has effectively improved the corrosion resistance properties in strong acid environments. However, recent studies [2, 29]

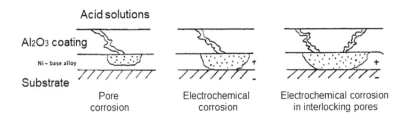

FIGURE 5.12

Schematic illustration displaying the corrosion of plasma-sprayed Al₂O₃ coating in the hydrochloric acid (HCl) environment [30].

corroborated that the control of intrinsic coating properties, including thickness and porosity features, can lead to considerable improvements in the corrosion resistance properties which apparently further show the importance of ceramic-processing science behind the selection of a coating system based on the engineering considerations.

5.5.3 Thermal Barrier Coatings

Thermal barrier coatings are an important class of ceramic coatings primarily used to shield the engineering materials operating under high-temperature service environments (e.g., hot sections of a turbine engine) in order to improve the strength and durability by remarkably reducing the working temperature of the substrate (>300°C) and protecting against the in-service wear attacks that include environmental oxidation, abrasion and corrosion caused by the rapidly moving hot fluids and gases [2, 33]. The requisite properties of a thermal barrier coating material are high-temperature melting point, low density, low thermal conductivity, smaller thermal expansion coefficient and excellent resistance to the aforementioned in-service wear. The properties of several important thermal barrier coating materials have been listed in Table 5.5; however, it is the most common choice to select 8 wt. % yttria-stabilized zirconia (ZrO₂–8 wt. % Y₂O₃ or 8YSZ) ceramics in accordance with the degree of ranking [33].

Electron beam-assisted plasma vapor deposition (EB–PVD) process is the recommended practice to develop reliable coated structures that have a columnar microstructure and good adherence characteristics [34]. This module details the processing and failure analysis of the typical bi-layer ceramic thermal barrier coatings.

The preparation procedure of a bi-layered thermal barrier coating material (Figure 5.13a) with the conventional composition comprising a ceramic top coat of 8 wt. % yttria-stabilized zirconia (ZrO₂–8 wt. % Y₂O₃ or 8YSZ) and a bond coat of NiCrAlY deposited on the IC6 superalloy (cast Ni₃Al base alloy) using the electron beam-assisted physical vapor deposition process can be demonstrated as follows. Primarily, the smooth surface of the substrate material, IC6 superalloy (15 mm × 10 mm × 3 mm), has been prepared by grinding it with different grades of abrasive papers in order to ensure the surface roughness (Ra) of about 1–3 μm prior to the deposition process and cleaning of the grounded surface with the use of ethanol or acetone in the ultrasonicate bath to obtain good adhesion characteristics. The substrates are kept in a sample holder and placed inside the PVD deposition chamber, which is further evacuated to a pressure of about 0.001 Pa in order to avoid oxidation, uniformly heated to a temperature range between 800°C and 900°C to provide better interfacial adhesion strength properties and a bond coat (NiCrAlY) is deposited with uniform thickness of about 50 μm at a deposition rate of 1 μm min⁻¹. These bond coat-deposited samples are then annealed at a temperature of 1050°C for a period of two hours

TABLE 5.5

Key Characteristics of the Widely Practiced Thermal Barrier Coating Materials [33]

Materials	Advantages	Disadvantages
7–8 YSZ	(1) high thermal expansion coefficient (2) low thermal conductivity (3) high thermal shock resistance	(1) sintering above 1473 K (2) phase transformation (1443 K) (3) corrosion (4) oxygen-transparent
Mullite	(1) high corrosion resistance (2) low thermal conductivity (3) good thermal-shock resistance below 1273 K (4) not oxygen-transparent	(1) crystallization (1023–1273 K) (2) very low thermal expansion coefficient
Alumina	(1) high corrosion resistance (2) high hardness (3) not oxygen-transparent	(1) phase transformation (1273 K) (2) high thermal conductivity (3) very low thermal expansion coefficient
YSZ + CeO$_2$	(1) high thermal-expansion coefficient (2) low thermal conductivity (3) high corrosion resistance (4) less phase transformation between m and t than YSZ (5) high thermal-shock resistance	(1) increased sintering rate (2) CeO$_2$ precipitation (>1373 K) (3) CeO$_2$-loss during spraying
La$_2$Zr$_2$O$_7$	(1) very high thermal stability (2) low thermal conductivity (3) low sintering (4) not oxygen-transparent	(1) relatively low thermal expansion coefficient
Silicates	(1) Cheap, readily available (2) high corrosion resistance	(1) decomposition into ZrO$_2$ and SiO$_2$ during thermal spraying (2) very low thermal expansion coefficient

to strengthen the interfacial adhesion between the bond coat and substrate (NiCrAlY/IC6) by the formation of a metallic bond due to the interdiffusion of elemental species.

Shot blasting of these annealed structures has been performed with a blast ball size of about ~ 250 μm diameter and blast pressure of nearly about 3 atm. to obtain a more compact crystal structure, improve the oxidation resistance by reducing the size of grains grown under annealing and the closure of porosity to limit oxygen ingress. The shot-blasted coated structures are then subjected to a pre-oxidation step in order to form a pre-oxide layer (Al$_2$O$_3$) that is closely ~1 μm thick for further improvement in oxidation resistance (Figure 5.13b). Further, a ~120 μm thick top ceramic coat (8YSZ) has been deposited onto the pre-oxidized bond coat of the substrate material under the deposition temperature range and vacuum pressure of about 800°C–900°C and 0.01 Pa, respectively. The microstructural features of the prepared coated structure have exhibited columnar grains with good adhesion properties that would be a marked reflection for the spalling resistant ceramic thermal barrier coatings of improved performance [4].

A series of research investigations concerned about the degradation behavior of thermal barrier coatings have strongly insisted that the oxidation and thermal shock are primitive wear types responsible for failure, and furthermore suggested that the aggravation of this degradation process is typically controlled by the thickness of a thermally grown oxide layer resulting from the growth of the pre-oxide alumina layer under service [2, 4].

Isothermal oxidation behavior of the superalloy (IC6) with and without coatings in air atmosphere at a temperature of 1373 K for a period of 300 h has been presented in

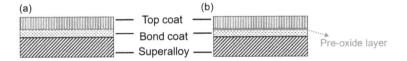

FIGURE 5.13
Schematic illustration showing the structural change in bi-layer thermal barrier coating, (a) before pre-oxidation step and (b) after pre-oxidation step during PVD processing. The formation of a pre-oxide layer (Al_2O_3) under the pre-oxidation stage can be observed.

Figure 5.14. It is clearly observed that the oxidative weight gain is significantly higher for the superalloy without coating (~3.25 mg cm^{-2}) and considerably lower for the superalloy with thermal barrier coating (~0.36 mg cm^{-2}).

Therefore, the presented experimental observations notably reflect that the current ceramic thermal barrier coating material (8YSZ) forms an excellent oxidation protective layer and credibility of the use of ceramic coating material can be further identified with the notice of relatively moderate oxidative weight gain (~0.75–1.5 mg cm^{-2}) in the superalloy coated with the only bond coat (MCrAlYSi/MCrAlY). It can be inferred from the results of this oxidation test that the relatively small weight gain in the specimen with thermal barrier coating is indicative of the lower thickness of the TGO layer. However, the effect of TGO layer thickness on spalling behavior under thermal cycling has been presented in Figure 5.15, which clearly displays that the tendency of spalling increases with the increasing thickness of the thermally grown oxide layer.

In accordance with these earlier studies, the basic failure mechanism of the bi-layered thermal barrier coating under thermal spalling can be demonstrated using microstructure analysis as presented in Figure 5.16 as follows [4, 34]:

- Accumulation of residual stress in the coated structure occurred mainly due to sintering of the ceramic top coat, formation of determinantal interdiffusion products via oxidation, growth of the thermally grown oxide (TGO) layer to a certain extent and fatigue of the TGO layer during temperature cycling.
- An increase in residual stress, caused by the aforementioned typical factors, induces and enhances the brittleness of the coated structure and gradually weakens the interfacial adhesive strength between the top ceramic coat (8YSZ) and the bond coat (NiCrAlY).
- Exceeding the residual stress in the coated structure above a certain value considerably weakens the interfacial adhesion and causes the formation of cracks between bond coat and TGO layer, consequently leading to failure under thermal spalling.

Therefore, the aforementioned case study has clearly elucidated that the ceramic processing science, in the perspective of choice of coating system, plays a decisive role in determining the properties of the coated structure, such as thickness of the TGO layer that typically controls the performance and service life of the ceramic thermal barrier coating systems.

5.5.4 Emissive Refractory Coatings

Emissive refractory coatings are a special class of ceramic coatings which are usually applied to the hot face lining of the several high temperature furnaces working in ferrous and non-ferrous metallurgy divisions in order to primarily provide the high thermal

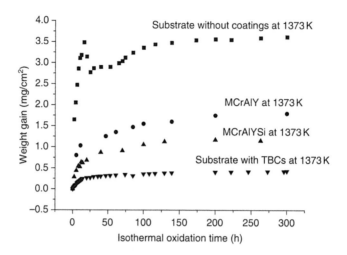

FIGURE 5.14
Isothermal oxidation behavior of superalloy (IC6) with and without TBC coatings at a temperature of 1373 K [34].

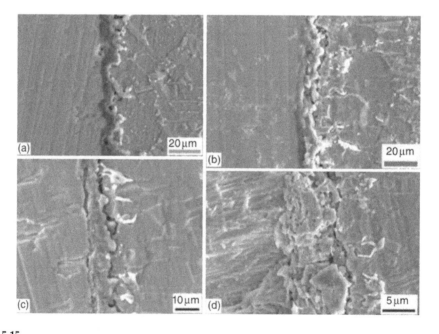

FIGURE 5.15
FE–SEM cross-sectional images showing the important effects of pre-oxide layer thickness on the spalling behavior of bi-layered thermal barrier coating after thermal shock at a temperature of 1323 K. Pre-oxide layer thickness of displayed TBC images are (a) 1 μm, (b) 2.5 μm, (c) 3.12 μm and (d) 4.25 μm, respectively [34].

efficiency by considerably improving the ability of a refractory lining to reradiate the absorbed thermal energy, just like a emissive body [35, 36]. Further, these emissive ceramic coatings are being chemically and mechanically bonded to the hot face refractory lining and they develop dense, non-wettable and reaction-constrained (chemically inert) thick surfaces that strongly protect the refractory against the integral action of disparate in-service wear attack and greatly improve the service life of the refractory. A concise idea on

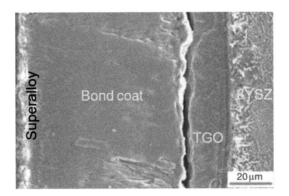

FIGURE 5.16
FE–SEM microstructural features showing the thermal shock failure behavior of bi-layered TBC coating [34].

the general processing methodology and responsible conditions for the failure of emissive refractory ceramic coatings have been presented as follows.

A typical composition of emissive refractory coatings usually constitutes a refractory pigment (e.g., ZrO_2, $ZrSiO_4$, Al_2O_3), a high emissive agent (e.g., Cr_2O_3, CoO, Fe_2O_3, NiO) that is conventionally selected in accordance to the earlier-stated engineering considerations and a binder or suspension agent, which are usually aqueous suspensions of silicates or phosphates, that act as a glue to form the bonding between the coating and hot-wall refractory lining. The deposition or installation procedures involve the preparation of a coating suspension comprised of a mixture of the aforementioned chemical ingredients in appropriate proportions, spraying of this mixture onto the surfaces of the refractory wall to a definite thickness generally ranging from 10–250 µm and drying of this mixture usually at a temperature of nearly about ~500°C to release the residual organic decomposable materials and form a strongly adhered emissive refractory ceramic coating with the intrinsic properties of designated measures [37]. Failure of these coatings usually occurred due to aggravating in-service high-temperature wear processes most commonly including abrasion, erosion, thermal shock, oxidation, and chemical corrosion by molten liquids and gases [38] Detailed experimental investigations concerning the decisive role of emissive material properties on the performance and durability of versatile refractory components have been described elsewhere [39].

5.5.5 Electric Insulation Coatings

Electric insulative coatings are ceramic coatings that have a primary requisite property of high dielectric strength in order to effectively insulate the conductive media or materials from the surrounding environment and should be capable to guard the substrate component against miscellaneous mechanical loads and corrosive environments [40]. Recently, sol-gel-based SiO_2–Al_2O_3 electrically insulative ceramic composite coatings have been attained greater importance ascribed to their high dielectric strength properties resulting from the comparable microstructural features relative to the traditional electrical porcelains. This module demonstrates sol-gel processing and failure analysis of the electrically insulative SiO_2–Al_2O_3 composite coatings [41, 42].

Sol-gel processing technology used for the preparation of electrically insulative SiO_2–Al_2O_3 ceramic composite coatings is discussed as follows. Primarily, Al_2O_3 powder particles (d ≈ 3.6µm) are added to the alcohol-based polymer silica solby keeping the solid loading ratio between silica solto alumina powder of about 2:1 in terms of weight. The prepared mixture is subsequently ball milled to produce a homogeneous mix of coating formulation, which is then spray deposited on stainless steel substrates to a thickness of about ~300 µm. The sprayed deposits are further dried at a temperature of ~343 K to form gel by the release of residual organic matter, firing at a temperature of ~723 K to pyrolyze the remaining organic constituents to develop ceramic coatings with the intrinsic properties of good adhesion, high density and low surface porosity. It is noteworthy to mention that the alumina-based composite coatings are prepared using aqueous sol-gel suspensions and the corresponding processing details can be found elsewhere [42].

High dielectric strength is the primary requisite property of an electric insulative ceramic coating system, particularly for applications concerning high temperatures such as low-profile integrated heating elements used in kettles, coffee warmers and so on. However, it is apparent that the dielectric strength property is significantly influenced by the operating temperature profiles and intrinsic properties of the coated structure, including thickness, surface porosity, degree of thermal expansion coefficient mismatch and microcrack flaw population resulting from the processing defects; these properties have a key role in determining the degree of performance and consequent in-service failure behavior of the coating.

Figure 5.17 shows the comparative temperature-dependent dielectric behavior plots of sol-gel processed, ~300 µm-thick silica–alumina composite coatings and the ~200 µm-thick alumina–alumina composite coatings determined using a *hi-pot* test where the operating temperatures range up to 673 K. The dielectric strength is found to be linearly decreases with increasing temperature, and larger dielectric strength estimates have been exhibited by the silica–alumina composite coatings relative to the alumina–alumina composite coatings. Therefore, the considerable increase in dielectric strength of the silica–alumina composite coatings is attributed to the beneficial features like an increase in density, appreciable sealing of surface porosity and optimal range of thickness. However, the plausible in-service failure analysis for the electric insulative ceramic coatings prepared by sol-gel processing can be explained as follows:

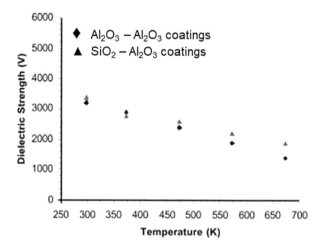

FIGURE 5.17
Temperature-dependent dielectric strength behavior of sol-gel processed SiO_2–Al_2O_3 (▲) and Al_2O_3–Al_2O_3 (♦) composite coatings [42].

- Formation of microcracks in the electrically insulative coated structures solely attributed to the processing flaws originated by the release of residual organics during the drying and firing stages of the fabrication process.
- Origin of residual stress due to increase and coalescence of the flaw population triggered by the establishment of considerable mismatch in thermal expansion coefficients among the substrate (steel or aluminum) component, matrix phase (SiO_2) and particle phase (Al_2O_3) of the composite coating material (SiO_2–Al_2O_3) under cycling of temperatures in service.
- Exceeding the residual stress after a certain extent leading to failure of the coating by interface delamination cracking when the temperature cycling meets a critical measure.

Therefore, the current discussion depicts that the choice of coating material (SiO_2–Al_2O_3) and process control has considerable importance in controlling the microstructural features to obtain the desirable intrinsic properties of the coated structure for the development of high-performance electric insulating ceramic coatings with improved dielectric strength properties even at high operating temperatures.

5.5.6 Biomedical Coatings

Biomedical coatings are the other important division of ceramic coatings that must be capable of bonding the implant with the bone and should possess the requisite properties, primarily chemical similarity, bioactive nature, porosity, permeability, chemical inertness and good abrasion resistance. A series of investigations have firmly confirmed that the plasma-sprayed hydroxyapatite coatings have such vital properties [2, 43]. However, the scientific and technological efforts required to produce plasma-sprayed hydroxyapatite (HAp) coatings with improved adherence characteristics is still a challenging task [44, 45]. This module discusses the processing and failure analysis of hydroxyapatite coatings deposited on titanium alloy substrate [46].

High-velocity oxy fuel (HVOF) technology (c.f. section 5.4.3.1) has been employed to deposit the hydroxyapatite coatings on titanium alloy (Ti–6 Al–4V) substrate specimens with the use of spray-dried HAp powders. Identification details of prepared specimens can be discussed in the following section.

Recent research investigations related to the adherence or bond strength characteristics of the hydroxyapatite coating (HAp) to the titanium alloy (Ti–6Al–4V) have received predominance over the conventional materials in terms of *in vitro* and *in vivo* wear evaluations since the loosening of the implantation is the most usual failure in clinical applications. A series of studies have reported that the morphological features and the corresponding intrinsic properties of hydroxyapatite coatings (HAp) are typically dependent on the process parameters of thermal spray technology [2, 45]. Two kinds of hydroxyapatite coatings are deposited on titanium alloy (Ti–6 Al–4V) substrate specimens through the earlier-stated process by varying the flow rates of gases. These specimens are identified as `C1′ (flow rates of $O_2$ gas/$H_2$ gas = 700 scfh/1,200 scfh) and `C6′ (flow rates of O_2 gas/H_2 gas = 600 scfh/1,200 scfh), respectively. The bond strength characteristics of these specimens have been evaluated under tensile loading conditions (ASTM C633) and observed as the specimen 'C1′ has a higher bond strength (~31 MPa) relative to the specimen 'C6′ (~25 MPa). The increase in bond strength characteristics is attributed to the beneficial features of applying high flow rates of gases which lead to the formation of a strong mechanical interlocking bond between the coating material and the substrate component.

The microstructure features of the specimen 'C1' (Figure 5.18a) reveals that the fracture is located within the splats as illustrated in the schematic whereas specimen C6 (Figure 5.18b) displays the fracture in the interface region of splats resulting from the rapid solidification of fully molten particles. The deviation in fracture tendencies caused by the considerable changes obtained in microstructural features are typically dependent upon the melting state of powders and leading to the important and interesting observations. These observations strongly insist that the improved adhesion properties of the specimen C1 are attributed to the evaluation of a bimodal microstructure comprising fully molten (FM) weak regions together with the partially molten (PM) tough regions It is noteworthy to mention that the failure behavior of these thermally sprayed hydroxyapatite (HAp) coatings follow the laws of brittle fracture [47].

According to the above discussion, the effective control of process variables is found to have a decisive role in determining the considerable improvement in important intrinsic properties like good adherence characteristics of the HVOF-sprayed hydroxyapatite (HAp) coatings to develop sustainable bioimplants with improved clinical performance.

5.5.7 Aesthetic Appearance Coatings

Implant aesthetics is the one leading field of bio-ceramics wherein bio-implants are being coated with aesthetic appearance ceramic coatings in order to impart aesthetic qualities, including smoothness, color, luster and gloss, together with greater resistance to disparate wear processes against abrasion and chemical corrosion. In recent times, research investigations pertaining to the processing and failure analysis of aesthetic appearance ceramic

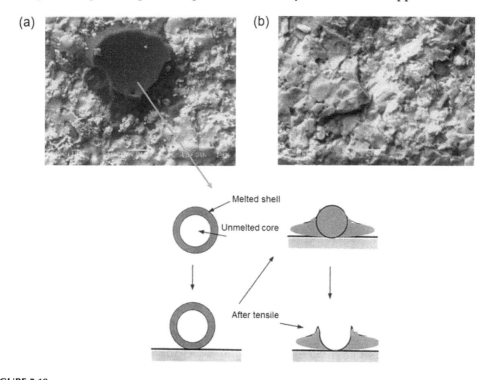

FIGURE 5.18
Scanning electron micrographs displaying the tensile fractured surfaces of (a) specimen 'C1' and (b) specimen 'C6'. Schematic illustration showing fracture behavior of *splat* [46].

coatings received tremendous interest particularly porcelain materials, which are commonly fused and deposited onto either metal or ceramic dental crowns in order to provide the aforesaid aesthetic qualities [48–50]. The basic mechanical properties of the various dental materials have been listed in Table 5.6. A series of studies have been reviewed in the perspective of clinical experiences, and they strongly insisted that the porcelain–fused–metal (PFM) dental crowns are still considered as the gold standard [51] for the refurbishment of damaged teeth against the brittle ceramic dental crowns which have unacceptable failure rates and limited durability (ref. Table 5.6). This module discusses the idea of processing and failure analysis of the porcelain aesthetic appearance ceramic coatings deposited onto the metal crowns.

The method of preparation for the porcelain–fused–metal (PFM) dental crowns is detailed as follows. The casted metal block substrates have been primarily ground and polished in order to ensure the top and bottom surfaces are flat and parallel. The polished specimens are then degassed at a high temperature (T ≈ 980°C for Au alloy) to avoid the bubble formation in the interface region between porcelain ceramic and metal substrate. The degassed metal surfaces are then sand blasted to control the surface roughness characteristics to provide good adhesion properties for the coating to a metal substrate. Following this, porcelain is applied in the form of slip onto the metal surface, fired at a temperatures typically in the range of about ~900°C to sinter each layer of porcelain in a sequential fashion for the development of a porcelain coat with definite thickness and good adherence characteristics. The resultant porcelain–fused–metal (PFM) dental crowns are further ground and polished to obtain the aforementioned aesthetic qualities [52].

Failure of the aesthetic porcelain ceramic coating deposited on the metal dental crown has usually been demonstrated by the Hertzian indentation measurements that notify the

TABLE 5.6

Mechanical Properties of the Various Important Dental Materials [49]

Material	Name	Supplier	Modulus E (GPa)	Hardness H (GPa)	Toughness T (MPa m$^{1/2}$)	Strength S (MPa)
Core ceramic						
Glass-ceramic	Empress II	Ivoclar	104	5.5	2.9	420
Alumina(infiltr)	InCeram	Vita Zahnfabrik	270	12.3	3.0	550
Zirconia(infiltr)	InCeram	Vita Zahnfabrik	245	13.1	3.5	440
Zirconia(Y-TZP)	Prozyr	Norton	205	12.0	5.4	1400
Alumina-matrix composite(AMC)	DC25	CeramTec	350	19.3	8.5	1150
Veneer ceramic						
Porcelain	Mark II	Vita Zahnfabrik	68	6.4	0.92	130
	Empress I	Ivoclar	67	5.6	1.4	160
Glass	Soda-lime	Fisher Scientific	73	5.2	0.67	110
Core metal						
Au-allow	Argident 88	Argen	92	1.2		
Co-alloy	Novarex	Jeneric/Pentron	231	3.0		
Substrate						
Polycarbonate	Hyzod	AIN Plastic	2.3			
Epoxy	RT Cure	Master Bond	3.5			
Tooth						
Denatl cement	Various		2–8			
Dentin			16			

key role of a metal substrate in fracture behavior of the ceramic-coated structure. The yielding behavior of the metal component under loading beneath the contact region promotes the initiation of radial cracks particularly in thin structures at the interfacial zone (lower surface) between the ceramic coating and metal substrate. These radial cracks are considered seriously detrimental since they grow relatively at lower loads and are extended to the subsurface region of the ceramic-coated structure and lead to consequent failure [52, 53]. It has been strongly insisted by many clinical research investigations that the effective control of intrinsic properties of the ceramic coating based on the engineering considerations is the only tool to limit the tendency of radial cracking and to develop the long service life of aesthetic porcelain ceramic coatings for the gold standard metal crowns.

5.6 Conclusions

Ceramic coatings are 2D-layered structures often deposited on the surface of the substrate components by various deposition procedures (e.g., plasma vapor deposition, sol-gel deposition and electroplating) to improve service performance and sustainability. In such incidents, these are strongly recommended as promising coating materials for miscellaneous engineering applications due to their unique properties, including extreme hardness coupled with good toughness in wear-resistant ceramic coatings for tool materials; electrical insulation in electrical insulating coatings for microelectronic circuits; high temperature resistance together with greater oxidation resistance and excellent chemical stability in thermal barrier coatings for diverse regions of nuclear reactors and gas turbine engines and so on. Failure of the ceramic coatings most commonly occurred in diverse engineering application areas by the combined attack of disparate wear processes, primarily involving regular abrasion and erosion to the aggressive high-temperature oxidation, corrosion, creep and thermo-mechanical fatigue modes of wear. However, the extent of action of a wear process typically varies with the kind of service environment and substrate material. A concise idea on the fundamentals of processing science primarily involving principles of coating processes, engineering considerations to select a coating system and the requisite intrinsic properties of a coated structure to develop endurable ceramic coatings of enhanced performance has been presented. The review presented on processing and failure analysis of miscellaneous engineered ceramic coatings have strongly corroborated that a coating system comprises the coating material and the coating process, which should be selected on the basis of distinct engineering considerations in order to develop high-performance and durable ceramic coatings for versatile engineering applications.

References

1. Chan, K. K. 1987. Expanding world of ceramic coatings. *Ceramic Industria* 129(3):24.
2. Basu, B., and Balani, K. 2011. *Advanced Structural Ceramics*. John Wiley & Sons: New York.
3. Holmberg, K., and Matthews, A. 2009. *Coatings Tribology—Properties, Mechanisms, Techniques and Applications in Surface Engineering*. Tribology and Interface Engineering Series, No. 56, Elsevier, UK.

4. National Research Council. 1996. *Coatings for high-temperature structural materials: trends and opportunities*. National Academies Press: Washington D.C.

5. Dahotre, N. B., Kadolkar, P., and Shah, S. 2001. Refractory ceramic coatings: Processes, systems and wettability/adhesion. *Surface and Interface Analysis: An International Journal Devoted to the Development and Application of Techniques for the Analysis of Surfaces, Interfaces and Thin Films* 31(7):659–672.

6. Dahlquist, Carl A. 2006. The theory of adhesion. In Arthur A. Tracton (Ed.), *Coatings Technology: Fundamentals, Testing, and Processing Techniques*. CRC Press: Boca Raton, FL.

7. McColm, I. J., and Clark, N. J. 1988. *Forming, Shaping, and Working of High-Performance Ceramics*, p. 345. Blackie and Son Ltd: Glasgow, Scotland.

8. Loehman, R. E. 2010. *Characterization of Ceramics*. Momentum Press: New York.

9. Matthews, A., and Rickerby, D. S. (Eds.). 1991. *Advanced Surface Coatings: A Handbook of Surface Engineering*. Blackie and Son LTD.: Glasgow, Scotland.

10. Park, J.-H., and Sudarshan, T. S. (Eds.). 2001. *Chemical Vapor Deposition*, Vol. 2. ASM International: Materials Park, OH.

11. Pauleau, Y., and Péter, B. B. (Eds.). 1996. *Protective Coatings and Thin Films: Synthesis, Characterization and Applications*, Vol. 21. Springer Science & Business Media, the netherlands.

12. McHargue, C. J., Kossowsky, R., and Hofer, W. O. (Eds.). 2012. *Structure-Property Relationships in Surface-Modified Ceramics*, Vol. 170. Springer Science & Business Media, the netherlands.

13. Kutz, M. (Ed.). 2015. *Mechanical Engineers' Handbook*. Vol. 3: Manufacturing and Management. John Wiley & Sons, New York.

14. Cao, G. 2004. *Nanostructures & Nanomaterials: Synthesis, Properties & Applications*. Imperial College Press, UK.

15. Herman, H., and Sampath, S. 1996. Thermal spray coatings. In K. H. Stern (Ed.), *Metallurgical and Ceramic Protective Coatings*, pp. 261–289. Springer: Dordrecht, the netherlands.

16. Pawlowski, L. 2008. *The Science and Engineering of Thermal Spray Coatings*. John Wiley & Sons, New York.

17. Oguchi, H., Ishikawa, K., Ojima, S., Hirayama, Y., Seto, K., and Eguchi, G. 1992. Evaluation of a high-velocity flame-spraying technique for hydroxyapatite. *Biomaterials* 13(7):471–477.

18. Babu, P. S., Basu, B., and Sundararajan, G. 2008. Processing–structure–property correlation and decarburization phenomenon in detonation sprayed WC–12Co coatings. *Acta Materialia* 56(18):5012–5026.

19. Fang, J. J., Li, Z.-X., and Shi, Y. W. 2008. Microstructure and properties of TiB2-containing coatings prepared by arc spraying. *Applied Surface Science* 254(13):3849–3858.

20 .Yandouzi, M., Sansoucy, E., Ajdelsztajn, L., and Jodoin, B. 2007. WC-based cermet coatings produced by cold gas dynamic and pulsed gas dynamic spraying processes. *Surface and Coatings Technology* 202(2):382–390.

21. Guo, Z., and Tan, L. 2009. *Fundamentals and Applications of Nanomaterials*. Artech House: Norwood, MA.

22. Callister, W. D., and Rethwisch, D. G. 2011. *Materials Science and Engineering*, Vol. 5. John Wiley & Sons: New York.

23. Van Vlack, L. H. 1989. *Elements of Materials Science and Engineering*. Addison-Wesley: Reading, MA.

24. Ohring, M. 2001. *Materials Science of Thin Films*. Elsevier: New York.

25. Herman, H., Sampath, S., and McCune, R. 2000. Thermal spray: Current status and future trends. *MRS Bulletin* 25(7):17–25.

26. Heimann, R. B. 1996. Applications of plasma-sprayed ceramic coatings. *Key Engineering Materials* 122–124:399–442, Trans Tech Publications: Zurich, Switzerland.

27. Bansal, P., Padture, N. P., and Vasiliev, A. 2003. Improved interfacial mechanical properties of Al2O3-13wt% TiO2 plasma-sprayed coatings derived from nanocrystalline powders. *Acta Materialia* 51(10):2959–2970.

28. Shaw, L. L., Goberman, D., Ren, R., Gell, M., Jiang, S., Wang, Y., Xiao, T. D., and Strutt, P. R. 2000. The dependency of microstructure and properties of nanostructured coatings on plasma spray conditions. *Surface and Coatings Technology* 130(1):1–8.

29. Fauchais, P., and Vardelle, A. 2012. Thermal sprayed coatings used against corrosion and corrosive wear. In H. Jazi (Ed.), *Advanced Plasma Spray Applications*. InTech Open, UK.

30. Dianran, Y., Jining, H., Jianjun, W., Wanqi, Q., and Jing, M. 1997. The corrosion behaviour of a plasma spraying Al2O3 ceramic coating in dilute HC1 solution. *Surface and Coatings Technology* 89(1–2):191–195.

31. Yan, D., He, J., Li, X., Dong, Y., Liu, Y., and Zhang, J. 2004. Corrosion behavior in boiling dilute HCI solution of different ceramic coatings fabricated by plasma spraying. *Journal of Thermal Spray Technology* 13(4):503–507.

32. Cai, J. P., and Li, B. 2000. Thermal spraying ceramic coatings. *Materials for Mechanical Engineering* 24(1):5–7.

33. Cao, X. Q., Vassen, R., and Stoever, D. 2004. Ceramic materials for thermal barrier coatings. *Journal of the European Ceramic Society* 24(1):1–10.

34. Zhang, D. 2011. Thermal barrier coatings prepared by electron beam physical vapor deposition (EB–PVD). In H. Xu and H. Guo (Eds.), *Thermal Barrier Coatings* (Part I). Woodhead Publishing Ltd, UK.

35. Hellander, J. 1991. How high emissivity ceramic coatings function advantageously in furnace applications. *Materials & Equipment/Whitewares Ceramic Engineering and Science Proceedings* 162–169.

36. Burford, R. A., and Fowler, J. B. 1995. Ceramic, black-body, high emissivity coatings for improved efficiency of fired heaters. No. Conference-950908-. *ASM International, Materials Park, OH (United States)*.

37. Holcombe Jr. C. E., and Chapman, L. R. 1999. High emissivity coating composition and method of use. U.S. Patent No. 6,007,873.

38. Fisher, G. 1986. Ceramic coatings enhance performance engineering. *American Ceramic Society Bulletin* 65(2):283.

39. Hove, J. E. 1965. *Ceramics for Advanced Technologies*, Vol. 7. John Wiley & Sons: New York.

40. Tushinsky, L., Kovensky, I., Plokhov, A., Sindeyev, V. and Reshedko, P., 2013. *Coated Metal: Structure and Properties of Metal-Coating Compositions*. Springer Science & Business Media: New York.

41. Moulson, A. J., and Herbert, J. M. 1990. *Electroceramics*. pp. 154–156. John Wiley & Sons Ltd: New York.

42. Olding, T., Sayer, M., and Barrow, D. 2001. Ceramic sol–gel composite coatings for electrical insulation. *Thin Solid Films* 398–399:581–586.

43. Mucalo, M. (Ed.). 2015. *Hydroxyapatite (HAp) for Biomedical Applications*. Elsevier, UK.

44. Sergo, V., Sbaizero, O., and Clarke, D. R. 1997. Mechanical and chemical consequences of the residual stresses in plasma sprayed hydroxyapatite coatings. *Biomaterials* 18(6):477–482.

45. Yang, C. Y., Wang, B. C., Chang, E., and Wu, J. D. 1995. The influences of plasma spraying parameters on the characteristics of hydroxyapatite coatings: A quantitative study. *Journal of Materials Science: Materials in Medicine* 6(5):249–257.

46. Li, H., Khor, K. A., and Cheang, P. 2000. Effect of the powders' melting state on the properties of HVOF sprayed hydroxyapatite coatings. *Materials Science and Engineering: A* 293(1–2):71–80.

47. Kingery, W. D., Kent Bowen, H., and Uhlmann, D. R. 1976. *Introduction to Ceramics*, Vol. 183. John Wiley & Sons: New York.

48. Kuchinski, F. A. 1993. Corrosion resistant thick films by enamelling. In J. B. Watchman and Richard A. Haber (Ed.), *Ceramic Films and Coatings*, pp. 77–130. Noyes Publications: New Jersey.

49. Lawn, B. R., Pajares, A., Zhang, Y., Deng, Y., Polack, M. A., Lloyd, I. K., Rekow, E. D., and Thompson, V. P. 2004. Materials design in the performance of all-ceramic crowns. *Biomaterials* 25(14):2885–2892.

50. Kelly, J. R. 1997. Ceramics in restorative and prosthetic dentistry. *Annual Review of Materials Science* 27(1):443–468.

51. Canadian Agency for Drugs and Technologies in Health. 2015. *Porcelain-Fused-To-Metal Crowns versus All-Ceramic Crowns: A Review of the Clinical and Cost-Effectiveness*. Canadian Agency for Drugs and Technologies in Health.

52. Zhao, H., Hu, X., Bush, M. B., and Lawn, B. R. 2001. Cracking of porcelain coatings bonded to metal substrates of different modulus and hardness. *Journal of Materials Research* 16(5):1471–1478.

53. Deng, Y., Lawn, B. R., and Lloyd, I. K. 2002. Characterization of damage modes in dental ceramic bilayer structures. *Journal of Biomedical Materials Research* 63(2):137–145.

6

Refractories and Failures

K. Sarath Chandra and Debasish Sarkar

CONTENTS

6.1 Introduction

Refractories are a special class of structural ceramics and can be defined as typical non-metallic inorganic materials that constitute a unique combination of properties such as high refractoriness under high load; higher hardness; excellent wear resistance in terms of impact and erosion; exceptional corrosion resistance against slag, metal, and hot flue gases; able to withstand a high degree of thermal shock loadings; and so on. These unusual characteristics have recommend refractories as potential candidates to serve as lining materials for high temperature vessels where the process environments greatly involved with distinct types of mechanical, chemical, and thermal wear. Figure 6.1a shows that the largest consumers of refractory are integrated iron-making and steel-making plants (73%) around the globe and the second largest consumers are cement and lime industries (13%) [1]. Thus, throughout this chapter, more emphasis is placed on the ferrous metallurgy and allied processes. Refractories have been categorized into several classes, as shown in Figure 6.1b, to ensure a decisive path for the selection of suitable refractory material with desirable properties for the particular area of application. Classification based on the chemical nature is of prime importance since the rightful

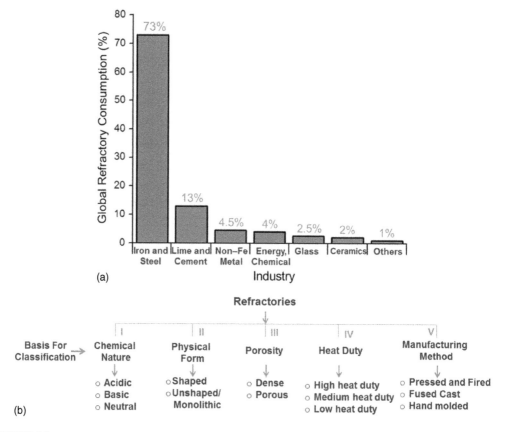

FIGURE 6.1

(a) Global refractory consumption of various ferrous and non-ferrous industries in terms of percentage and (b) distinct classification approaches of refractories on the basis of the chemical nature, physical form, porosity, heat duty, and manufacturing method.

selection of a refractory for any application typically depends on the chemistry of the high temperature process environments.

Acidic refractories have superior corrosion-resistant characteristics against acidic process environments, which makes the SiO_2 refractory an exceptional choice for glass melting furnaces. Basic refractories such as MgO possess excellent resistance against alkaline (basic) slags and highly recommended for basic oxygen furnaces (BOF) in steel making. Further, neutral refractories including carbon and silicon carbide (SiC) have extreme chemical stability against both the acidic and basic types of corrosive slag environments render them as rightful selection for blast furnaces in iron-making industries [2].

The failure of a refractory often occurs in metallurgical process environments due to a combined wear attack involving the most common types of mechanical, chemical, and thermal wear processes; however, the predominant wear mechanism varies with the nature of the process chemistry in distinct service atmospheres. Mechanical wear includes degradation by impact, abrasion, and erosion types of forces; chemical wear involves oxidative damage primarily in carbon-bonded refractories by air containing oxygen and corrosive damage by the combined action of liquid melts together with rapidly moving hot flue gases; and thermal shock is responsible for the cracking and failure of a refractory. A series of studies have concluded that considerable divergence in locally induced thermal gradients has a dominant influence on aggravating the factors responsible for the failure of a refractory and this argument can be further corroborated by undertaking conventional batch-type manufacturing processes where refractories are subject to frequent temperature cycling. However, the extent of divergence in the induced temperature gradients is primarily dependent on the composition, properties, size, shape, and installation procedure of the structural refractory components [3, 4]. Therefore, it requires a great deal of serious effort for a ceramic engineer to fabricate refractories with engineered wear-resistant properties by correlating the wear factors responsible for failure as a function of ceramic process variables such as composition and properties.

Therefore, this chapter is designed in such a way as to provide the foundation for a comprehensive understanding on the following important elements:

- A concise idea on the properties and selection criteria of refractories for various applications.
- A detailed view on mechanism, characterization, and control of the different wear processes encountered by the refractories under distinct service environments.
- The decisive effects of ceramic processing variables (e.g., composition) on failure behavior of the various refractories depending upon the application.
- A short description on the installation procedures for shaped and unshaped (monolithics) refractories.

6.2 Properties and Testing Standards

Properties reflect the characteristic features of a refractory and unambiguously show its durability under diverse and extreme high temperature metallurgical process environments. The various types of refractory properties are categorized into distinct classes including physical, mechanical, corrosive, thermal, and thermo-mechanical properties.

Testing refractories is indispensable to improve and optimize a definite set of properties by tailoring elements of the composition with the aid of ceramic processing science to meet the demands of a particular application environment, which may be solely distinctive for every application [5]. For example, a low cement castable can withstand a high spout temperature compared to a conventional castable, LD converter MgO–C brick, achieving a high density at 800 ton press compared to 150 ton press. This section discusses a distinct set of refractory properties and outlines the testing procedures and standards, as listed in Table 6.1.

6.3　Selection Criteria for Miscellaneous Applications

The high temperature metallurgical process environments of industrial refractories in a wide variety of applications primarily deal with integrated wear attacks in terms of thermal, mechanical, and chemical wear caused by the instantaneous charging of feedstock material, turbulent liquid melts, rapidly moving hot flue gases, and so on. The selection of refractory ceramics for such an extreme purpose requires a set of property criteria to manage the demands of specific application areas. For example, dense refractories of improved mechanical strength with greater resistance against abrasion, corrosion, oxidation, and thermal shock wear are an excellent choice for the most common metallurgical service environments operated under either acidic or basic conditions at elevated temperatures. However, selecting porous refractories containing a porosity level above ~45 vol.% as a primary feature is a solution for high temperature insulation [1, 20]. This section deals with the selection criteria of several classes of refractories for miscellaneous industrial applications based on their elementary properties, as listed in Table 6.2 [20].

6.4　Wear

Wear is a degenerative process responsible for the deterioration and failure of installed refractory structures due to an integrated attack by several degradation mechanisms under service. Thermal shock, mechanical wear, and chemical wear are the three predominant wear mechanisms encountered by refractories in iron-making and steelmaking process environments. Degradation by thermal shock involves the effects of differential thermal stresses; mechanical wear comprises the actions of impact, abrasion, and erosion processes, and chemical wear includes oxidative as well as corrosive damage by oxygen containing air and liquid melt at elevated temperatures. To design and fabricate refractories with improved wear-resistant properties by manipulating the process variables, it is imperative to analyze the wear mechanism and characterization of the particular wear process to recognize its severity [1, 2]. In this section, the mechanism, characterization, and control of the three major wear processes including thermal shock, mechanical wear, and chemical wear are discussed in detail. In addition, a phenemenonlogical model of failure analysis is presented to distinguish the effect of distinct types of wear on the aggravating factors accountable for refractory failure.

TABLE 6.1

Properties and Testing Standards of Refractories

Class	Properties	Definition/Significance/Formulae	Testing Method	Testing Standard
I	Physical			
	1. Density, porosity, and specific gravity	Bulk density (BD) is the ratio of the mass to the bulk volume of material.	Boiling method	ASTM C 20-00 [6]
		Apparent porosity (AP) is the ratio of the volume of surface pores to the bulk volume of material.	Vacuum method	
		Apparent specific gravity (ASG) refers to the specific gravity of the refractory where the permeable voids are excluded from the bulk volume.		
		Formulae:		
		$$\text{Bulk density (BD)} = \left(\frac{w_d}{w_{so} - w_{su}} \right)$$		
		$$\text{Apparent porosity (AP) (\%)} = \left(\frac{w_{so} - w_d}{w_{so} - w_{su}} \right) \times 100$$		
		$$\text{Apparent specific gravity (ASG)} = \left(\frac{w_d}{w_d - w_{su}} \right) \times 100 \text{ where } w_d$$		
		(g) is the dry weight, w_{so} (g) is the soaked weight, w_{su} (g) is the suspended weight.		
	2. Firing shrinkage	Change in linear dimension or shrinkage under firing signifies the relative densification of a refractory. It is determined by	Dilatometry	ASTM E 228-17 [7]
		$$\text{Linear shrinkage } (\epsilon) (\%) = \left(\frac{L_2 - L_1}{L_1} \right) \times 100$$		
		where L_2 (mm) is the length of the refractory specimen after firing and L_1 (mm) is the length of the refractory specimen before firing.		
	3. Permeability	Permeability signifies the rate of the fluid (gas/liquid) penetration through the porous structure of a material. Refractories comprising larger open or interconnecting porosity have a high permeability.	Standard testing method for permeability	ASTM C 577 [8]
	4. Elastic modulus	Elastic modulus (E) signifies the stiffness parameter. Refractories with a larger elastic modulus value are more susceptible to thermal shock due to increased inherent brittle tendency.	Sonic test	ASTM C 885-87 [9]

(Continued)

TABLE 6.1 (CONTINUED)

Properties and Testing Standards of Refractories

Class	Properties	Definition/Significance/Formulae	Testing Method	Testing Standard
	5. Permanent linear change on reheating (PLCR)	This is the linear dimensional change that occurs in a refractory under second firing or reheating. Using this measurement, the structural integrity of the refractory lining can be ensured. It can be determined by $$PLCR(\varepsilon_r)\,(\%) = \left(\frac{L_3 - L_2}{L_2}\right) \times 100$$ where L_3 (mm) is the length of the refractory specimen after reheating and L_2 (mm) is the length of the refractory specimen before reheating.	Standard testing method for permanent linear change measurement	ASTM C 113-14 [10]
II	Mechanical			
	1. Cold crushing strength (CCS)	This is the compressive strength of the refractory. A high CCS value is an indication of good resistance to abrasion and corrosion. This property measures the bond rupture strength of the refractory under compressive loading. It can be determined as Cold crushing strength $(\alpha_c) = \left(\dfrac{P_c}{A_C}\right)(MPa)$ where P_c (N) is the compressive load responsible for the disintegration of the refractory and A_c (cm^2) is the loading area or cross-sectional area of the specimen.	Standard testing method for cold crushing strength measurement	ASTM C 133-97 [11]
	2. Cold modulus of rupture (CMOR)	Also known as flexural or bending strength, it is an important property for brittle materials. This property measures the ability of the material to withstand deformation under bending loads. The formulae to measure the CMOR of the refractory are given as Flexural strength under three-point bending test: $(\sigma_{f\text{-}III}) = \left(\dfrac{3P_b l}{2bd^2}\right)(MPa)$ Flexural strength under four-point bending test: $(\sigma_{f\text{-}IV}) = \left(\dfrac{3P_b l}{4bd^2}\right)(MPa)$ where P_b (N) is the bending load responsible for the rupture, b (mm) is the breadth of the refractory specimen, d (cm) is the length of the refractory specimen, and l (cm) is the span length.	Standard testing method for cold modulus of rupture measurement	ASTM C133-97 [11]

(Continued)

TABLE 6.1 (CONTINUED)

Properties and Testing Standards of Refractories

Class	Properties	Definition/Significance/Formulae	Testing Method	Testing Standard
	3. Abrasion resistance	The surface property related to the mechanical wear and is defined as the resistance of the refractory material exhibited against the abrasive action of the moving solid: $$\text{Abrasion loss}(A) = \left(\frac{\Delta m}{\rho_{bd}}\right)(cm^3)$$ where Δm (g) is the weight loss of the specimen due to abrasive wear and ρ_{bd} (g/cm³) is the bulk density of the specimen.	Standard testing method for abrasion resistance measurement	ASTM C 704 [12]
III	Thermal	1. Coefficient of thermal expansion	It refers to the extent of change in the linear dimension or the linear shrinkage of the refractory per unit temperature difference and can be determined as Linear coefficient of thermal expansion $$(\alpha) = \left(\frac{L_2 - L_1}{\Delta T L_1}\right) \times 100$$ where L_2 (mm) is the length of the refractory specimen after firing, L_1 (mm) is the length of the refractory specimen before firing, ΔT (°C) is the temperature difference.	Standard testing method for thermal expansion coefficient measurement	ASTM C 113-14 [10]
	2. Thermal conductivity	Intrinsic property exhibits the ability to conduct heat. High thermal shock resistance results from the good thermal conductivity of a refractory.	Hot wire method	ASTMC 1113/ C1113M-09 [13]
	3. Refractoriness (or) pyrometric cone equivalent temperature (PCE)	Softening temperature at which a refractory begins to fuse or deform under the action of heat. Refractoriness decreases with the increase in impurity content.	Standard testing method for PCE measurement	ASTM C 24-09 [14]
IV	Thermo-mechanical	1. Hot modulus of rupture (HMOR)	Can also be referred to as hot strength, which is the bending strength at elevated temperature and is the most desirable property for iron-making and steelmaking refractories. The formulae are similar to those specified for CMOR.	Standard testing method for HMOR measurement	ASTM C 583 [15]

(Continued)

TABLE 6.1 (CONTINUED)

Properties and Testing Standards of Refractories

Class	Properties	Definition/Significance/Formulae	Testing Method	Testing Standard
	2. Creep	A crucial property that elucidates the structural integrity of the refractory linings and can be defined as high temperature deformation against time under particular load and temperature conditions.	Standard testing method for creep measurement	ASTM C 832 [16]
	3. Refractoriness under load (RUL)	Key selection property indicates the safe temperature to use a refractory under the integrated action of load and heat. For the RUL measurement, the specific load is considered as 0.2 MPa.	Standard testing method for RUL measurement	ASTM C 16-03 [17]
	4. Thermal shock resistance	This is the ability of a refractory to exhibit resistance against a spalling type of failure caused by the action of induced thermal stresses promoted under frequent temperature cycling. A high thermal shock–resistant refractory should have the characteristic features of a low thermal expansion coefficient, good thermal conductivity, low elastic modulus, and optimum strength combined with controlled porosity.	Wedge-splitting test	ASTM C 1171 [18]
V	Corrosion	1. Slag/metal corrosion resistance	This property is the ability of a refractory to show resistance against the penetration of molten slag or metal, degrading the refractory by dissolution and erosion mechanisms.	Rotary slag testing	ASTM C 874-11a [19]

TABLE 6.2

Selection Criteria of Several Classes of Refractories for Miscellaneous Industrial Applications

	Refractory Type	Mineral Phases	Elementary Properties	Area of Application
I. Oxide	1. Silica (SiO_2)	Cristobalite Tridymite Quartz	1. Low specific gravity 2. Low specific heat 3. Low thermal conductivity 4. thermal expansion coefficient 5. High thermal shock resistance	High duty silica refractories used in coke ovens, hot stoves, soaking pits, and glass tank crown applications.
	2. Alumina (Al_2O_3)	Corundum β–Al_2O_3	1. High specific gravity 2. High mechanical strength 3. High thermal conductivity 4. High refractoriness 5. High slag corrosion resistance	Fired refractories used in blast furnace hot stoves, stopper heads, sleeves, soaking pit covers, and reheating furnaces.
	3. Fire clay (SiO_2–Al_2O_3)	Mullite Cristobalite	1. Low specific gravity 2. Low specific heat 3. Low thermal conductivity 4. Low thermal expansion coefficient 5. Low strength at high temperature 6. Less slag penetration	High duty refractories used in coke ovens, reheating furnaces, and cement kilns.
	4. Zircon ($ZrSiO_4$)	Zirconium silicate	1. High specific gravity 2. High slag resistance 3. High thermal shock resistance	Fired refractories used in ladles, nozzles, stopper heads, and sleeves.
	5. Magnesia (MgO)	Periclase	1. Low durability at high humidity 2. Relatively low strength at high temperatures 3. Low thermal shock resistance 4. High refractoriness 5. High basic slag resistance	Fired refractories used in hot metal mixers, secondary refining vessels, rotary kilns, checker chamber of glass tanks, and electric arc furnace walls
	6. Chrome (Cr_2O_3)	Chromite	1. Low strength at high temperatures 2. Low thermal shock resistance 3. High refractoriness	Buffer brick between acidic and basic refractory bricks

(Continued)

TABLE 6.2 (CONTINUED)

Selection Criteria of Several Classes of Refractories for Miscellaneous Industrial Applications

	Refractory Type	Mineral Phases	Elementary Properties	Area of Application
	7. Magnesia–chrome (MgO–Cr$_2$O$_3$)	Periclase Spinel	1. Good thermal shock resistance at low MgO content 2. High hot strength 3. High refractoriness 4. High RUL value 5. High basic slag resistance	Fired refractories used in rotary cement kilns, lime kilns, hot metal mixers, secondary refining vessels, and electric arc furnaces
	8. Magnesium aluminate spinel (MgAl$_2$O$_4$)	Periclase Spinel Corundum	1. High slag resistance 2. High hot strength 3. High thermal shock resistance	Fired bricks used in ladles and rotary cement kilns
II. Oxide–carbon	1. Magnesia–carbon (MgO–C)	Periclase Graphite	1. High thermal shock resistance 2. High slag resistance	Unfired bricks used in ladles, basic oxygen steelmaking furnaces, and electric arc furnaces
	2. Alumina–carbon (Al$_2$O$_3$–C)	Corundum Graphite	1. High thermal shock resistance 2. High refractoriness 3. High corrosion resistance	Fired/unfired structures used in continuous casting refractories primarily in slide gate plates and submerged entry nozzles (SEN)
	3. Alumina–silica–carbon–silicon carbide (Al$_2$O$_3$–SiO$_2$–C–SiC)	Mullite Cristobalite Silicon carbide graphite	1. High hot strength 2. High thermal shock resistance	Unfired refractories used in blast furnaces, iron ladles, torpedo ladles, and electric arc furnaces
III. Non-oxide	1. Carbon	Graphite Carbon	1. Low oxidation resistance 2. High refractoriness 3. High slag resistance	Fired refractories used in the hearth of blast furnaces (BF) and electric arc furnaces (EAF)
	2. Silicon carbide (SiC)	Silicon carbide	1. Low oxidation resistance at high temperatures 2. High thermal conductivity 3. High refractoriness 4. High hot strength 5. High thermal shock resistance 6. High slag resistance	Fired refractories used in blast furnaces (BF), incinerators, and kiln furniture

(Continued)

TABLE 6.2 (CONTINUED)

Selection Criteria of Several Classes of Refractories for Miscellaneous Industrial Applications

Refractory Type	Mineral Phases	Elementary Properties	Area of Application
3. Silicon carbide–carbon (SiC–C)	α–SiC Graphite	1. High thermal conductivity 2. High refractoriness 3. High hot strength 4. High thermal shock resistance	Fired refractories used in incinerators
4. Silicon carbide–silicon nitride (SiC–Si$_3$N$_4$)	α–SiC α–Si$_3$N$_4$	1. Relatively high oxidation resistance 2. High hot strength 3. High thermal shock resistance	Fired refractories used in blast furnaces (BF) and kiln furniture

6.4.1 Thermal Shock–Assisted Wear

Thermal shock is a serious wear mechanism that leads to severe cracking and eventual failure of iron-making and steelmaking refractory components by the action of thermal stresses engendered by induced incompatible changes in dimensions due to frequent rapid cycling of temperatures. Refractory spalling, peeling, flaking, and slabbing are the different forms of thermal fatigue or thermal shock. The key factors that aggravate thermal shock are low thermal conductivity, high and nonlinear thermal expansion coefficients, low strain to failure, large component size, external mechanical loading, and uneven and sudden changes in temperatures. Thermal shock or fatigue resistance is the material property that signifies the ability of the refractory to resist fracture from thermal stresses. The widely practiced experimental protocols to determine the thermal shock resistance of refractories are the wedge-splitting method, the normalized statistical method, the water quenching method, and the small prism method. These test procedures have been extensively detailed in earlier studies [3, 21]. This section comprehensively reviews the mechanism, characterization, and control of the thermal shock wear.

6.4.1.1 Mechanism and Characterization of Thermal Shock

The mechanism of refractory wear by thermal shock is illustrated for one of the most thermal shock–prone areas: the bottom portion of a basic oxygen steelmaking furnace (BOF). Thermal shock cracking and failure of BOF bottom zone refractories usually occur due to pinch spalling, which involves a greater expansion of the hot face compared to the cold face of the refractory due to restrained loadings under service. Figure 6.2a shows the process environment in a BOF and the subsequent thermal shock cracking pattern. This process comprises injecting natural gas and oxygen at a temperature of 200°C through a series of bottom tuyeres into a molten steel bath at 1650°C. Thus, the existing temperature difference between the area right beside the tuyere zone and the molten steel bath induces a huge thermal gradient of nearly $\Delta T/\Delta y \approx 7.25$°C/mm across a 200 mm thick wall and thermal stress, which cause cracking and failure of the refractory [3]. This failure mechanism at the microstructure level can be demonstrated in the following five stages [22, 23], as shown in Figure 6.2b.

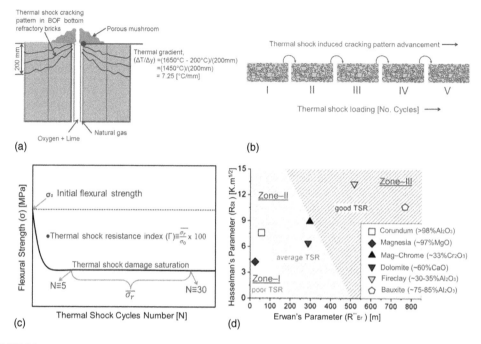

FIGURE 6.2

Mechanism and characterization of thermal shock failure in refractories. (a) Thermal shock cracking pattern in bottom zone refractory bricks of a basic oxygen furnace (BOF); (b) stepwise mechanism of thermal shock cracking to failure; (c) characterization of thermal shock resistance using a normalizing method; and (d) plot of thermal shock resistance parameters "R_{st}" against "$R^{||||}_{Er}$," characterizing the degree of thermal shock resistance for miscellaneous refractories.

Stage 1: Refractory structure prior to a thermal shock attack where the flaw population is solely attributed to the processing defects.

Stage 2: Generation of microcracks due to the action of localized thermal stresses at the regions of processing flaws under early stages of thermal shock.

Stage 3: Elastic strain energy assists the growth of microcracks to larger size and propagation with the advancement of thermal shock cycling.

Stage 4: Accumulation and intersection of multiple cracks trigger localized damage with the further progression of thermal shock cycling.

Stage 5: Refractory rupture caused by the extension of intersecting cracks to the surface region when thermal shock loads exceed a critical measure.

The thermal shock behavior of refractory ceramics can be characterized in two different ways, which determine the degradation in strength property as a function of progressive wear damage [24, 25]:

1. Estimating the thermal shock resistance index (TSRI, Γ) by a normalized method (Figure 6.2c).

2. Evaluating the thermal shock resistance parameters R_{st} and $R^{||||}_{Er}$ (Figure 6.2d).

Figure 6.2c shows a schematic plot of the change in the flexural strength of refractory ceramics corresponding to the number of thermal shock cycles (N). It has been observed that the strength retained in the refractory remains an approximate constant value after a certain degree of thermal shock loading in terms of the number of cycles and exhibits a thermal shock damage saturation plateau. This saturation level or plateau is shown in Figure 6.2c for thermal shock loading cycles (N) between N=5 and N=30. The strength of the refractory at this saturation level is known as the mean retained strength σ_r (MPa). Therefore, the thermal shock resistance of a refractory can be estimated by determining the thermal shock resistance index (TSRI, Γ) and is defined as

$$\Gamma \equiv \frac{\overline{\sigma_r}}{\sigma_o} \times 100 \tag{6.1}$$

where σ_o (MPa) is the initial flexural strength of the refractory prior to thermal shock; this evaluation approach is referred to as the normalized method. The higher the TSRI (Γ) value, the better the thermal shock resistance of any refractory.

Evaluating the thermal shock resistance parameter is another approach to estimate the thermal shock behavior of industrial refractories. Many types of thermal shock resistance parameters have been derived, but Hasselman's thermal shock resistance parameter, R_{st} (K m$^{1/2}$), and Erwan's thermal shock resistance parameter, $R^{||||}_{Er}$ (m), are considered in the present case to study failure under thermal shock crack propagation because a refractory brick holds a large population of processing flaws [22, 25]. The aforesaid parameters can be mathematically expressed as

$$R_{st} = \left(\frac{\gamma_f}{\alpha^2 E (1-\vartheta)} \right)^{1/2} \tag{6.2}$$

$$R^{||}_{Er} = \left(\frac{E\gamma_f}{2\sigma_m^2 (1-\vartheta)} \right) \tag{6.3}$$

where γ_f (J/m) is denoted as the specific fracture energy or work of fracture; σ_m (MPa) is the integral mean stress; α (K^{-1}) is the thermal expansion coefficient; υ (–) is Poisson's ratio; and E (MPa) is Young's modulus or the modulus of elasticity. The thermal shock resistance parameter (R) scales inversely with the thermal gradient and designing the R parameter to a higher value signifies a notable improvement in resistance against thermal shock failure in a refractory. Figure 6.2d presents a plot of Hasselman's thermal shock resistance parameter (R_{st}) versus Erwan's thermal shock resistance parameter ($R^{||||}_{Er}$), and classifies the various refractories into three different zones attributed to their degree of thermal shock resistance estimated by correlating their distinct properties, as specified in Table 6.3.

The distinct thermal shock resistance zones defined for various refractories are

Zone I: Poor thermal shock resistance

Zone II: Average thermal shock resistance

Zone III: Good thermal shock resistance

The poor resistance to thermal shock for Zone I refractories such as magnesia (97% MgO) is due to the large coefficients of thermal expansion and brittle oxide bonding. Whereas the good thermal shock resistance of Zone III refractories such as fireclay (30%–35% Al$_2$O$_3$) and bauxite (75%–85% Al$_2$O$_3$) is primarily attributed to the smaller thermal

TABLE 6.3

Correlating the Miscellaneous Properties of Refractory Ceramics with Thermal Shock Resistance

Refractory	$\rho \times 10^{-3}$ (kg/m³)	ϕ (%)	υ (-)	σ_c (MPa)	$E \times 10^{-2}$ (MPa)	γ_f (J/m²)	$\alpha \times 10^6$ (/K)	k (W/m/K)	C_p (J/kg/K)	R^{III}_{Er} (m)	#TSR
Magnesia (~97% MgO)	3.1	12–18	0.22	80	400	200	13.5	15	1300	24.7	Low
Corundum (>98% Al₂O₃)	3.3	20	0.22	150	350	200	8	6	1050	56.8	Moderate
Mag–chrome (~33% Cr₂O₃)	3.1	15–20	0.22	50	250	250	9	6	1000	296	Good
Dolomite (~60% CaO)	2.9	15	0.22	60	300	275	12	3	1050	288	Average
Fireclay (30%–35% Al₂O₃)	2.1	15–20	0.22	40	150	150	6	2	1100	516	Good
Bauxite (75%–85% Al₂O₃)	2.8	20	0.22	80	350	300	7	3	1050	767	Very good

Note: ρ (kg/m³) is the bulk density of the refractory; ϕ (%) is the apparent porosity; υ (–) is Poisson's ratio; σ_c (MPa) is the compressive strength; E (MPa) is Young's modulus or elastic modulus; γ_f (J/m²) is the work of fracture or specific fracture energy; α (/K) is the coefficient of thermal expansion, k (W/m/K) is the thermal conductivity; C_p (J/kg/K) is the specific heat capacity; R^{III}_{Er} (m) is Erwin's thermal shock resistance parameter; and #TSR is the degree of thermal shock resistance.

expansion coefficient and the absorption of thermally induced stresses by viscous glassy phases formed in situ via melting impurities within the refractory structure at elevated temperatures.

6.4.1.2 Control of Thermal Shock

The key steps involved in the control of thermal shock wear process to develop the refractories with improved thermal shock resistance property are summarized as follows [3, 22]:

1. The thinner refractory lining practice and the choice of raw materials with large thermal conductivity, good thermal diffusivity, low heat capacity, and smaller thermal expansion coefficient characteristics markedly reduce the thermal gradient and thermal stresses, leading to a significant improvement in the thermal shock resistance property. The high thermal shock resistance of graphite-containing refractory oxides (e.g., MgO–C) over brittle refractories (e.g., MgO) is a classic example of this case and is explained by correlating the divergence in crack propagation behavior with failure mode using stress (σ)–strain (ε) diagrams, as shown in Figure 6.3. The stable and gradual cessation of crack growth owing to crack meandering around the flaky structure of graphite results in ductile failure whereas unstable crack propagation is responsible for the catastrophic failure of brittle refractories.

2. Low strength refractories with the high work of fracture (γ) significantly arrest crack propagation, thereby substantially increasing the value of the thermal shock resistance parameter, notably improving the thermal shock resistance property. The high work of fracture (γ_f) in a refractory can be obtained by optimum porosity with a controlled pore size, selecting microdefects containing raw materials such as tabular alumina and stainless steel needles, incorporating nanoscaled additives, and reducing brittle oxide bonding by using higher amounts of carbon together with the limited additions of antioxidants.

3. The use of thermal expansion allowance (TEA) and mortar joints considerably limits the thermal stresses originating from the restrained thermal expansion phenomenon. The newly modified Tom's rule for thermal expansion allowance (TEA) is given as

$$TEA = 0.5 \times (\% \varepsilon_{RLC} \times T) \times (1 + \% \varepsilon_{PLC}) \times \Delta y \tag{6.4}$$

where ε_{PLC} (%) is the permanent linear change; ε_{RLC} (%) is the reversible linear change; T (°C or K) is the temperature; and Δy (mm) is the thickness of the refractory lining.

6.4.2 Mechanical Wear in Vessel

The mechanical wear of refractory materials is particularly detrimental for charge pad and bottom portions of several iron-making and steelmaking vessels and occurs by three different mechanisms: impact, abrasion, and erosion wear primarily caused by various process and service factors such as feedstock material, hot metal, and rapidly moving hot flue gases. The degradation of refractory material by the impact wear mechanism involves the action of a larger force in a shorter time, whereas abrasion and erosion types of wear gradually lead to extreme damage and limit the service life of the refractory. The severity of these

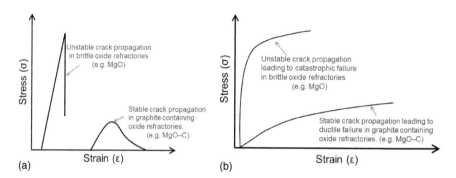

FIGURE 6.3
Schematic stress (σ)–strain (ε) curves for (a) crack propagation and (b) failure behavior in brittle oxide-bonded (e.g., MgO) and graphite-containing oxide (e.g., MgO–C) refractory ceramics under thermal shock.

types of wear mechanism is typically influenced by key factors including the physical characteristics of the feedstock, such as size, shape, hardness, and specific gravity; the overall weight of the burden; the angle of impingement; the height of the furnace; and the velocity and diameter of the hot metal impact stream. The wear resistance property of a refractory specifies the ability of a material to withstand the wearing down of its surface due to the action of comprehensive mechanical forces. The solid particle erosion measurement (ASTM G76-04) is a widely used characterization method to determine the wear resistance of a refractory where the impact and erosion wear mechanisms are dominant. In this section, the mechanism and characterization of the erosive wear process are explained and details on process variable considerations for the wear control of refractory ceramics are provided.

6.4.2.1 Mechanism and Characterization of the Mechanical Wear

A brittle erosion mechanism is proposed as the dominant material removal process for several classes of fired and unfired refractories because the ingredient composition constitutes millimeter-sized brittle coarse aggregates together with a micron-sized secondary phase. For fired refractories, the secondary phase is a binder (e.g., SiO_2 in an Al_2O_3 refractory) and for unfired refractories such as carbon-containing oxides (e.g., Al_2O_3–C), graphite is considered as the major secondary phase. Figure 6.4a shows schematic and microstructural evidence for the brittle erosion mechanism of a refractory by the solid particle erosion wear process where the refractory or target surface is worn out by the rapidly impinging erodent particles of greater hardness than the target [3, 26].

The key elements of the wear process are

1. The formation of radial and lateral cracks in coarse aggregates and the secondary phase of the refractory due to rapid impingement of hard erodent particles.

2. The growth of the generated cracks attributed to the accumulated impact of energy during the impingement process at the surface and subsurfaces of the target.

3. The wear of the refractory aggregate surface as chips ascribed to the growth of radial cracks.

4. The formation of erosion pits by dislodging the millimeter-sized aggregates due to the growth of lateral cracks.

5. The failure of the refractory caused by the gradual fall in strength attributed to progressive cracking and aggregate pull-out from the matrix.

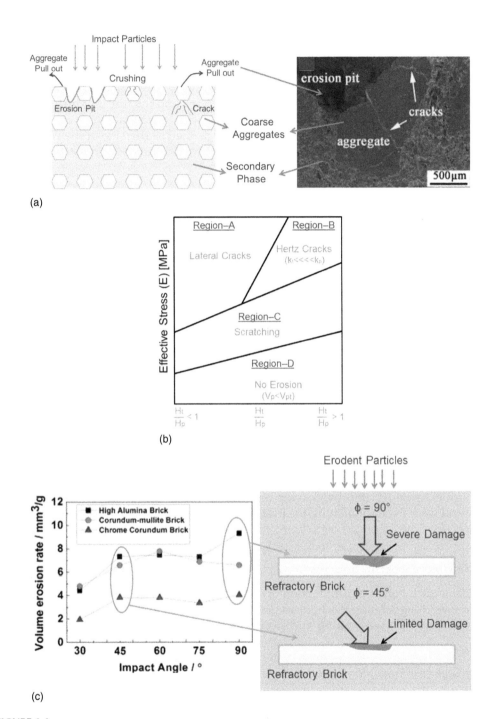

FIGURE 6.4

Mechanism and characterization of the wear process. (a) Schematic of the brittle erosion wear process including characteristic microstructure features; (b) characterization and estimation of refractory wear behavior using wear map; and (c) effect of impingement angle on the erosive wear rate. Inset shows the extreme damage that occurs at an impingement angle of $\phi = 90°$.

The characteristic microstructural features of the eroded region, such as aggregate surface cracking and erosion pit formation, strongly indicate that the brittle erosion mechanism is the leading process in the mechanical wear of a refractory. The wear behavior of refractories is characterized using a graphical plot known as a "wear map" [27, 28]. This plot correlates the erosive wear rate (E_r) with the hardness (H) and fracture toughness (K) characteristics of the target or refractory and eroded particles to estimate the wear condition. The erosive wear rate, E_r (mm^3/g), of a refractory is defined as the ratio of the volume loss of the target material to the total weight of the eroded particles and is calculated as follows:

$$E_r = \frac{\Delta W}{\rho * t * \dot{m}} \tag{6.5}$$

where ΔW (g) is the average weight loss of the target; t (s) is the test period; $\dot{m}$ (g/s) is the eroded particle flux; and ρ (g/cc) is the density of the target. Figure 6.4b shows a schematic wear map of refractories displaying diverse erosive wear mechanisms in four distinct regions; details are listed in Table 6.4.

A series of studies concluded that the impingement angle is a serious damaging factor for mechanical wear in refractories. Figure 6.4c shows the room temperature erosive wear for three different kinds of fired Al_2O_3-based refractories as a function of the impingement angle (ϕ). The wear rate is found to be markedly increased with an increasing impingement angle and reaches maximum at $\phi = 90°$ attributed to the larger impact force; this wear trend further characterizes that the claimed erosive wear mechanism is brittle in nature. The reduction in the degree of erosive wear damage with a decreasing impingement angle from $\phi = 90°$ to $\phi = 45°$ is schematically presented. The greater erosive wear resistance of chrome-corundum refractories is attributed to their high density, low porosity, strong binder phases, and hard aggregates.

6.4.2.2 Control of Mechanical Wear

The considerations for improved wear resistance of refractory ceramics via control of mechanical wear by manipulating the process and service variables are listed in Table 6.5 [3, 29].

6.4.3 Chemical Wear in Contact

Chemical wear involves a combination of simultaneous oxidation and corrosion damage and is a severe degradation mechanism mainly in carbon-containing oxide and non-oxide refractories used in iron-making and steelmaking furnaces where the chemical process

TABLE 6.4

Discrete Regions of Wear Map Indicating the Diversity in Wear Mechanism [27]

Region	Wear Condition	Wear Mechanism
A	$V_p > V_{pt}$; $H_t/H_p > 1$	Fracture guided by lateral cracking
B	$V_p > V_{pt}$; $H_t/H_p > 1$; $K_t <<< K_p$	Fracture guided by hertz cracking
C	$V_p > V_{pt}$; $H_t/H_p < 1$	Scratching
D	$V_p < V_{pt}$	No damage, because wear process is not fully introduced

Note: V_p (m/s) is the erodent particle velocity; V_{pt} (m/s) is the critical erodent particle velocity; H_t (GPa) is the hardness of the target; H_p (GPa) is the hardness of the erodent particle; K_t (MPa $m^{1/2}$) is the fracture toughness of the target; and K_p (MPa $m^{1/2}$) is the fracture toughness of the erodent particle.

TABLE 6.5

Process Considerations for Improved Mechanical Wear Resistance in Refractories

Process/Service Variables		Considerations for Improved Wear Resistance of the Refractory
1. Feed stock particles (or) burden	Size and shape	Smaller size particle fractions having smooth edges rather than sharp corners exert limited impact force and result in improved wear resistance.
	Specific gravity and hardness	Particle characteristics with low specific gravity and smaller hardness value reduce the impact force and enhance the wear resistance.
	Weight	Decrease in weight of the burden considerably minimizes the impact force and improves the wear resistance properties.
2. Hot metal impact stream	Diameter	Reduced diameter of the hot metal impact stream significantly drops the impact force and upgrades the wear resistance properties.
	Weight	Decrease in the weight of the hot metal markedly lowers the impact force and increases the wear resistance properties.
3. Furnace	Height	Impact force increases with the height of the furnace. Therefore, the lower the height the greater the wear resistance.
	Angle	Charging burden or hot metal at lower impingement angles significantly controls the mechanical wear and improves the wear resistance.
	Stir rate	Controlled hot metal stir rates result in enhanced wear resistance since the extent of wear is proportional to the stir rate.
4. Refractory	Work of fracture	Refractories with high work of fracture possess good wear-resistant properties due to the considerable activity of stress-absorption mechanisms.
	Hardness and density	Dense refractories with considerable hardness properties is a characteristic feature of wear-resistant materials attributed to the marked reduction in the impact forces triggered by the erodent media.

mainly deals with service environments enriched with oxidative and corrosive media under high temperature and pressure conditions. The oxidative wear process is composed of depleting carbon in the refractory which is said to occur by means of direct oxidation in oxygen-enriched air atmosphere and indirect oxidation in FeO-containing slags at elevated temperatures. The process of corrosive wear constitutes a deterioration of the refractory due to both chemical and mechanical interaction with the hot flue gases and molten liquids such as slag and metal. This latter form of corrosive damage is known as "melt corrosion" and is considered the dominant attack. The relative resistance against chemical wear can be characterized by estimating the oxidation and slag corrosion resistance properties of a refractory. The standard test to measure the oxidation resistance of a refractory is ASTM C-863. The static slag corrosion cup or brick test (ASTM C-621) and the dynamic rotary slag corrosion tests (ASTM C-874) are the commonly used methods to determine the corrosion resistance properties [30, 31]. In this context, the mechanism and the characterization of the chemical wear in terms of both oxidation and melt corrosion processes are comprehensively discussed together with the process variable manipulation to control and improve the chemical wear resistance of carbon-containing oxide refractories.

6.4.3.1 Mechanism of Chemical Wear

Chemical wear of carbon-containing oxide refractories is a complex process that involves the simultaneous attack of several degradation mechanisms, leading to disparate damage of a refractory structure, as shown schematically in Figure 6.5a. A general scheme of the mechanism for the chemical wear process of oxide–carbon refractories at the microstructure level is presented in Figure 6.5b and is explained in the following five stages [30].

Stage 1: Exposure of a refractory to the wear environment. A typical oxide–C refractory is subjected to the chemical wear environment and the microstructure features clearly show that the oxide grains are bonded together by homogeneously distributed adjacent graphite flakes. At this stage, the defect population in the refractory is solely ascribed to the processing flaws.

Stage 2: Formation of a decarburized layer. Direct and indirect oxidation of carbon in the form of crystalline graphite or pyrolitic carbon primarily by oxygen containing air at high temperatures and iron oxide (FeO) in molten slag respectively, and leads to the formation of decarburized layers. The corresponding reactions are [32]

a. $\text{Direct oxidation reaction}: 2C_{(s)} + O_{2(g)} \rightarrow CO_{2(g)}(T > 500°C)$ [Rxn. 6.1]

b. $\text{Indirect oxidation reaction}: C_{(s)} + FeO_{(L)} \rightarrow Fe_{(L)} + CO_{(g)}(T > 1400°C)$ [Rxn. 6.2]

These formed decarburized layers increase the flaw population, which further acts as slag veins and allows the penetration of liquid slag into the refractory matrix to aggravate the next level of melt corrosion.

Stage 3: Slag infiltration and dissolution. The infiltrated slag penetrates through the decarburized layers creating refractory–slag interfaces that act as reaction sites to form new compounds. The local chemical potential gradient across the interface drives the ionic transport and leads to the dissolution of oxide grains into slag media (inset, Figure 6.5b). The formation of forsterite (Mg_2SiO_4) at temperatures above 1400°C in MgO–C refractories strongly encourages the considerable dissolution of MgO grains in steelmaking slags and the corresponding slag dissolution reaction is given as

$$2MgO_{(s)} + SiO_{2(L)} = Mg_2SiO_{4(s)}\left[T > 1400°C\right]$$ [Rxn. 6.3]

Stage 4: Erosion. The continuous action of the slag dissolution process and the reduction in oxide grains by carbon at elevated temperatures are the two primary factors promoting refractory erosion. The occurrence of an MgO–C reaction in carbon-bonded magnesia systems is a typical example of this case and is given as

$$MgO_{(s)} + C_{(s)} = Mg_{(g)} + CO_{(g)}\left[T \approx 1600°C\right]$$ [Rxn. 6.4]

Stage 5: Failure. Cyclic and simultaneous activity of decarburization, slag dissolution, and erosion processes render these refractories porous with considerably reduced strength, leading to structural failure under the effects of thermomechanical loads.

6.4.3.2 *Characterization of Chemical Wear*

As has been detailed earlier in this section, the chemical wear of a refractory is the corrosive damage, primarily caused by the simultaneous action of air oxidation and hot metal or slag corrosion processes in the service environment. In order to characterize the ability

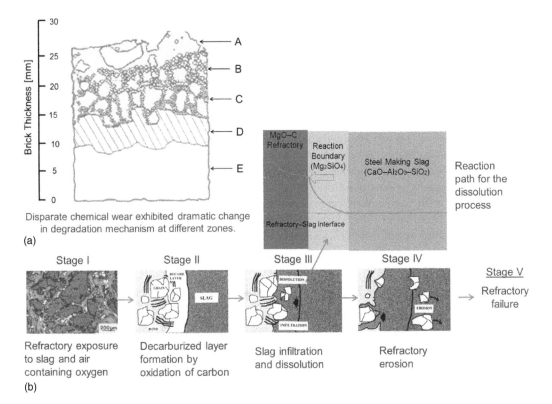

(a) Schematic presenting the disparate chemical wear attack and (b) a mechanism showing the distinct stages involved in the chemical wear process of oxide–C refractories. Inset shows the reaction path for the dissolution process.

FIGURE 6.5

of a refractory to withstand the considerable effects of chemical wear, the recommended practice is to evaluate the oxidation resistance against air and the corrosion resistance against slag or metal as follows.

A series of studies have reported that the oxidation process of carbon in a refractory is initiated by a reaction between carbon and oxygen containing air at elevated temperatures, forming oxides of carbon around the exterior boundary of a refractory. This reaction gradually progresses toward the interior of the refractory structure, oxidizing the carbon layer by layer, and remained the partially oxidized refractory with a clearly marked interface between the porous-reacted layer as the oxidized region (yellow–brown) and the unoxidized core (black), as shown in Figure 6.6a. The characterization tools used to quantify the oxidation resistance of carbon-containing refractories are an estimate of the areal oxidation index (OI) on the basis of the oxidized area in the oxidized area method and a rate constant (k_e) measurement based on the oxidized volume in a kinetic model [33, 34].

These procedures require the use of model specimen geometry representing a partially oxidized cubic refractory block, which is schematically presented in both cross-section and isometric views (Figure 6.6b). The expressions provided for estimating the oxidation resistance are

1.
$$\text{Areal OI} = \left(\frac{Y^2 - y^2}{Y^2} \right) \qquad (6.6)$$

2. $\quad$ Rate constant $(k_e) = \dfrac{d^2}{t} = \left(\dfrac{Y^3 - y^3}{6Y^2}\right)^2 \times \dfrac{1}{t} \quad \left[\rightarrow d = \left(\dfrac{Y^3 - y^3}{6Y^2}\right)\right] \qquad (6.7)$

where d (cm) is the diffusion depth as a linear dimension; t (s) is the time of oxidation; Y (cm) is the length of each side of the cubic refractory block; and y (cm) is the length of the unoxidized cubic core.

Figure 6.6c shows the relative oxidation damage of MgO–C refractories at 1400°C for 10 hours, estimated in terms of both the oxidized area method and the rate constant measurement model using Equations 6.6 and 6.7, respectively. Prior to the oxidation test, the refractory specimens were coked at 1000°C. The compositional details of the refractory series together with their coked properties are listed in Table 6.6. It is observed that the oxidation susceptibility of these carbon-containing refractories diminishes with increasing expanded graphite (EG) content. Areal oxidation index (OI) calculations show that the

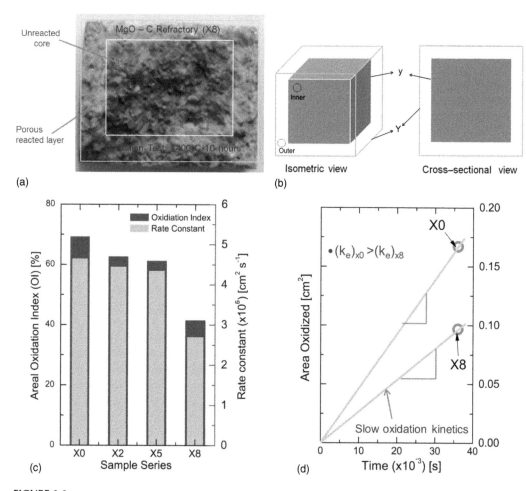

FIGURE 6.6

Characterization of the oxidation wear process in carbon-bonded MgO refractories. (a) Digital image of a partially oxidized refractory sample (X8); (b) schematic view of specimen model geometry; (c) estimation of relative oxidation damage using both oxidized area method and rate constant measurement model; and (d) kinetics of oxidation process.

TABLE 6.6

Composition and Properties of MgO–C refractory series

| Identity | Ingredient Composition (wt.%) | | | | | Refractory Properties Coked at 1000°C/4 hours | | | | |
	Fused MgO	Graphite	Expanded Graphite (EG)	Antioxidants	Phenolic Resin	BD (g/cc)	AP (%)	CCS (kg/cm²)	HMOR (kg/cm²)
X0	93	5	0	2	4	2.84	13.6	206	40
X2	93	4.8	0.2	2	4	2.86	11.9	174	37
X5	93	4.5	0.5	2	4	2.89	10.8	190	50
X8	93	4.2	0.8	2	4	2.91	10.1	239	90

extent of oxidation damage for refractory specimens without EG (X0) is significant and approximately 30% higher than the refractory compositions with 0.8 wt.% EG (X8), indicating that the X8 composition has superior oxidation resistance among the listed series of refractory compositions.

The oxidation resistance characteristics evaluated on the basis of rate constant calculations are found to be in good agreement with the previously claimed results. The kinetics of the oxidation process for these carbon-containing magnesia (MgO–C) refractory compositions is shown in Figure 6.6d. These plots further corroborate the preceding assertions and deduce that the oxidation process is kinetically slower by 40% in the X8 refractory composition containing 0.8 wt.% EG with a rate constant of about $k_e \approx 2.7 \times 10^{-6}$ cm^2/s than conventional graphitic compositions (X0). The excellent improvement in the oxidation resistance characteristics of the refractory compositions with EG (X8) is attributed to the high coked density, low coked porosity, enhanced cold and hot strength properties engendered by the hierarchical packing of aggregates with varying size fractions, and the homogeneous formation of hot strength in situ structures such as Al_4C_3, AlN, and $MgAl_2O_4$ cubic entities in the refractory matrix.

The melt corrosion of a refractory by slag under dynamic conditions is a severe wear process in most steelmaking vessels and predominantly occurs due to erosion–dissolution mechanisms. It is imperative to characterize the corrosion process as kinetically controlled because the dissolution reaction rate is limited by the mass transportation of refractory aggregates into liquid slag through a refractory–slag interface [30]. Determining the corrosion resistance of a refractory by means of a corrosion rate is a kinetic measurement and is tested using the dynamic slag corrosion method where the corrosion behavior of a refractory is studied as a function of several experimental variables such as temperature (°C), rotation rate (rpm), rotation speed (cm/s), and immersion period (s). However, the corrosion rate under dynamic slag corrosion conditions can be estimated on the basis of both the erosive wear and slag dissolution characteristics of the refractory using equation 6.8 and is expressed as; [35]:

$$\vartheta_{de} = -\frac{\Delta D}{\Delta t} = 0.01 k \frac{(n_s \rho_s - n_b \rho_b)}{\rho} = A_0 U^b \qquad (6.8)$$

where d_i (cm) and d_f (cm) are the initial and final diameters of a test refractory cylinder, respectively; t_i (cm) and t_f (cm) are the initial and final immersion periods, respectively; k (cm/s) is the mass transfer coefficient; n_s (%) and n_b (%) are the oxide contents of slag at the interface and bulk of the refractory, respectively; ρ (g/cm^3) is the bulk density of the refractory, ρ_s (g/cm^3) and ρ_b (g/cm) are the density of slag at the interface and bulk of the refractory, respectively; U (cm/s) is the rotating speed; b (–) is the power law exponent; and A_0 (–) is a constant.

Controlled atmosphere dynamic slag corrosion experiments were performed for MgO–C (97 wt.% MgO, 5 wt.% C), doloma (38 wt.% MgO, 59 wt.% CaO), and magnesia–doloma (62 wt.% MgO, 36 wt.% CaO) cylindrical refractory rods against the CAS steelmaking slag with composition (56 wt.% CaO, 33 wt.% Al_2O_3, 11 wt.% SiO_2) under different test conditions. The results for a typical case are shown in Table 6.7.

The estimated corrosion rates (υ_{de}) based on both the erosion and dissolution characteristics using Equation 6.8 have been found to be considerably low for graphite-containing magnesia refractories.

The excellent corrosion resistance of magnesia–graphite (MgO–C) refractories is attributed to the high thermal conductivity and the large wetting angle characteristics of

TABLE 6.7

Slag Corrosion Rate Results of Oxide Refractories with and without Carbon

Refractory	MgO–C	Dol–Mag	Doloma
Temperature (T) (°C)	1650	1650	1650
Rotation rate (R) (rpm)	200	200	200
Corrosion rate (υ_{de}) (cm/s)	4.21×10^{-6}	45.5×10^{-6}	83.6×10^{-6}
Mass transfer coefficient (k) (cm/s)	1.71×10^{-4}	2.54×10^{-4}	4.60×10^{-4}

graphite [3]. High thermal conductivity offers rapid dissipation of established thermal gradients, decreases the temperature of the refractory, increases the viscosity of slag media, and reduces the slag penetration rate. It is a well-known fact that larger wetting angles decelerate the slag infiltration process by reducing the capillary forces.

The microstructural evidence classically displayed in Figure 6.7 further corroborates the effects of a carbon-enriched matrix on limiting the dissolution process of magnesia grains in MgO–C refractories.

6.4.3.3 Control of Chemical Wear

Important considerations established by process variable manipulation to control and improve the chemical wear resistance of carbon-bonded oxide refractories are listed in Table 6.8.

The earlier reviewed discussions on wear processes (c.f. 6.4) elucidated the action of a particular wear mechanism on the degradation behavior of a refractory; however, installed refractories in service encounter or experience integrated wear attacks.

A model analysis is portrayed in Figure 6.8 to specify the definite role of each wear mechanism responsible for the failure of a refractory. Therefore, the degradation under service is aggravated by mechanical (impact, abrasion, and erosion) and chemical wear (air oxidation, slag corrosion, and erosion), but failure of the refractory is stimulated by a type of wear called "Thermal Shock".

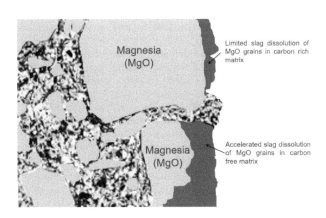

FIGURE 6.7

SEM image showing the diversity in the dissolution process of MgO grain in the presence of a carbon-rich refractory matrix.

TABLE 6.8

Process Considerations for Improved Chemical Wear Resistance of Refractory [3, 30]

Process Variables		Considerations for Improved Oxidation and Corrosion Resistance of the Refractory
1. Oxide grain	Source and purity	Fused oxide grains have been chosen to obtain improved corrosion resistance properties because of their large crystallite size and low grain porosity. In the case of MgO–C refractories, selecting fused magnesia grains with 97.5% purity and an impurity content of CaO/SiO_2 ratio >2 with minimal B_2O_3 concentration is common practice to impart enhanced corrosion resistance and hot strength by the formation of high temperature C_2S and C_3MS_2 phases in the refractory matrix [36].
	Crystallite size	Large crystallite sizes of oxide grains possess a low surface area that limits the rate of slag dissolution as well as high temperature reduction reactions, resulting in improved corrosion resistance.
	Grain porosity	Low grain porosity of oxide raw materials is favorable for better corrosion resistance because of their reduced interaction with molten slags.
2. Carbon	Source	Commonly used carbon sources are graphite, resin or pitch as binder, and nanosized high surface area carbons to partially replace graphite in low carbon-containing refractories.
		1. Improved corrosion resistance properties can be obtained using graphite because of large wetting angles and high thermal conductivity.
		2. Greater extent of dense MgO layer formation in resin-bonded bricks results in the enhanced slag penetration resistance characteristics that has strongly recommended to select resin as a binder instead of pitch.
		3. Use of high surface area carbons such as nanocarbon, expanded graphite, and reduced graphene oxide is often desirable in low carbon refractories, which uniquely offer the catalytic formation of hot strength structures in the inter-refractory pore space binding the oxide grains, leading to an increase in density and a reduction in porosity, improving corrosion resistance as well as oxidation resistance.
	Purity	Highly pure carbons have a greater tendency to oxidation, which is a destructive aspect for refractory applications. However, graphite with fixed carbon content levels in the range of about 94%–97% have been selected.
	Amount	Extent of graphite addition has a decisive role on the chemical wear of a refractory and the considered amount is in the range of about 10–35 wt.%. Exceeding this concentration of graphite (>35 wt.%) causes a greater reduction in slag penetration resistance.
3. Antioxidants	Source	Commonly used antioxidant sources are metal/alloy additives (Al, Si, Mg, and Al–Mg) and ceramic powders (SiC and B_4C).
		1. Metal additives improve the oxidation resistance by promoting the following activities: casing the graphite with local dense layers such as MgO and Mg_2SiO_4; depositing condensed species in the inter-refractory pore space; and playing the role of CO reducing agents; which considerably limits the oxidation of graphite.
		2. Boron-based compounds enhance corrosion as well as oxidation resistance by forming low-melting liquid phases that fill the interconnecting open porosity and limit the ingress of slag and oxygen containing air.
	Amount	Optimum additions are beneficial for effective chemical wear resistance.
4. Slag	Composition	High basicity with increasing amounts of MgO content (for the case of MgO–C refractories) and diminishing amounts of MnO, Cr_2O_3, and Fe_2O_3 in the slag composition is beneficial for improved corrosion resistance.
	Viscosity	High viscosity limits the slag penetration depth due to poor wetting characteristics and enhance the corrosion resistance.

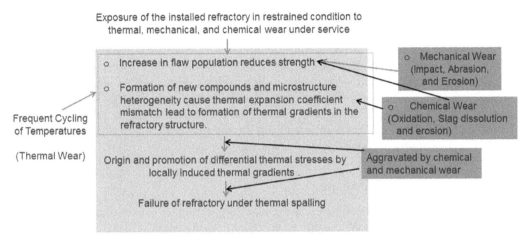

FIGURE 6.8
Failure analysis protocol for installed refractories under service.

6.5 Thermal Gradient Divergence

Refractories are the structural components that constitute several processing flaws including non-spherical cavities and micro and macro cracks. Refractory-operating conditions in most high temperature environments are often batch-type models where refractories are subjected to frequent temperature cycling combined with integrated wear attack by thermal shock mechanical wear and chemical wear processes during service. This wear phenomenon suggests that the flaw population is compounded and refractories are severely degraded with thermal shock wear where the remaining wear mechanisms aggravate the damage. The preceding assertions have clearly stated that the failure of a refractory under service often occurs by spalling or strainingof the mechanically and chemically worn regions due to the origin of locally induced thermal gradients. However, the extent of the strain developedin the worn regions of refractory and the consequent failure behavior are typically controlled by the divergence in the magnitude of the induced thermal gradients. A series of studies have reported that the process and installation factors of a refractory can exhibit a significant influence on thermal gradient divergence [2, 37]. They are

- Composition and properties
- Size and shape of the brick
- Installation process parameters

6.5.1 Composition and Properties

Two different kinds of silica refractories fabricated using quartzite and fused silica as the chief SiO_2 source have been considered to illustrate the influence of composition and properties on thermal gradient divergence and the consequent failure [38]. The evaluated refractories have similar residual quartz content but differ in thermal expansion coefficients. The high temperature stress–strain relationships of the examined refractories at a temperature of 1000°C, shown in Figure 6.9, displayed substantial dissimilarity in their failure behavior.

The catastrophic failure in conventional quartzite-based SiO_2 refractories indicates unstable crack propagation, but the ductile failure behavior of the new fused silica-based SiO_2 refractories reflects stable crack propagation characteristics. The low coefficient of thermal expansion and the homogeneous distribution of microcracks in fused silica-based SiO_2 refractories are the primary responsible factors that allow rapid dissipation of the induced thermal gradients, minimizing the microstructure damage, and markedly depressing the consequent spalling and cracking.

6.5.2 Brick Dimension and Shape

Failure behavior under thermal spalling significantly varies with a change in the size and dimension of the refractory bricks due to considerable alteration of the thermal gradient. A series of standard sizes and shapes of miscellaneous industrial refractory bricks is presented in Figure 6.10 [39].

In order to explore the effect of refractory size on temperature gradient discrepancy and the subsequent spalling type of failure, three different compositions of refractory bricks with two dimensions have been considered and designated brick sizes B_1 ($150 \times 25 \times 25$ mm) and B_2 ($160 \times 40 \times 40$ mm), respectively. The compositions of refractories based on their main ingredient are identified as A2 (electrofused alumina, ~2.3 mm aggregate), A8 (electrofused alumina, ~8 mm aggregate), M2 (electrofused eutectic mullite–zirconia, ~2.3 mm aggregate), M8 (electrofused eutectic mullite–zirconia, ~8 mm aggregate), and Z2 (electrofused eutectic alumina–zirconia, ~2.3 mm aggregate). The change in retained strength (CMOR) after a thermal shock test conducted at 800°C as a function of the dimension and composition of the refractory is presented in Figure 6.11a.

The decrease in the strength of the ingredient composition irrespective of the aggregate size follows such a trend and clearly exhibits that refractories with a larger dimension (B_2; W=40 mm) are more susceptible to thermal shock damage compared to smaller dimension refractories (B_1; W=25 mm). Further, experimental results corroborated that the low work of fracture, thermal expansion anisotropy, mechanical size effects, and the enhancement

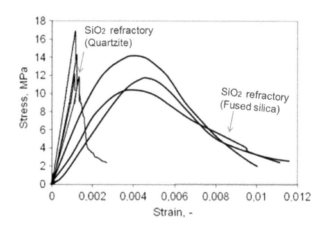

FIGURE 6.9
Stress–strain curves showing the high temperature failure behavior of conventional quartzite-based SiO_2 and fused SiO_2 refractories at 1000°C.

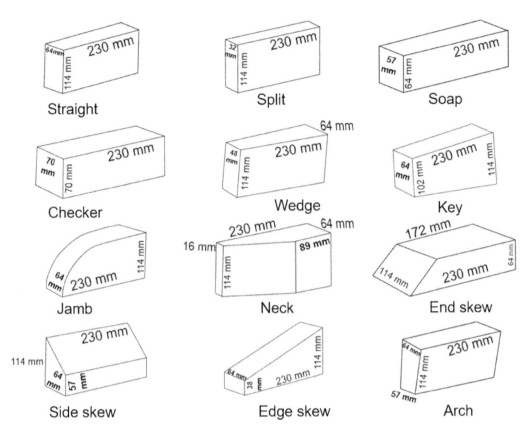

FIGURE 6.10
Sketches of the standard size and shape of industrial refractory bricks.

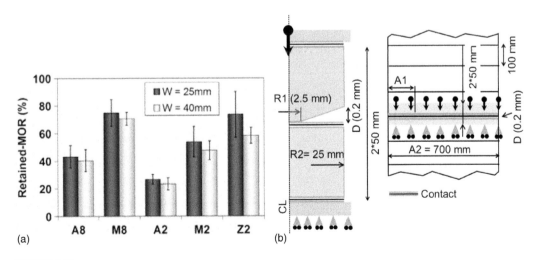

FIGURE 6.11
Discrepancy in the induced temperature gradient and consequent thermal shock damage as a function of (a) size or dimension and (b) shape of the refractory brick.

of the inherent brittleness with increasing brick volume are the contributing factors for the origin of locally induced thermal gradients and consequent thermal shock failure [40]. Figure 6.11b displays the origin of the compressive thermal stresses responsible for thermal spalling in basic oxygen furnace (BOF) linings in a non-parallel contact surface between standard bricks and banana-shaped bricks due to a marked deviation in the thickness parameter [41]. Therefore, it can be deduced that the size and shape of the refractory structure have a considerable effect on thermal shock damage.

6.5.3 Installation Parameters

In the masonry refractory lining, mortar is a crucial element that binds the refractories, stabilizes the structural brickwork, and prevents joint openings. Its softening behavior upon heating provides thermal expansion allowance (TEA) for the installed refractories laid under a constrained state and considerably limits the generation of thermally induced stresses accountable for spalling under service. Numerous research investigations on the thermo-mechanical considerations for refractory installation processes have reported that mortar thickness is a typical process parameter that considerably manages the ability of mortar to provide expansion allowance to a certain extent and has a substantial influence on the control of compressive thermal stresses responsible for discrepancy in temperature gradient and consequent thermal shock failure [38, 42].

Figure 6.12 is a schematic illustration of the effect of increasing mortar joint thickness (d) on compressive strains generated in the mortar joint (Δd) under a certain compressive force upon heating for two different types of mortars, designated mortar "A" and mortar

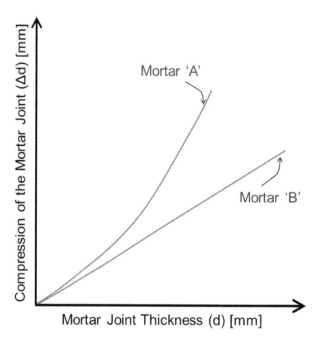

FIGURE 6.12
Compressive strain behavior of a mortar joint as a function of the joint thickness and mortar particle interactions at a working temperature of 800°C.

"B," respectively. The plots clearly show that the generation of compressive strain increases with the increase in mortar thickness. However, the higher value of the compressed mortar joint thickness (Δd) in mortar "A" is indicative of the high degree of thermal expansion allowance (TEA) for the refractory brick structure, which is attributed to the low frictional forces between the interacting particles of the mortar. Therefore, it can be inferred that the induced stress should be low in mortar "A" as a function of compressive thermal loads, which is a marked feature exhibiting the effective dissipation of locally induced thermal gradients and control over the consequent spalling failure compared to the joint types of mortar "B".

With reference to experimental observations and theoretical arguments presented in this section, there would be significant divergence in locally established thermal gradients in various regions of the refractory, which can cause a spalling type of failure in a disparate way. However, the role of several processing variables (e.g., composition) on the divergence in thermal gradient and consequent failure behavior of various refractories upon the combined action of the wear damage in distinct service environments are reviewed in the following section.

6.6 Processing and Failure Analysis

This section discusses processing and failure analysis of different classes of shaped and unshaped refractories depending upon the kind of application and hostile service environment.

6.6.1 Silica (SiO_2) Refractory

Silica (SiO_2) refractories are a class of fired or burnt ceramics that contain a significant amount of SiO_2 >93 wt.% and minor amounts of common impurities or fluxing oxides such as lime (CaO), alumina (Al_2O_3), and iron oxide (Fe_2O_3). Super duty SiO_2 bricks (flux oxides <0.5 wt.%) and high duty SiO_2 bricks (flux oxides >0.5 wt.%) are two important categories of these refractories depending on their flux content. The prominent application areas of SiO_2 refractories are the roof of a glass tank furnace, coke oven batteries, and hot blast stoves in iron-making and steelmaking industries because of their relatively low specific gravity, excellent acidic melt corrosion resistance, high refractoriness under load (RUL) value, and good volume stability at high temperatures. Silica exhibits polymorphism and exists in three polymorphic forms, quartz, tridymite, and cristobalite, whereas each polymorphic form has a low and a high temperature transition known as "inversions" and "conversions," respectively. These transitions are provided in the following assembled reaction [43, 44]. These temperature-dependent transitions are associated with significant volume changes that may result in cracking and failure of the refractory. Thus, instantaneous polymorphic transitions in silica refractories are detrimental to the aforementioned leading properties and these transitions are typically controlled by processing variables such as firing temperature, residual quartz content, and choice of mineralizer, which have been explicitly detailed elsewhere [1]. In this section, a processing and failure analysis of SiO_2 bricks used in coke oven batteries are discussed.

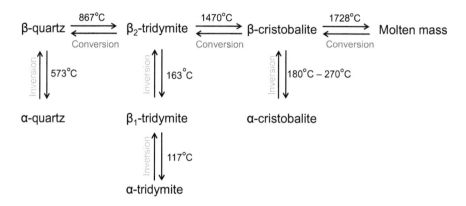

The principal raw material used in the fabrication of SiO_2 bricks is quartzite with a purity level of nearly $\approx 95\%$. Initially, quartzite grains of different size fractions are prepared by a series of operations: washing, crushing, grinding, and screening. These graded quartzite fractions are combined in predetermined proportions, mixed with a binder and plasticizer in a muller, pan, or counterflow-type mixer to obtain a homogeneous mix. The most commonly used plasticizer is water that provides and retains the shape, whereas hydrated lime of about 3% is used as a binder as well as a mineralizer that catalyzes the formation of desirable tridymite or cristobalite polymorphic phases. This friable or aggregate mix is shaped on a hydraulic or friction screw press to green blanks and is further dried to remove moisture at a maximum temperature of 150°C–200°C for a period of 24–30 hours. Firing of the dried bricks is mostly performed in batch-type kilns with very slow heating rates especially during the inversion of α-quartz to β-quartz, which occurs in the temperature range 550°C–600°C. The selection of a firing temperature in the range of 1420°C–1430°C favors the maximum formation of the tridymite phase, whereas the cristobalite phase formation is dominant in firing temperatures of between 1480°C and 1500°C [20]. The process flow sheet for the manufacturing of silica bricks is presented in Figure 6.13.

The operation of coke oven batteries is a batch-type process that involves charging coal into the coking chambers and heating it in the absence of air in the temperature range 1050°C–1400°C to form metallurgical coke by releasing the volatile constituents of coal. This method of coke preparation is known as "carbonization." The heat transfer between the heating chamber and the coking chamber to perform the carbonization process occurs through the SiO_2 brick-lined refractory wall interface. Therefore, coke oven SiO_2 bricks are subjected to extreme wear attack under service and their failure analysis can be detailed as follows [45, 46]:

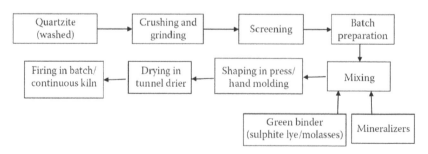

FIGURE 6.13
Process flow sheet for the fabrication of SiO_2 refractories.

1. Mechanical wear in terms of impact and erosion caused by the charging and discharging of hard, angular, and sharp-edged coal and coke, respectively. This wear effect leads to the chipping and dislodging of SiO_2 grains, which enhances the flaw population, extends microstructural heterogeneity, and reduces the strength of the brick.

2. Chemical wear caused by the formation of an indistinct mixture of calcium alumina silicate due to the deposition of carbon on the SiO_2 grains in the form of coal or coke, which contains considerable amounts of lime (CaO) and alumina (Al_2O_3) as impurities.

3. Thermal shock of the degraded refractory caused by frequent cycling of temperatures due to batch-type processing.

4. Cracking and the failure of the refractory ascribed to the thermal spalling of heterogeneous regions of microstructures under progressive integral wear attack. These microstructure heterogeneities vary in terms of chemical (corroded and uncorroded regions) and mechanical (differential strength) characteristics.

Numerous industrial research reports have explicitly demonstrated that the failure of SiO_2 bricks can be significantly controlled by strictly maintaining the processing variables such as low residual quartz content combined with large amounts of desirable cristobalite or tridymite phase formations in the sintered microstructures of the fired brick.

6.6.2 Alumina (Al_2O_3) Brick

Alumina (Al_2O_3) refractories, an important class of burnt or fired ceramics, are of paramount importance since alumina plays a leading role in a wide variety of refractory compositions because of its unusual properties. Refractories classified in the high alumina category should contain Al_2O_3 in the range 45–99 wt.% and these materials are further divided depending on the specific purity level of Al_2O_3. The typical properties of the 60 wt.% class high alumina refractories include low porosity, high heat strength, creep resistance with excellent volume stability, and load-bearing capacity; these refractories are recommended as potential candidates for applications primarily involving blast furnace stove checkers, blast furnace lining, preheating zones of cement rotary kilns, and so on. This section details the processing and failure analysis of 60 wt.% class high alumina refractories used in blast furnace stove checkers.

The major raw materials used to fabricate this class of high alumina refractory are calcined bauxitic clay and pure kaolin. The manufacturing process involves mixing a predetermined proportion of varying refractory aggregate size fractions with the green binder and plasticizer in either a pan, muller, or countercurrent mixer to obtain a homogeneous mix with adequate binding and plastic characteristics. Hydraulic or friction screw–type presses are utilized for uniaxial compaction of the prepared homogeneous mix into green refractory blocks, which are further fired using continuous or batch types of kilns in the temperature range 1450°C–1750°C depending on the alumina content [1].

Hot blast stoves are generally cylindrical shaft kilns used for preheating air that is blown into a blast furnace to increase the process efficiency. These stoves constitute two types of chambers: the combustion chamber where the furnace gas is allowed to burn and the checker chamber where high alumina refractories are installed to store the generated heat in a whole volume of checker bricks, which can be further utilized to heat up the cold blast at any particular temperature. Therefore, it is obvious that high alumina refractories

(60% class) intimately interact with the blast furnace gas to ensure complete heat transfer. The various steps involved in the failure mechanism of 60% class high alumina refractories are as follows [20, 47]:

1. Blast furnace gas carries dust that contains alkali matter including Fe_2O_3, SiO_2, Al_2O_3, P_2O_5, and CaO.
2. The deposition of dust onto the surface of a refractory during the period of interaction leads to clogging of checker bricks holes.
3. The penetration of the fused alkali matter into the open pores of the refractory under service temperatures in the range 1200°C–1250°C leads to different high temperature chemical reactions and the formation of undesirable compounds such as lucite and kaliophilite. These reactions are accompanied by sudden volume changes in the refractory matrix that reduces the creep strength and causes cracking.
4. The failure tendency of blast stove refractories is commonly aggravated by key wear factors including an increase in the hot blast temperature, a prolonged dome temperature of 1250°C, the high dust content of blast furnace gas, and a greater amount of clogging of checker bricks holes.

The experimental results corroborated that selecting 60% class high alumina refractories exhibited considerable resistance to creep failure compared to refractories with compositions containing a lower content of alumina that have undergone a catastrophic failure.

6.6.3 Magnesia–Chrome (MgO–Cr$_2$O$_3$) Brick

Magnesia–chrome (MgO–Cr$_2$O$_3$) refractories are a class of composite-type oxide refractories with a concentration of MgO typically above 50 wt.%. There are several categories of magnesia–chrome refractories depending on the nature of the bonding, which is designated as chemically bonded, silicate bonded, direct bonded, rebonded or co-burnt, and fused grain types. The various bonded refractories find important applications mostly in electric arc furnace (EAF) steelmaking furnaces and degassers in ferrous metallurgy, cement kilns in the ceramic industry, and many non-ferrous metallurgical furnaces because of their excellent hot strength, high volume stability, good thermal shock resistance, and greater corrosion and erosion resistance characteristics; however, the amount of these properties differs for each type of bonding in a refractory. In this section, the processing and failure analysis of direct bonded magnesia–chrome refractories used in the burning zone of rotary cement kilns is discussed.

In this category of magnesia–chrome refractories, magnesia and chromia grains are bonded together to form a direct or sintered bond under firing. The manufacturing process involves sizing highly pure magnesia and chromite grains into discrete fractions using a series of comminution and screening operations, and mixing these size-graded grains into predetermined proportions with a binder in a pan or counterflow mixer to obtain a homogeneous mix that is further uniaxially pressed into refractory blocks followed by firing at temperatures above 1700°C. The extent of direct bond formation increases with the increase in firing temperature together with prolonged firing schedules due to the formation of higher amounts of primary and secondary spinels ($FeCr_2O_4$, $MgCr_2O_4$, $MgAl_2O_4$, and $FeAl_2O_4$) in the refractory matrix, which is credible for the aforementioned distinctive properties [1, 20].

The operation performed in the burning zone of a rotary cement kiln involves the preparation of a sintered clinker by heating the raw materials of cement to a temperature of about 1450°C. At this condition, sintering of these materials occurs by a chemical reaction that involves melting, and formation of a high temperature coat on the surface of refractory and is enriched with Ca_3SiO_5 and Ca_2SiO_4 phase chemistry. Therefore, the direct bonded magnesia–chrome bricks installed in the burning zone undergo a combined wear attack during service. The failure analysis of these refractories can be described as follows [48, 49]:

1. The mechanical wear of the refractory caused by the abrasion and erosion effects of sintered clinker due to rotational movement of the cement kiln renders the brick weaker in strength.

2. The chemical wear is dominant and extends the heterogeneity of the microstructure. The typical features of the chemical wear are (i) the disappearance of forsterite, primary and secondary spinel phases from the hot face of the brick; (ii) the penetration of the clinker phases present in the coat including Ca_3SiO_5, Ca_2SiO_4, and $Ca_3Al_2O_6$ through the hot face; and (iii) the formation of hexavalent chromium in the coated clinker layer.

3. The thermal shock of a refractory due to the establishment of large thermal gradients across the kiln wall because of the rotational movement of the cement kiln.

4. A progressive integrated wear attack leads to cracking and failure of a refractory by thermal spalling of the heterogeneous regions of the microstructure, as discussed earlier (c.f. 6.6.1).

The chemical wear can be considerably reduced by promoting enriched formation of corrosion-resistant primary and secondary spinels in the microstructure of sintered bricks by manipulating the firing schedules.

6.6.4 Magnesia–Carbon (MgO–C) Refractory

A magnesia–carbon (MgO–C) refractory is one of the most important unfired class of "oxide–C" composite refractories containing about 8–30 wt.% total carbon content with graphite as the chief carbon source. These refractories are primarily used as the working linings for several high temperature steelmaking vessels, mostly electric arc furnace (EAF), basic oxygen furnace (BOF), and ladle. The exploitation of such challenging service environments has been attributed to the generic features of MgO–C refractories, which include high refractoriness, excellent slag or metal corrosion resistance, and good thermal shock resistance. In this section, the processing and failure analysis of MgO–C bricks used in BOFs is discussed.

The major raw materials used for processing or fabricating MgO–C refractory bricks are magnesia; graphite as the chief carbon source; antioxidants such as Al, Si, Mg, and B_4C as additives to protect the carbon as well as for hot strength improvement; and phenolic resins such as resol and novalac as organic binders. The most commonly used magnesia sources are fused magnesia, seawater magnesia, and sintered magnesia with characteristics including high purity, large periclase crystal size, CaO/SiO_2 ratio above 2, and minimum B_2O_3 content, which is the globally considered choice for operations under severe steelmaking wear environments. The processing of carbon-containing oxide refractories involves three main steps as shown in Figure 6.14: preparing a homogeneous mix of raw

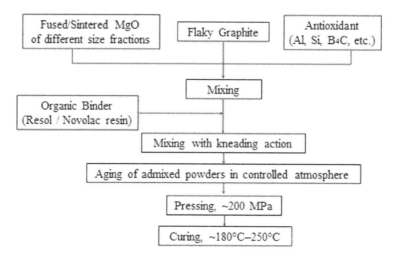

FIGURE 6.14
Process flow sheet for the fabrication of MgO–C bricks.

materials; shaping the mix by compaction; and subjecting the shaped refractory to a heat treatment step known as "curing" or "tempering."

Therefore, to fabricate MgO–C refractory bricks, a homogeneous mix of uniformly resin (binder)-coated aggregate particles is produced by mixing the raw materials including varying size fractions of sintered and fused MgO grains, graphite, and additives such as antioxidants with the resin type of binder ($\mu \approx 6{,}000$–$8{,}000$ cps) in a pan or a Hobart mixer using a kneading action for about 0.5 hours followed by a few hours of aging [1, 29].

The aged mix is subsequently fed into a steel die and uniaxially compacted to a brick shape by a hydraulic or friction screw press without forming any lamination cracks. The shaped bricks are then tempered or cured where volatiles of the organic binder have been burnt off and adequate strength is imparted to the brick by the formation of a three-dimensional interlocking carbon–carbon chain network via the polymerization of resin. The commonly used industrial compaction pressure and curing temperatures are in the ranges 150–200 MPa and 180–250°C, respectively.

A BOF operation involves the oxidation of impurities (C, S, P, Mn, etc.) by pure oxygen blown at high velocity resulting in the formation of molten steel and slag at temperatures of approximately 1650°C. The refractories at the charge pad zone of a BOF (Figure 6.15a) undergo a simultaneous wear attack that is sequentially analyzed and explained as follows [3]:

1. **Mechanical wear:** Impact and erosion types of wear caused by the charging of room temperature steel scrap and high temperature hot metal at 1300°C.

2. **Chemical wear:** This type of wear is ascribed to the decarburization, dissolution, and erosion of MgO grains in a refractory by low viscous and highly penetrable steelmaking slags.

3. **Thermal wear:** This kind of wear mainly involves localized thermal shock damage to the charge pad refractory due to severe cycling of temperatures frequently caused by the concurrent charging of steel scrap at room temperature and hot metal at 1300°C.

4. **Failure:** A progressive integrated wear attack develops a marked discrepancy in the induced temperature gradients in the regions with considerable microstructure heterogeneity in terms of chemical composition and flaw population, causing differential thermal spalling, leading to the failure of the refractory. Therefore, the chemical wear in terms of corrosion and oxidation is considered the dominant wear mechanism.

Figure 6.15b shows the chemical wear parameters [4] in terms of slag penetration depth and areal oxidation index (O.I.) of the cured MgO–C refractory compositions, as listed in Table 6.9. It is evident that the T4 composition containing 0.9 wt.% nanocarbon exhibiting decreased oxidation index (O.I.) is an indication of higher oxidation resistance and competitive corrosion resistance compared to the conventional TC composition containing 10 wt% graphite.

The beneficial features of the improved properties of nanocarbon-containing compositions are attributed to a highly reactive nanocarbon that promotes the catalytic formation of carbide phases on reaction with metal powders, coats the surfaces of carbon particles, strongly binds the refractory aggregates, and manipulates the porosity characteristics by reducing the pore volume. Therefore, a reduction in the graphite content and the use of small amounts of nanocarbon exhibit marked resistance against chemical wear, reflecting an explicit way to exploit ceramic processing science.

6.6.5 Alumina–Carbon (Al$_2$O$_3$–C)

Alumina–carbon (Al$_2$O$_3$–C) refractories are an important class of carbon-containing oxide (oxide–C) composite materials that find important applications in continuous steel-casting processes, specifically in flow control devices such as slide gate plates, submerged entry nozzles (SEN), and monoblock stoppers because of their high hot strength, good thermal shock resistance, and excellent melt corrosion resistance. In this section, processing and failure analysis of slide gate plate refractories is discussed.

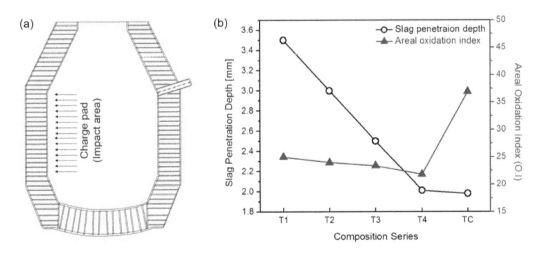

FIGURE 6.15
(a) MgO–C bricks installed at the charge pad zone of a basic oxygen furnace (BOF) and (b) chemical wear behavior of an MgO–C refractory composition series.

TABLE 6.9

Composition and Properties of the Nanocarbon-Containing MgO–C Refractory Series

| Identity | Ingredient Composition (wt.%) | | | | | Refractory Properties | | | |
	Fused MgO	Graphite	Nanocarbon	Antioxidants	Resin + Pitch Powder	BD (g/cc)	AP (%)	CCS (MPa)	HMOR (MPa)
T1	96.0	3	0	1	3.75	3.05	4.7	30	2.5
T2	95.7	3	0.3	1	3.75	3.07	4.6	40	2.9
T3	95.4	3	0.6	1	3.75	3.09	4.5	47	4.1
T4	95.1	3	0.9	1	3.75	3.13	4.3	54	4.5
TC	89.0	10	0	1	3.75	3.05	3.8	37.5	3.9

The raw materials used in the manufacturing of Al_2O_3–C slide gate plate refractories are tabular alumina primarily for high thermal shock resistance, graphite as the chief carbon source, metallic and non-metallic additives as antioxidants, and phenolic resin as a binder. The process flow is similar to the fabrication of MgO–C refractories as previously discussed (Section 6.6.4). Refractory aggregates of varying size fractions, graphite, and additives are mixed with the organic binder in a pan mixer to obtain a homogeneous mix that is further aged for 24 hours. The aged mix is then uniaxially shaped under a compaction pressure of about 200 MPa followed by curing or tempering at 200°C for 24 hours to impart adequate strength to the shaped slide gate plate refractories through the formation of an interlocking carbon–carbon network in three dimensions of the refractory microstructure. These cured plates are fired under a non-oxidizing atmosphere in the temperature range 800°C–1400°C [20, 50].

Al_2O_3–C slide gate plates are used to control the molten steel flow rate where the refractory is exposed to a severe integrated wear attack. Figure 6.16a shows a scanning electron microscope (SEM) image of microcrack formation together with macrocracking and spalling type of failure for slide gate plate refractories with various geometries. The failure analysis of these refractories by the action of various wear mechanisms can be explained as follows [3, 51]:

1. **Mechanical wear:** Damage involving impact and erosion types of wear at the bore portion of a slide gate plate is caused by the throughput rate of molten steel (tonnes/minute) at a temperature of 1650°C. However, the extent of erosive damage varies considerably with the size of the bore.

2. **Chemical wear:** This kind of degradation is dominated by an aggressive Ca–vapor attack on alumina slide gate plates, which usually occurs during the flow of liquid steel containing calcium at a concentration level higher than 35 ppm. Further, decarburization and slag/metallic corrosion are primitive wear types.

3. **Thermal wear:** Extreme thermal shock damages the lower nozzle of the slide gate plate refractory caused by induced temperature gradients due to the sudden exposure of the system to an ambient atmosphere and a high temperature during the liquid steel flow through the valve.

4. **Failure:** The advancement in integrated wear attack by means of the aforesaid types of wear cause s differential thermal spalling and cracking at the bore portion, leading to rapid failure of slide gate plate refractories. Thermal shock is found to be the dominant wear mechanism.

Figure 6.16b shows the thermal shock behavior of a coked Al_2O_3–C refractory composed of a carbon source in terms of 1 wt.% graphite and varying expanded graphite (EG) in the range 0–1.0 wt.%. It can be clearly observed that for the residual strength before and after five thermal shock cycles, the expanded graphite (EG)-containing refractory compositions exhibited a dominance over conventional graphite systems (0 wt.% EG). The advantageous properties of the new compositions are attributed to the generic microstructural features of the expanded graphite (EG), including the presence of nano and micro-pores between graphene laminae, which offers high compressibility and flexibility during thermal cycling, promotes a stress absorption–release mechanism, and improves thermal shock resistance [52]. Therefore, from the viewpoint of ceramic processing, the choice and amount of carbon source considerably influences the control of the thermal shock wear of slide gate plate refractory compositions.

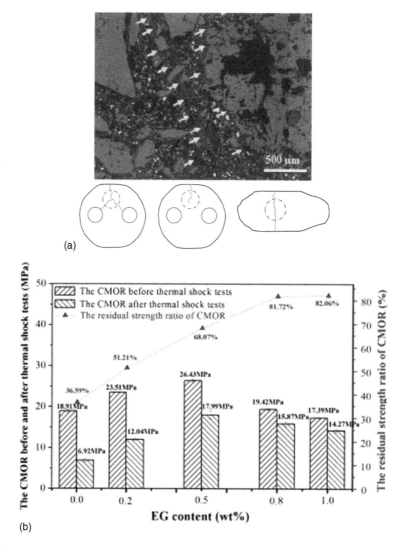

FIGURE 6.16
(a) Thermal shock cracking and spalling of slide gate plates with different geometries. The formation of microcracks can be clearly seen from the SEM image. (b) The effects of carbon source and amount on the thermal shock behavior of refractories with Al_2O_3 and C compositions.

6.6.6 Castables

Castable refractories are an important category of monolithic refractories because of their enhanced performance and ease of installation over shaped bricks. Castables are defined as materials that constitute a homogeneous mixture of graded refractory aggregates and fines blended together with bonding agents and property improvement additives in a predetermined fixed ratio based on a complex and technical formulation to impart specific and tailored properties to the casted shapes, recommending these materials for various critical applications. Castables have been categorized into many different types but classification based on the type and amount of bonding material (cement, sol, and phosphate) has a key role in selecting the particular type of castable with certain properties for a typical application. For example, the excellent properties of the Al_2O_3–SiC–C castable composition, including superior corrosion resistance, high hot strength, greater thermal shock resistance, and high load-bearing capacity, endorse these castable materials as suitable for blast furnace trough applications where the wear environment mainly deals with thermal shock, oxidation, erosion, and slag and metallic corrosion [1, 20]. This section demonstrates the processing and failure analysis of an Al_2O_3–SiC–C blast furnace trough castable.

The raw materials used to fabricate Al_2O_3–SiC–C trough castables are tabular- and reactive-type alumina as the alumina source, silicon carbide, graphite as the chief carbon source, cement or silica sol as a binder, and property improvement additives such as metal antioxidant powders and magnesium aluminate spinel. Graded refractory aggregates of different size fractions combined with the additives are blended in a predetermined proportion, which is further mixed with a liquid binder (water is used in the case of cement bonding systems) in a planetary mixer until a homogeneous mixture with the proper consistency is obtained. The desired shape of the castable refractory can be achieved using the vibrocasting technique that involves pouring of castable mix into a mold under vibration where the mixture undergoes a curing process, leading to the formation of rigid and hard mass due to the hydraulic or chemical setting phenomenon. These cured cast articles are demolded, dried in an ambient air atmosphere and further by a hot air oven for 24 hours, and then fired to form the sintered bond.

The function of a blast furnace trough is to transport the tapped hot metal and slag from the blast furnace where the Al_2O_3–SiC–C refractory experiences severe integrated wear attack but chemical wear is found to be the dominant process. The concise idea of the chemical wear process involves the formation of a decarburized layer on the hot face of the lining due to extreme erosion and oxide dissolution processes caused by hot metal and slag. This dissolution process aggravates the chemical wear and follows the Marangoni effect. The rate of decarburized layer formation on the hot face lining determines the wear rate and degradation of Al_2O_3–SiC–C castables. A recent study found that trough castables bonded with silica sol showed improved slag corrosion resistance over low cement castables attributed to cement-free compositions that don't exhibit the regular deleterious effects caused by lime-containing liquid phases. Therefore, it is corroborated with regards to the science of ceramic processing that the selection of an appropriate binding agent has a decisive role in considerable control of the chemical wear effects [20, 53].

6.6.7 Tundish Spraying Masses

Spraying masses are low dense monolithic materials that find important applications as hot face coating for the permanent working linings of tundish steelmaking practices where this coating primarily acts as a protective layer against the integral wear attack encountered in high temperature process environments. The additional advantages of these spray masses include minimum rebound loss, flexibility for use in cold or hot conditions, better performance against wear attack offered by the minimal coating thickness, less skull formation, and easy deskulling. The standard spraying masses are mostly made of MgO-based materials and the composition ingredients are similar to castables but with a further reduced particle size. The installation method includes preparing the homogeneous free-flowing slurry by mixing the dry composition (spray mass) with 15%–25% of water and pumping this fully wetted slurry to the spray nozzle where it is atomized with air and sprayed effectively onto the permanent refractory lining of a tundish. The installed spray coat is heated for predetermined curing cycles to completely remove the water, becoming drier and harder due to the formation of a chemical bond [1]. The integral wear attack encountered by this hot face lining involves chemical wear caused by molten metal/slag, mechanical wear exerted by the deskulling process, and thermal stresses resulting from frequent cycling of temperatures, leading to cracking and degradation. A detailed failure analysis of the tundish spray mass coatings is explicitly presented elsewhere [3].

6.6.8 Insulating Refractories

Insulating refractories are porous ceramics with porosity levels above 45% and most commonly serve as potential members for backup linings of high temperature furnaces such as glass tanks and tunnel kilns to reduce the heat losses from the process environment to the open atmosphere. The control of heat loss is solely attributed to the structural configuration of the insulating refractory where dissipation of the thermal energy occurs by entrapped air in the porous zones of the refractory because air is not a conductor of heat. Designing the matrix of an insulating refractory with the uniform distribution of small-sized pores is a characteristic approach to offer maximum insulation under service. A widely used practice to generate porosity within the refractory structure is firing the shaped refractory mix containing combustible organic materials that burn out during firing and create the pore space. However, the use of foaming agents, chemical bloating techniques, and porous textured materials such as vermiculite are the typical protocols employed to create a high level of porosity within the refractory structure for special purposes. Insulating refractories have been classified into four different categories based on their capacity to withstand heat or their application temperature, as listed in Table 6.10.

Ceramic fibers are an important division of insulating materials and attained marked industrial importance because of their direct usage of hot face refractory lining where melt corrosion resistance and hot strength are not the chief requirements. Alumina, alumina-silica, and zirconia are the most commonly used ceramic fibrous systems. These materials can be spun and fabricated to any kind of desired shape such as boards, textiles, blankets, felt, and so on. However, prolonged service periods and excessive temperatures are the typical common factors in the deterioration and subsequent failure of ceramic fibers. These factors primarily cause a change in the chemical composition of the ceramic fiber with the formation of new ceramic phases that are associated with noticeable structural changes in volume, leading to sagging of the fibrous structure and consequent degradation [1, 36].

TABLE 6.10

Classification of Insulating Refractories Depending on the Application Temperature

Type of Insulating Refractory	Application Temperature	Refractory Compositions
1. Insulating refractory (Class I)	1000°C	Calcium silicate, vermiculite, and siliceous earth materials
2. Insulating refractory (Class II)	1400°C	Ceramic fibers, lightweight fire clays, lightweight castables, and aluminosilicates
3. High-temperature insulating refractory	1600°C	Alumina fibers, bubble alumina, and lightweight mullite and alumina
4. Ultrahigh-temperature insulating refractory	1800°C	Porous zirconia and non-oxide compounds

6.7 Installation Protocols

The installation of refractories in the construction of different types of furnaces is a specialized area of masonry work. There are five principal categories of furnace structural geometries including flat walls, cones, cylinders, spherical domes, and arches. A perfect construction with good structural integrity requires the proper alignment of different sizes and shapes of refractory bricks, providing adequate expansion joints, and eliminating wider mortar joints. The refractory installation process in any kind of furnace architecture construction involves the tapping of shaped refractory bricks to a place where they are laid and tightly bonded together using an air-setting mortar. These mortars are a class of unshaped refractories or monolithics [39]. The different installation procedures for shaped and unshaped refractories in various furnace architectures are detailed in the following sections.

6.7.1 Shaped Refractories

The installation procedure for shaped refractory bricks in the construction of furnace walls is described in Figure 6.17a [54].

The laying or installation method is observed as a combination of alternative stretcher and header courses, which is the most common type of arrangement to build wall structures with strongly tightened and bonded brickwork. However, the choice of bonding type used to construct a particular furnace architecture mainly depends on factors such as furnace design, wall thickness, operation conditions, maintenance services, and so on. The different types of bonds used for construction purposes are as follows:

1. Complete headers
2. Entire stretchers
3. Combination of alternate headers and stretchers
4. Composite walls

Newly designed refractory bricks (e.g., wedge, skew, arch, jamb, and key shapes; as shown in Figure 6.10) , are used in combination to construct distinct architectures such as cylindrical shaped linings, cones, and arches (Figure 6.17b) of various iron-making and steel-making furnaces, which have been detailed elsewhere [2].

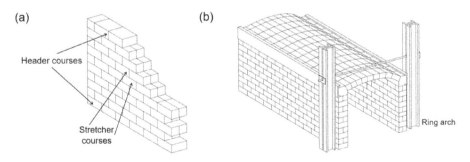

FIGURE 6.17
(a) Combination of alternate header and stretcher course type of bonding used for furnace wall construction and (b) schematic geometry of ring arch structure.

6.7.2 Unshaped Refractories

Different installation procedures for various unshaped or monolithic refractories are detailed in Table 6.11.

TABLE 6.11

Methods of Installation for Various Unshaped/Monolithic Refractories [1, 3]

Unshaped/Monolithic Refractory	Method of Installation	Description
1. Castables	Casting or pouring	A class of monolithic refractory comprising a dry mix of aggregates with varying size fractions. This dry material is usually mixed with water and installation is carried out by pouring this wet mass with or without vibration in the gap space where the lining is required.
2. Gunning masses	Gunning	These are repair masses comprising a dry mix of aggregates with finer-sized fractions compared to castables. A wet gunning type of installation technique is used to carry out the repair work where this process involves the pumping of wet masses prepared by mixing the former dry material with the water through the hose using a piston pump that further disperses with compressed air through the nozzle.
3. Ramming masses	Ramming	These are granular masses used for repair work, which constitute a dry mix of finer-sized aggregate fractions compared to the castable mix. The common ramming techniques, including shovel tamping and hand or pneumatic ramming, are used for this refractory installation provided that the installed ramming layer thickness is kept in the range 100–150 mm.
4. Plastic masses	Pasting	These are wet masses used for repair works and are available in a premixed condition with liquid. Installation can be carried out immediately using the pneumatic ramming technique. The quality and consistency of these repair masses vary with the type of application.
5. Spray masses/shotcrete	Spraying/shotcreting	These are the repair masses of finer-sized aggregates and installation can be done by directly spraying on the repair location of the refractory lining. This spraying method is similar to the gunning technique. High-end performance of the refractory lining can be obtained with the shotcrete.
6. Mortar	Vibration	These are finer-sized aggregates compared to castables and may be of heat-setting or air-setting type. Installation can be done by a vibration technique to fill the refractory joints for construction purposes as well as to protect the linings against liquid metal/slag penetration.
7. Dry vibratable masses	Vibration	These are the premixed dry masses containing a thermosetting type of binder. Refractory lining installation is carried out using a vibration technique and further cured by a hot air blowing system.

6.8 Conclusions

Integrated high temperature characteristic features in terms of physical, chemical, thermal, and thermo-mechanical properties strongly suggest that refractories are promising materials to serve as linings for metallurgical furnaces operated under extreme wear process environments at elevated temperatures. The failure of various refractories working under distinct service environments is found to occur due to the action of three of the most common types of wear processes: mechanical, chemical, and thermal wear. The developed failure analysis concept corroborated that mechanical wear including degradation caused by impact and erosion effects and chemical wear including oxidative damage caused by air containing oxygen as well as corrosive damage by molten melts are considered typical wear processes that aggravate the factors credible for failure; however, thermal shock has been found to be the wear process solely responsible for the failure of refractories. The key process parameters including composition, properties, brick size and shape, and installation mortar joint thickness have exhibited a marked influence on the divergence in locally induced thermal gradients that aggravate the failure activity of a refractory. Failure analysis studies have corroborated that the effect of composition variables on the change in failure behavior of refractories is remarkable. Further, a brief outline on the installation procedures of shaped and unshaped refractories offers a concise idea on the brick masonry work and refurbishment practices.

References

1. Sarkar, R. 2016. *Refractory Technology: Fundamentals and Applications*. CRC Press: Boca Raton, FL.
2. Schacht, C. 2004. *Refractories Handbook*. CRC Press: Boca Raton, FL.
3. Vert, T., and Smith, J. 2016. *Refractory Material Selection for Steelmaking*. John Wiley & Sons: New York.
4. Bag, M., Adak, S., and Sarkar, R. 2012. Study on low carbon containing MgO–C refractory: Use of nano carbon. *Ceramics International* 38(3):2339–2346.
5. Chesters, J. H. 1973. *Refractories: Production and Properties*. Iron and Steel Institute: London.
6. ASTM. 2015. Standard test methods for apparent porosity, water absorption, apparent specific gravity, and bulk density of burned refractory brick and shapes by boiling water. ASTM C20-00: West Conshohocken, PA.
7. ASTM. 2006. Standard test method for linear thermal expansion of solid materials by thermo-mechanical analysis. ASTM E228-17: West Conshohocken, PA.
8. ASTM. 1999. Standard test method for permeability of refractories. ASTM C577: West Conshohocken, PA.
9. ASTM. 2012. Standard test method for Young's modulus of refractory shapes by sonic resonance. ASTM C885-87: West Conshohocken, PA.
10. ASTM. 2014. Standard test method for reheat change of refractory brick. ASTM C113-14: West Conshohocken, PA.
11. ASTM. 2015. Standard test methods for cold crushing strength and modulus of rupture of refractories. ASTM C133-97: West Conshohocken, PA.
12. ASTM. 2015. Standard test method for abrasion resistance of refractory materials at room temperature. ASTM C704/C704M-15: West Conshohocken, PA.
13. ASTM. 2013. Standard test method for thermal conductivity of refractories by hot wire (platinum resistance thermometer technique). ASTM C1113/C1113M-09: West Conshohocken, PA.

14. ASTM. 2013. Standard test method for pyrometric cone equivalent (PCE) of fireclay and high alumina refractory materials. ASTM C24-09: West Conshohocken, PA.
15. ASTM. 2015. Standard test method for modulus of rupture of refractory materials at elevated temperatures. ASTM C583-15: West Conshohocken, PA.
16. ASTM. 2015. Standard test method for measuring thermal expansion and creep of refractories under load. ASTM C832-00: West Conshohocken, PA.
17. ASTM. 2018. Standard test method for load testing refractory shapes at high temperatures. ASTM C16-03: West Conshohocken, PA.
18. ASTM. 2016. Standard test method for quantitatively measuring the effect of thermal shock and thermal cycling on refractories. ASTM C1171-16: West Conshohocken, PA.
19. ASTM. 2011. Standard test method for rotary slag testing of refractory materials. ASTM C874-11a: West Conshohocken, PA.
20. The Technical Association of Refractories, Japan (Ed.) 1998. *Refractories Handbook*. The Technical Association of Refractories: Tokoyo, Japan.
21. Surendranathan, A. O. 2014. *An Introduction to Ceramics and Refractories*. CRC Press: Boca Raton, FL.
22. Kelly, A., and Zweben, C. H. 2000. *Comprehensive Composite Materials*. Elsevier: Oxford, UK.
23. Honggang, S., Hongxia, L., Pengtao, L., and Shixian, Z. 2007. Research on slag resistance of Al_2O_3-Cr_2O_3 refractories under temperature fluctuation condition. In: *Proceedings of the Unified International Technical Conference on Refractories (UNITECR)*. Santiago, Chile.
24. Li, K., Wang, D., Chen, H., and Guo, L. 2014. Normalized evaluation of thermal shock resistance for ceramic materials. *Journal of Advanced Ceramics* 3(3):250–258.
25. Brochen, E., Pötschke, J., and Aneziris, C. G. 2014. Improved thermal stress resistance parameters considering temperature gradients for bricks in refractory linings. *International Journal of Applied Ceramic Technology* 11(2):371–383.
26. Yang, J. Z., Fang, M. H., Huang, Z. H., Hu, X. Z., Liu, Y. G., Sun, H. R., Huang, J. T., and Li, X. C. 2012. Solid particle impact erosion of alumina-based refractories at elevated temperatures. *Journal of the European Ceramic Society* 32(2):283–289.
27. Wada, S. 1996. Effects of hardness and fracture toughness of target ceramics and abrasive particles on wear rate by abrasive water jet. *Journal of the Ceramic Society of Japan* 104(1208):247–252.
28. Momber, A. W., and Kovacevic, R. 2003. Hydro-abrasive erosion of refractory ceramics. *Journal of Materials Science* 38(13): 2861–2874.
29. Caniglia, S., and Barna, G. L. 1992. *Handbook of Industrial Refractories Technology: Principles, Types, Properties and Applications*. William Andrew: Norwich, NY.
30. Lee, W. E., and Zhang, S. 1999. Melt corrosion of oxide and oxide–carbon refractories. *International Materials Reviews* 44(3):77–104.
31. Lee, W. E., and Moore, R. E. 1998. Evolution of in situ refractories in the 20th Century. *Journal of the American Ceramic Society* 81(6): 1385–1410.
32. Behera, S., and Sarkar, R. 2016. Effect of different metal powder anti-oxidants on N220 nano carbon containing low carbon MgO–C refractory: An in-depth investigation. *Ceramics International* 42(16):18484–18494.
33. Mahato, S., and Behera, S. K. 2016. Oxidation resistance and microstructural evolution in MgO–C refractories with expanded graphite. *Ceramics International* 42(6):7611–7619.
34. Sadrnezhaad, S. K., Mahshid, S., Hashemi, B., and Nemati, Z. A. 2006. Oxidation mechanism of C in MgO–C refractory bricks. *Journal of the American Ceramic Society* 89(4):1308–1316.
35. Jansson, S., Brabie, V., and Bohlin, L. 2004. Corrosion mechanism and kinetic behaviour of refractory materials in contact with CaO-Al_2O_3–MgO–SiO_2 slags. In: *VII International Conference on Molten Slags Fluxes and Salts* (pp. 341–347). The Southern African Institute of Mining and Metallurgy: Johannesburg, South Africa.
36. Lee, W. E., and Rainforth, M. 1994. *Ceramic Microstructures: Property Control by Processing*. Springer Science & Business Media: New York.

37. Andreev, K., Boursin, M., Laurent, A., Zinngrebe, E., Put, P., and Sinnema, S. 2014. Compressive fatigue behaviour of refractories with carbonaceous binders. *Journal of the European Ceramic Society* 34(2):523–531.

38. Andreev, K., Wijngaarden, M. V., Put, P., Tadaion, V., and Oerlemans, O. 2017. Refractories for coke oven wall: Operator's perspective. *BHM Berg- und Hüttenmännische Monatshefte* 162(1):20–27.

39. Norton, F. H. 1968. *Refractories*. McGraw-Hill: New York.

40. Miyaji, D. Y., Otofuji, C. Z., Pereira, A. H. A., and Rodrigues, J. D. A. 2015. Effect of specimen size on the resistance to thermal shock of refractory castables containing eutectic aggregates. *Materials Research* 18(2):250–257.

41. Andreev, K., and Harmuth, H. 2003. FEM simulation of the thermo-mechanical behaviour and failure of refractories: A case study. *Journal of Materials Processing Technology* 143:72–77.

42. Gregorová, E., Černý, M., Pabst, W., Esposito, L., Zanelli, C., Hamáček, J., and Kutzendörfer, J. 2015. Temperature dependence of Young's modulus of silica refractories. *Ceramics International* 41(1):1129–1138.

43. Kingery, W. D., Bowen, H. K., and Uhlmann, D. R. 1976. *Introduction to Ceramics*. John Wiley & Sons: New York.

44. Harrison, D. 1990. The selection of materials for use in coke ovens. *Materials and Design* 11(4):197–205.

45. Kasai, K., and Tsutsui, Y. 2008. Recent technology of coke oven refractories. *Age* 42:37.

46. Zublev, D. G., Barsky, V. D., and Kravchenko, A. V. 2017. Operation of the extreme heating channels in coke batteries. *Coke and Chemistry* 60(6):231–233.

47. Ghosh, B. N., and Rao, B. 1970. Alkali attack on blast furnace stove refractories. *Transactions of the Indian Ceramic Society* 29(1):1N–7N.

48. Qotaibi, Z., Diouri, A., Boukhari, A., Taibi, M., and Aride, J. 1998. Analysis of magnesia chrome refractories weared in a rotary cement kiln. *Annales de Chimie Science des Matériaux* 23(1–2):169–172.

49. Villalba Weinberg, A. V., Varona, C., Chaucherie, X., Goeuriot, D., and Poirier, J. 2016. Extending refractory lifetime in rotary kilns for hazardous waste incineration. *Ceramics International* 42(15):17626–17634.

50. Behera, S. K., and Mishra, B. 2015. Strengthening of Al_2O_3–C slide gate plate refractories with expanded graphite. *Ceramics International* 41(3):4254–4259.

51. Grasset-Bourdel, R., Pascual, J., and Manhart, C. 2014. Thermal shock on the lower slide gate plate when closing: Test development and post mortem investigations. In: *Proceedings of the Unified International Technical Conference on Refractories (UNITECR 2013)* (pp. 73–78). John Wiley & Sons, Inc., Hoboken, NJ.

52. Wang, Q., Li, Y., Sang, S., and Jin, S. 2015. Effect of the reactivity and porous structure of expanded graphite (EG) on microstructure and properties of Al_2O_3–C refractories. *Journal of Alloys and Compounds* 645:388–397.

53. Pilli, V., and Sarkar, R. 2016. Effect of spinel content on the properties of Al_2O_3–SiC–C based trough castable. *Ceramics International* 42(2):2969–2982.

54. Nandi, D. N. 1987. *Handbook on Refractories*. Tata McGraw-Hill: New Delhi.

7

Whiteware and Glazes

Partha Saha and Debasish Sarkar

CONTENTS

7.1 Introduction

Ceramics are the most ancient material used by humankind, in the form of earthenware. Many factors contribute to the use of ceramic materials as whiteware, such as hygiene and aesthetic appeal, excellent compressive strength, incredible wear, abrasion and corrosion resistance, resistance to various chemicals, water penetration resistance, superior heat and electricity resistance, refractory behavior, and easy maintenance. The cost of handmade fabrication is relatively cheap compared to automated production, and the degree of dimension stability from product to product varied in the former processing protocol. In ancient times, people used both glazed and unglazed ceramic utensils to obtain corrosion-free containers. Whiteware covers a vast field of ceramics, in which clay, quartz, and feldspar are the basic raw materials for making whiteware bodies. Whiteware encompasses many subcategories, such as sanitary ware, porcelain ware, earthenware, and stoneware [1]. Variation in the composition of the three basic ingredients results in these various whiteware products. Starting from the selection of the raw material to the sintering of the final product, certain protocols have to be followed, where each step plays an important role in the fabrication of a defect-free product. Every process is briefly discussed in the following sections. Information is provided on the processing parameters to fabricate defect-free whiteware bodies.

7.2 Raw Materials

The main raw materials used in conventional triaxial body fabrication are clay, quartz, and feldspar. Clays used in the whiteware body are classified into two types: plastic clay and non-plastic clay. Generally, china clay has a medium degree of plasticity. Fire clay and ball clay have higher plasticity in comparison to china clay. Quartz is used as a filler in this composition, maintaining the required viscosity of the liquid formed during firing. Feldspar acts as a fluxing agent providing liquid that fills the pores present in the ware during firing.

7.2.1 Batch Composition

The raw materials used for the triaxial body preparation are primarily divided into three main categories: clay, flux, and filler. Among these materials, clay is the prime raw material used to achieve the plasticity and workability of the product's shape and mechanical strength and the vitrification of the body. Quartz is used as a filler in the body composition that provides viscous fluid during densification. Flux such as feldspars are used in the vitreous phase to facilitate mass transfer during densification, which also leads to the early maturing of the body composition. A minimum impurity content (usually <1%) leads to a defect-free and high strength body during continuous production. However, the batch composition sometime varies because of the different resources of input raw materials. Conventionally, the triaxial body composition is made up of 50% clay including plastic and non-plastic clays with varying percentages as per the required properties, 25% quartz, and 25% feldspar. However, in industrial applications, the composition varies based on the fundamental requirements such as the application temperature, specific quality, weather conditions, etc. Thus, a wide range of whiteware are manufactured and their basic compositions can be depicted on a phase diagram, as highlighted in Figure 7.1 [2].

The clay–quartz–feldspar phase diagram represents the composition of all types of whiteware bodies, such as porcelain stoneware, white stoneware, soft porcelain, hard porcelain, laboratory porcelain, stoneware, earthenware, and dental porcelain. To understand batch composition, let us consider the example of dental porcelain near the feldspar region in Figure 7.1 (right), showing dental porcelain containing >75% feldspar, <10% clay minerals, and the rest quartz. In order to pick up the apposite composition for the particular whiteware, the mentioned phase diagram can be employed. In Figure 7.1, a typical hard porcelain is marked "O" and the composition is determined and guided by the dotted line. Herein, the particular position shows that the composition contains clay (48 wt.%), quartz (28 wt.%), and feldspar (24 wt.%). Owing to the conventional approach, one can pick up any mass fraction for whiteware.

7.2.2 Various Clays

Both plastic and non-plastic clays have individual potential in the triaxial body. A brief discussion on both china clay and ball clay will provide a better insight into selecting the ratio of individuals properties in the cumulative clay content.

China clays are usually considered both primary and secondary clays depending on the location of their origin. Primary china clays are found at region of origin, having a

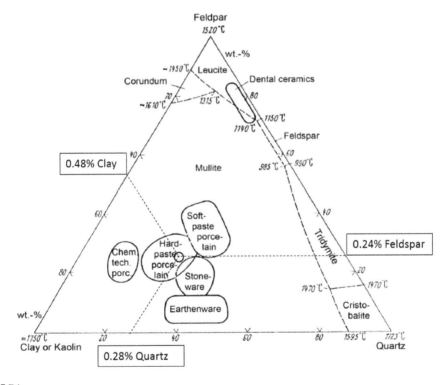

FIGURE 7.1
Ternary phase diagram of clay, quartz, and feldspar [2]. A typical hard porcelain is marked "O" and the composition is determined and guided by the dotted line, where clay = 48 wt.%, quartz = 28 wt.%, and feldspar = 24 wt.%, respectively.

coarse grain size, with secondary or sedimentary china clays transferred from their original place of origin by weathering, preferentially through wind, river, or rain, to a different location. Over years, interaction with these transferring agents makes these clays less pure as well as continuously reducing their particle size. The main component of these clays is kaolin ($Al_2O_3.2SiO_2.2H_2O$), which contains 39.8 wt.% Al_2O_3, 46.3 wt.% SiO_2, and 13.9 wt.% H_2O. China clay expedites few characteristic features of body ware, as follows;

1. Provides a good casting thickness.
2. Contributes medium plasticity as it possesses relatively large clay particles.
3. Provides white color to the fired body.
4. Exhibits low dry strength (preferably not less than 15 kg/cm²) and high fired shrinkage.
5. Resources of alumina to develop in situ mullite in the fired body that eventually enhances the mechanical strength of the body.

In order to fabricate a whiteware body, the china clay composition should have minimum impurities, for example, the content of Fe_2O_3, TiO_2, and CaO should be less than 2 wt.%, 2 wt.%, and 1 wt.%, respectively.

Ball clay is generally used in the body composition as it imparts plasticity and workability to the slip. As its name suggests, it is cut into ball-shaped structures during mining and possesses a high percentage of organic mass that leads to plasticity. It is composed of kaolin as its primary mineral, although quartz, feldspar, and mica are present in lesser quantities. Furthermore, the darker shade of ball clay in comparison to china clay is due to higher quantities of impurities such as TiO_2 and Fe_2O_3 (preferably <2.75 wt.%) and carbonaceous materials (should be <2 wt.%). Obviously, such impurities drastically affect the resultant physical properties of the clay. Even though they increase the green strength of the body, however, they decrease the casting thickness and fired strength. The addition of ball clay provides the following properties to the body [3]:

1. Higher plasticity and workability.
2. Higher content providing excessive dry shrinkage because of fine particles.
3. Higher green strength but lower firing strength.
4. Buff-burning ceramic bodies, low maturing and less translucent due to impurities.

Other plastic clays used in minute quantities are bentonite and pyrophyllite. Among clay minerals, not only kaolinite but also illite, halloysite, smectite, pyrophyllite, sericite, dickite, interstratified vermiculite, and chlorite can be used in whiteware bodies.

7.2.3 Fillers

SiO_2 is used as a filler in the whiteware body, having the mineralogical name quartz and its corresponding mineral rock is quartzite. Quartzite is white in color and contains various impurities and muscovite, biotite, and hematite. Crystalline silica is also available as silica sand and sand stone and is buff white or yellowish white in color. Quartz follows a polymorphic transformation during cooling at 573°C, leading to a sudden substantial volume change during transformation. Thus, the residual quartz may cause fracture of the whiteware body. Below 1200°C, it remains unreactive, preferring to form mullite above this temperature. Typical reasons for the addition of an inexpensive filler such as quartz are

1. It acts as a shrinkage controller during drying and firing.
2. It controls the viscosity of the liquid formed at high temperature by dissolving into it during body maturing.
3. It enhances the mechanical response of the fired body and it is acid resistant.

However, prior to the addition of such a filler, it is necessary to maintain the composition of raw material more than 98 wt.% SiO_2 content along with a maximum limit of 1, 0.25, and 0.5 wt.% of Al_2O_3, Fe_2O_3, and CaO, respectively. In addition, a carbonate-free, speck, and patch-free white mass without any fusion and having a low loss of ignition (LOI) value <0.5 wt.% is expected in whiteware fabrication.

7.2.4 Fluxes

Several raw materials are used as fluxes in the whiteware industry, such as potash feldspar ($K_2O.Al_2O_3.6SiO_2$, orthoclase/microcline), sodium feldspar ($Na_2O.Al_2O_3.6SiO_2$, albite), calcium feldspar ($CaO.Al_2O_3.6SiO_2$, anorthite), nepheline syenite ($Na_2O.Al_2O_3.2SiO_2$), and sericite ($K_2O.3Al_2O_3.6SiO_2.2H_2O$). Owing to the impurity content, natural flux material results in a low-melting phase. These low-melting compounds provide the liquid phase for mass diffusion and transportation during densification at comparatively lower temperatures. The type of alkali ion present and its percentage content in the composition play a vital role in the fluxing action. Reasons for the addition of such fluxes during whiteware body preparations are

1. Fluxing agents soften early and provide a liquid medium to the solid clay and quartz particles to form a vitrified dense body by liquid phase sintering. This leads to the formation of a body with low water absorption and chemical inertness.
2. During firing, the flux melts act as a surfactant to decrease the surface tension of the solid particles to bring them together leading to densification.
3. To enhance the green strength and provide low drying shrinkage.
4. To boost the mechanical responses of the fired body, such as strength, hardness, and optical properties including translucency and glossiness, and a variation in thermal behavior such as increased coefficient of thermal expansion (CTE).

It is preferable that the feldspar melts at a low temperature (1200°C max) to form a glassy structure. Containing a minimum of 9–10 wt.% of the major alkali oxide and other alkali oxides in the range of 3–4 wt.% is advantageous for large-scale production. At the same time, it is necessary to minimize Fe_2O_3 with a maximum allowable limit of 0.2 wt.% in the mixture of 17–18 wt.% Al_2O_3 and 64–67 wt.% SiO_2. In practice, both potash and soda feldspar are required to make whiteware; potash provides a high viscous liquid and a long-range firing temperature, whereas soda feldspar provides low viscosity and high fluxing action in glazes.

Apart from these raw materials, pyrophillite ($Al_2O_3.4SiO_2.H_2O$) and sericite ($K_2O.3Al_2O_3.6SiO_2.2H_2O$) are extensively used in floor tiles as they provide a relatively lower thermal expansion coefficient, higher thermal shock resistance, and low drying and firing shrinkage. These properties consist of its flaky and acicular structure that eventually provide higher fired strength. It is quite non-plastic in nature, but analogous to clay and can replace talc in the whiteware body.

7.3 Whiteware Body Preparation

Ceramic body preparation involves the intimate mixing of various raw materials without or with water for "dry" or "wet" processing consist of binders, plasticizers, flocculants, deflocculants, etc., before being shaped, dried, and fired to the desired temperature. Table 7.1 highlights the common dry and wet processing methods and other process parameter(s) with their plausible applications. Sometimes, a surface is covered with glaze for decorative purposes, which matures along with the body at a slightly lower or the same temperature of firing. The raw materials that primarily contain clay (china clay, ball clay), quartz, and feldspar are beneficiated before being thoroughly intermixed to form the body composition. This mixing is essential to provide a homogeneous distribution of composition and minimize any inconsistency in the fired properties.

7.3.1 Dry Processing

The dry processing of materials where no or a minimum amount of water (<2%) is required involves operations such as crushing and grinding raw materials to powder form before shaping or forming it to the required geometry. Dry mixing or blending of individual components is performed in revolving cylinders. Since dry processing involves no or trace amounts of water, the removal of water by spray drying or filter pressing is not mandatory, rendering it a cost-effective and time-saving process.

a. **Dry Mixing:** Dry mixing is primarily used for non-clay-bearing compositions. However, clay-containing body compositions can also be explored where a small amount of water can be added to the dry-mix body to obtain the consistency required for plastic deformation during the forming/shaping stage. Dry mixing has a clear advantage over wet mixing because it does not demand filter pressing, which is mandatory for slip or slurry mixing processes. However, dry mixing

TABLE 7.1

Various Dry and Wet Processing Routes

Forming Method	Water Content	Pressure Used	Applications/Uses
Slip casting	20%–50%	No 0.5–2 MPa for pressure casting	Whiteware industry
Semi-dry pressing	4%–10%	High	Bricks, tile industry
Dry processing	0%–4%	High	Non-plastic body fabrication
Isostatic pressing	0%–2%	Very high	Whiteware industry
Hot pressing	Negligible	Very high	Metal carbides, borides, nitrides, etc.
Reactive hot pressing	—	Very high	MgO, Al_2O_3, etc.

has the disadvantage that as soluble salts in the body cannot be removed, these may interfere with the subsequent manufacturing stages or with the deflocculant if required. For non-clay-based bodies, binders, plasticizers, and lubricants are added to the dry mix to increase its workability. Auxiliary materials aid in shaping and providing rigidity to the body before firing.

b. **Plastic Mixing:** Plastic mixing is mainly used for clay-based bodies where water is added to develop plasticity. Mixing blends the coarse, medium, and fine granular materials that are fed into a pugged mill for further mixing, consolidation, and extrusion. The pugging operation is often performed in a vacuum to remove entrapped air and improve plasticity and strength. Ideally, a body must develop good plasticity at low/optimum moisture content, which minimizes drying-related defects and drying shrinkage.

c. **Dry and Semi-Dry Pressing:** Dry and semi-dry pressing does not involve drying shrinkage, therefore good dimensional tolerance can be achieved. Dry pressing is mainly used for tiles, certain sanitary ware, and porcelain products. The body must have sufficient plasticity to allow the powders to flow smoothly under the application of load and fill the die cavities. Organic binders and lubricants are frequently used for non-clay-based bodies during dry pressing to provide sufficient strength before firing.

d. **Isostatic Pressing:** A rubber bag is filled with granulated powder and immersed in a liquid/oil; thereafter, hydrostatic pressure is applied. The liquid exerts uniform pressure from all directions on the powder, achieving uniform density. This method is suitable for use with non-plastic materials and complex shapes to achieve high density. The advantages of isostatic pressing are (1) powders can achieve a high density without using any additives (binders, plasticizers, lubricants, water, etc.); (2) the intrinsic high density of isostatic pressed parts reduces the sintering temperature and firing shrinkage; and (3) defects related to drying and pressing, e.g., lamination, warping, and cracking, can be avoided. This process enhances both the properties and the production costs compared to conventional dry pressing.

e. **Hot Pressing:** During hot pressing, heat and pressure are simultaneously applied to a powdered mass, thereby forming/shaping and firing/sintering the component, rendering a sufficiently lower time to achieve a fully dense body at a lower temperature than firing/sintering performed under atmospheric pressure. Generally, high strength and good thermal shock resistance are obtained during hot pressing due to minimum grain growth (large grains grow at the expense of small grains) and the removal of pores from the sintered mass. Reactive hot pressing has also been explored on the principle of polymorphic transformation or chemical transformation. For example, dense MgO and Al_2O_3 are produced at relatively lower temperatures using $Mg(OH)_2$ and boehmite as precursors, respectively. However, the use of such a pressing protocol is limited in whiteware processing.

7.3.2 Wet Processing

During wet processing, clay-based slurries are prepared with the addition of a deflocculant for a stable slip. Sometimes, slurries are aged for one to two days before adding non-plastic components (quartz, feldspar, alumina, etc.). The viscosity of the stable slurries is adjusted through the addition of water and conventional additives (Na-poly(acrylic acid), Na-poly(methacrylic acid), Na_2SiO_3 as deflocculants; NaCl, $CaCl_2$ as flocculants; poly(ethylene glycol) as plasticizers; carboxymethyl cellulose and poly(vinyl alcohol) as binders) [4].

a. **Slip Preparation:** Slip is mainly used for whiteware body preparation where a stable suspension of clay, quartz, and feldspar (usually for a triaxial body) in water is prepared along with additives (deflocculant, binder, etc.) known as "slurry." The clay-based slurry is called "slip." The slurry/slip should be stable over time and the suspended fine particles must not settle over time. During wet processing, water from the body is removed using a filter press where the slurry/slip is compressed onto cloths or nylon sheets to squeeze out the water and the solid left behind is in the form of a filter cake. The filter cake is generally pugged before plastic making or dispersed in water for slip casting operations. Sometimes, before the filter pressing operation, the slurry/slip is passed through a sieve to remove coarse grain materials that would otherwise undermine the properties of the finished products. A fixed volume of slip is usually prepared in consideration of Brogniart's equation: $W = (P - A)(S / S - 1)$, where W is the weight of the dry materials; P is the weight of any volume of slip; S is the specific gravity of the dry material; and P is the weight of the same volume of water [5].

b. **Slip casting:** A well-dispersed slurry/slip is poured into a porous plaster of paris mold. Over time, water from the slurry is absorbed into the porous mold. As time progresses, a semi-solid layer builds up on the inside wall of the plaster of paris mold. Generally, slip casting is continued or repeated for longer durations until the required casting thickness is achieved and the remaining slurry/slip is discarded. The semi-solid layer dries over time and shrinks in the mold. In order to avoid the slip level falling into the mold during water removal from the slip, a "ring" is usually incorporated into the mold to hold the slip level at a certain height. The slip casting process generally involves liquid flow through the porous mold wall, which is described by Darcy's law. In one dimension, Darcy's law can be written as [6]

$$J = \frac{K \frac{dp}{dx}}{\eta_L} \tag{7.1}$$

where J is the liquid flux; K is the permeability of the porous mold; dp/dx is the pressure gradient in the liquid; and η_L is the viscosity of the liquid. Neglecting the mold resistance to the liquid flow, Darcy's law leads to a parabolic rate law equation of casting thickness (L_c) with time (t), expressed as

$$L_c^2 = \frac{2K_c pt}{\eta_L \left(\frac{V_c}{V_s} - 1 \right)} \tag{7.2}$$

where K_c is the permeability; p is the pressure difference across the cast; V_c is the volume fraction of solids in the cast; and V_s is the volume fraction of solids in the slurry. The rate of casting follows a parabolic rate law where the casting thickness is proportional to the square root of time. For sound and efficient casting, a few factors must be strictly followed:

(1) the body composition of the slurry, (2) particle size of the raw materials while preparing a slurry, (3) concentration of deflocculant and water content, (4) specific gravity and viscosity of the slurry, and (5) solid content in the slurry. A good and consistent slip flow depends on the colloidal behavior of clay and this phenomenon is briefly highlighted in the next section.

7.3.3 Colloidal Behavior of Clay

Isomorphous substitution facilitates partial replacement of the cation of a higher valence by lower valence cations, which leads to the clay structure becoming negatively charged, which is further enhanced by the external adsorption of other cations (M). The resultant clay is called "M-clay." It is also possible to replace M-cation with other N-cations by treating the M-clay solution with a salt solution of N-cations.

$$\text{M-clay} + \text{N}^+\text{A}^- \rightleftarrows \text{N-clay} + \text{MA} \tag{7.3}$$

The amount of M^+ replaced by N^+ depends on the concentration of the NA solution, the size of the M^+ and N^+, the valence of the cations, and the solubility of the product NA. Generally, smaller cations of higher valence adsorb faster than larger cations and the order of the adsorption is given by the Hofmeister series: $H^+ > Al^{3+} > Ba^{2+} > Sr^{2+} > Ca^{2+} > Mg^{2+} > NH_4^+ > K^+ > Na^+ > Li^+$ [7].

Clay particles preferentially adsorb smaller cations such as H^+ or higher valence Ca^{2+} in preference to a large monovalent ion such as Na^+ and the replacement of H^+ or Ca^{2+} with Na^+ would require a high concentration of Na^+ solution (NaCl). In practice, the treatment of H-clay with sodium carbonate and sodium hydroxide exhibits two different phenomena:

$$\text{H-clay} + \text{Na}_2\text{CO}_3 \rightleftarrows \text{Na-clay} + \text{H}_2\text{CO}_3(\text{carbonic acid}) \rightleftarrows 2\text{H}^+ + \text{CO}_3^{2-} \tag{7.4}$$

$$\text{H-clay} + \text{NaOH} \rightleftarrows \text{Na-clay} + \text{H}_2\text{O} \tag{7.5}$$

The hydrogen ion produced from the carbonic acid remains ionized and can compete with sodium ions for the adsorption of clay. Due to the small size of a H^+ ion, it is easily adsorbed unless the Na^+ ion concentration is high. Therefore, the reaction as described in Equation 7.3 proceeds to the left and the reaction to the right occurs to a limited extent. However, for Equation 7.4, the water molecule produced as a by-product is weakly ionized and the concentration of the H^+ ion is low to compete for adsorption with the Na^+ ion. Therefore, the reaction proceeds to the right and the H^+ ion is replaced with the Na^+ ion.

The exchange of one cation with another is known as "cation exchange" or "base exchange" capacity (c.e.c. or b.e.c.) and is expressed as a milli-equivalent of cations adsorbed per 100 g of clay. The c.e.c. of the montmorillonite type of clay is high (~70–150 m.e./100 g of clay), whereas kaolin clay (~3–6 m.e./100 g of clay) and ball clay (~15–40 m.e./100 g of clay) imply lower values. In the kaolin group of clay, the broken bond theory plays an important role in ion exchange. Unsatisfied valence bonds at the edges of the clay particles that can be easily exchanged with ions and further exchanged in a water solution lead to the development of surface charges on the clay particles. It should be borne in mind that the broken bond theory is the dominant mechanism for ion exchange in kaolin, and isomorphous substitution is the key reason for surface charges on the montmorillonite. The b.e.c. of clay mainly occurs due to (1) the broken bond theory in kaolin clay, (2) the isomorphous substitution of residual charges present in montmorillonite clay, and (3) the ionization of basal hydroxyl ions (OH⁻) that are present and ionized to negatively charged oxygen and positively charged hydrogen; the positively charged hydrogen is exchanged and a negative charge develops on the clay surface to attract another cation.

The stability of a clay suspension depends on the pH and the concentration of counterions. In water, clay behaves like the anion of a weak acid and ionization takes place as per Equation 7.6:

$$M\text{-clay} \rightleftharpoons M^+ + \text{clay}^-$$ (7.6)

The extent of ionization depends on the electropositivity of the cation. Charged clay particles produced by ionization have a tendency to repel other particles, resulting in deflocculation. Repulsion between particles depends on the type of exchangeable cation adsorbed on the clay, with high repulsion when the exchangeable cation is highly electropositive (Na⁺) and low repulsion when it is less electropositive (H⁺ or Ca²⁺). The magnitude of the zeta potential is an imaginary layer at the clay surface where the counterions and water molecules rigidly hold, which determines whether the particles will repel one another or floc together to form an agglomerate [8]. If H-clay is replaced with a sodium salt (NaOH), the sodium clay exhibits a high zeta potential and is enhanced in the presence of NaOH. A maximum value for the zeta potential is achieved when the maximum b.e.c. of NaOH is reached. However, further addition of NaOH has no effect, rather the zeta potential decreases since the excess Na⁺ ions crowd around the double layer and suppress the Debye length (1/k) resulting in a decrease in the zeta potential. When the H-clay is treated with a small addition of NaOH, viscosity starts to increase to a maximum and then gradually drops (see Figure 7.2a). The Na⁺ ion first replaces the H⁺ ion on the faces only, increasing the negative potential without replacing the edges, resulting in increased attraction between edges and faces. Further additions of NaOH raise the pH, which reverses the edge charges and the clay becomes negatively charged both at the edges and faces and the system becomes deflocculated (see Figure 7.2b).

Generally, for the kaolin group of minerals, sodium salts favor stability and cause deflocculation while hydrogen and calcium ions cause flocculation. The zeta potential value of various counterions is given in Table 7.2.

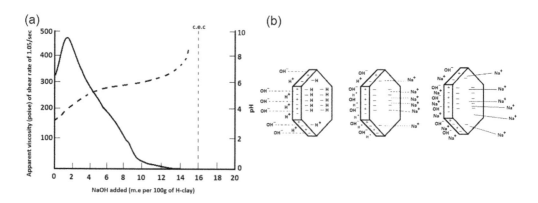

FIGURE 7.2
(a) pH and viscosity curves for H-ball clay (solid line = viscosity, broken line = pH) and (b) the probable cation and anion exchange mechanism during interaction in clay and NaOH [9].

TABLE 7.2

Zeta Potential of Various Clays

Nature of Clay	Zeta Potential (Millivolt)
Ca-clay	−10
H-clay	−20
Mg-clay	−40
Na-clay	−80
Natural clay	−30

It is generally accepted that the zeta potential must be above ±50 mV for a clay suspension to remain stable over time. A high zeta potential accompanied by deflocculation can occur when a small fraction of cation from clay is replaced with sodium ion. For example, in the kaolin group of materials, the Na ion favors stability and complete deflocculation while the hydrogen and calcium ions cause flocculation to occur. The Ca-clay may be deflocculated with sodium carbonate, sodium oxalate, and sodium phosphate, and the H-clay with NaOH. In addition, large anions (silicate, polyacrylate, phosphate, carbonate) are also adsorbed onto the clay surface and provide extra negative sites that are strongly ionized and the associated cations remain at a relatively greater distance from the surface.

Polyanions are preferably adsorbed on the edges either by anion exchange or by physical adsorption since the planar surface of clay particles is negative and repels polyanions. The dispersion of kaolinite is mainly governed by the stability of the system, which is governed by the type of medium anions and preferential adsorption. It has been found that SiO_3^{2-} ions preferentially adsorb into the kaolinite particles, and substantially increase the net negative charge on the particle (see Figure 7.3). It has also been observed that higher valence cations flocculate by suppressing the double layer resulting in coagulation.

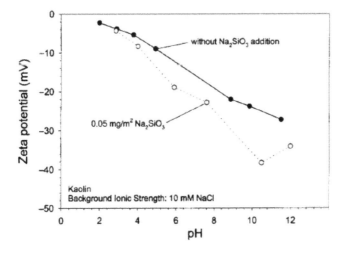

FIGURE 7.3

Effect of Na_2SiO_3 on the zeta potential of kaolin over a range of pH values against a background of ionic strength of 10 mM NaCl. An increase in the negative zeta potential indicates the specific adsorption of the SiO_3^{2-} ion on the clay particle surface [10].

7.4 Drying Whitewares

7.4.1 Classification of Driers

Drying plays a vital role in the fabrication of defect-free whiteware bodies. In order to fabricate whiteware bodies that have a critical shape, the maximum allowable moisture content has to be maintained before glazing. Despite glazing and firing faults and their consecutive defects, drying is an important parameter that is needed to maintain the proper thermal cycle in driers. Thus, various kinds of driers are used for industrial purposes, which can be categorized into two types: intermittent and continuous driers.

7.4.1.1 Intermittent or Batch-Type Driers

a. **Loft Driers:** Loft driers require large spaces or rooms heated by steam pipes that run through the walls or under the floors. These driers are for large refractories/whitwares such as sewer pipes, glass pots, etc. Drying in a loft drier is very slow and can take months. Additionally, the heavy ware needs to be adjusted or rotated regularly for uniform drying of the product. Hot floors are used extensively for handmade, complex refractory/whiteware shapes. A concrete or cement floor is heated by hot air or gases or by low-pressure steam pipes, and even though it is an inefficient way of drying, it somehow compensates for the problem of low thermal efficiency.

b. **Compartment Driers:** Compartment driers are mainly closed compartments where there is critical control over the dry and wet bulb temperature to efficiently maintain drying. The drying time of compartment driers is determined by the drying time of the largest ware, and these driers are more efficient than loft driers or hot floors.

c. **Humidity Cabinet:** A type of compartment drier, a humidity cabinet is built on a small scale and consists of insulation, equipped with an elaborate heat and humidity control apparatus to avoid heat leakage and provide high thermal efficiency.

In sum, the advantages of intermittent or batch-type driers are

1. Economical construction
2. Ease of operation
3. Opportunity to vary drying condition from one charge to another

Disadvantages of intermittent or batch-type driers are

1. Operation is not continuous so not suitable for continuous mass production unit
2. Thermal efficiency is low
3. It is difficult to provide the same drying conditions to all parts of the drier at the same time

7.4.1.2 Continuous Driers

a. **Tunnel Continuous Drier:** There are two types of tunnel continuous driers: parallel current and countercurrent. Tunnel driers are long tunnels (9–10 m) equipped with proper zone-wise heating elements, fans, and humidity and temperature controllers. A parallel current is when the drying gas moves in the tunnel in the same direction as the ware, and a countercurrent is when the drying gas moves in the opposite direction. In the absence of waste heat, the furnace underneath or the flue gases that pass through the duct can heat the tunnel. For drying complex-shaped sanitary ware, monorails are used and full control over humidity is maintained, which may take 24–26 hours. Small-shaped ceramics such as electrical porcelain are passed through the tunnel for about 4–6 hours. However, larger clayware requires approximately 2 weeks to dry in a semicontinuous tunnel chamber–type drier. This drier is divided into four compartments with different levels of humidity and circulation, where components spend 3–4 days in one compartment and then move to the next compartment to facilitate defect-free drying.

b. **Mangles:** Drier shelves are suspended on chains that ascend and descend tracks through a baffled chamber. The drying air flows in the opposite direction to the ware. The valves on steam coils that run through baffled walls control the gradual increase in temperature. This drier is used for mold release in potteries and for drying applied glaze over biscuit ware. During glazing, placing a dipping tank underneath the drying racks reduces the process time, and this approach is known as "glaze mangles."

c. **Rotary Driers:** Rotary driers are used to dry the raw materials that are usually in powdered or lump form. The materials are tumbled in a cylindrical drier containing projecting plates to provide intimate contact with the raw materials and drying air either in parallel current or countercurrent motion.

d. **Drum Driers:** Drum driers are conventionally operated in two modes: under atmospheric pressure and vacuum. These driers are used to dry fine materials in the slip, where the drum is heated and then partly dipped in the slip, so that a layer is retained on the drum, which dries quickly and can be removed by scraping. In sum, the advantages of a continuous drier are

1. Each piece of material undergoes the same drying schedule
2. It fits well in a continuous process

The disadvantages of a continuous drier are

1. High cost of installation and maintenance
2. Lack of flexibility

7.4.2 Drying Philosophies

Drying refers to the removal of water from solid, liquid, or gas, and it can be done using various methods. The main drying mechanisms relevant to the whiteware industry are summarized as follows:

a. **Precipitation by cooling:** In this method, either the water vapor is condensed into liquid or the water is frozen to ice and separated from the remaining solution or liquid.

b. **Absorption by a porous body:** This method is used in drying clay in plaster molds because of the capillary attraction of porous materials.

c. **Adsorption:** This is the removal of liquid (other than water) due to the presence of active surfaces on the substances.

d. **Evaporation:** This is the prime method for liquid removal from clay-based materials. It mainly depends on the vapor pressure of the liquid that is present in the atmosphere, as it enhances with increasing temperature and when the vapor pressure of the liquid is equal to the atmospheric pressure, the liquid starts to boil. Eventually, this liquid vaporizes below boiling point. The amount of water vapor present in the air also regulates the rate of evaporation, for example, the rate of water evaporation in a dry atmosphere is faster than in a humid atmosphere. Thus, humidity is an essential consideration in controlling the rate of evaporation and drying. The difference between the wet bulb and dry bulb temperature is used to calculate the humidity content in air, and such an apparatus is known as a "hygrometer."

7.4.3 Role of Humidity and Temperature in Drying

Humidity is controlled by varying the amount of water vapor present in air. Humidity in air is maintained by reusing air composed of a certain amount of humidity and fresh hot air in order to maintain the desired level of humidity. In consideration of humidity control, Wilson classified the drying system in three modes [11]:

a. The vapor wave experience high temperature and low percentage humidity over the entire drying period. In the first method, the attempt is only to regulate the drying cycle by maintaining the temperature of the air. As far as the operation of the drier is concerned, it is simple, but in terms of ware drying it is difficult because high temperature and low humidity lead to a high rate of surface evaporation and internal diffusion during the entire period.

The graph in Figure 7.4a shows rapid drying at the surface with large temperature differences between the interior and the surface. This large lag delays the increase in water pressure in the interior to a point where drying proceeds rapidly and safely, the distance BC representing the loss in weight that occurs before the minimum safe drying temperature has been reached in the interior; this safe drying temperature is known as the "critical drying temperature." This is the critical point because most stresses are developed beyond this critical temperature limit by the interior vapor pressure. As the drying proceeds, the rapid weight loss do not affect the ware, and the vapor pressure of the various masses increases and reaches maximum. Subsequently, the vapor pressure is further lowered to avoid increasing the attraction of water by clay particles and increasing the concentration of salts in the remaining water. This method is used to dry clays that have an open structure and a high rate of water diffusion.

b. A low percentage of humidity is maintained during the entire drying time and the temperature is kept maximum at the end. In the second method, as described in Figure 7.4b, the process is similar to the commercial drying of clay wares, but the surrounding air temperature is slowly increased until it reaches maximum

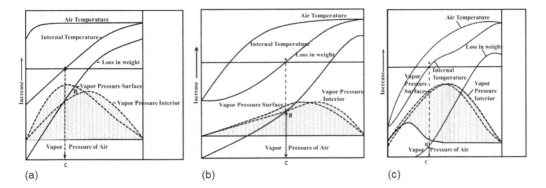

FIGURE 7.4
(a) First, (b) second, and (c) third type of drying system.

and is maintained for the remaining drying period. The maximum temperature is delayed until the interior has reached the critical drying temperature. Herein, the lag between the interior and exterior temperature declines. In both classes of drying, the humidity of the air remains the same throughout the system.

c. Herein, the temperature is raised during drying and the moisture content of the drying air is kept near saturation in the early part of the operation and then lowered in the later part. In the third system illustrated in Figure 7.4c, a high percentage of humidity is maintained in the initial stages of firing, allowing the interior temperature to cross the critical drying temperature without much weight loss, and as the temperature increases, slowly the lag between the interior and exterior is much less than the previous two drying systems. Owing to optimum control of the temperature and humidity, it provides maximum vapor pressure to all parts of the clay mass, preventing a lag in the vapor pressure of the exterior and interior and thus the interior temperature has passed the safe minimum. This drying system is safer than the previous two drying systems; however, it is slower than the first system but it is certainly faster than the second system.

7.4.4 Factors Affecting Drying

The mechanism of drying is controlled by the "diffusion of liquid through solid" and the "evaporation of liquid from surface." These two processes are briefly summarized as follows:

a. Diffusion of liquid through solid: Moisture begins to evaporate at the surface and once the moisture leaves the surface, water comes from the interior parts to the moisture-deficient part. Serious problems may occur when clay that undergoes shrinkage during drying follows different diffusion rates. Rapid surface evaporation causes a distinct moisture content in comparison with the bulk interior, which eventually leads to high surface shrinkage. Thus, the rate of diffusion is a critical concern during the drying of components. In common, the porosity gradient in a dense body is different from a porous body, which facilitates a high rate of diffusion in later systems. Similarly, it is necessary to avoid a large concentration gradient between the surface and the interior of the ware. This microstructure facilitates a low water evaporation rate rather than a diffusion rate from the surface.

b. Evaporation of liquid from surface: The retaining liquid film thickness depends on the surrounding air velocity, for example, the film thickness is thin in the presence of high air velocity and vice versa. Water that evaporates from a solid and diffuses into the surrounding air must pass through the film. The part of the film that is adjacent to the solid–liquid interface is always saturated, while the outer edge always contains the same amount of water vapor as the surrounding environment, thus a low humidity percentage enhances the rate of evaporation. The rate of drying for wet clay depends on the nature and the properties of the clay. Clays containing high colloidal matter content have very small capillaries that retard water movement and retain the water more efficiently. However, coarse-grained clays have large pores that allow the water to diffuse freely. High temperatures accelerate water mobility and reduce water attraction to solid particles. Thus, heat aids in shrinkage as it allows a readjustment in the shape and size by increasing the mobility of a plastic mass, resulting in material with less defects such as strain or distortion. Hence, the temperature of both the water and the clay mixture is also an important criteria to control the rate of drying.

7.4.5 Drying-Related Defects

All defects due to drying are mainly caused by shrinkage. The main defects are summarized as follows:

a. Strains: Until and unless a clay mass is dried at the same time and at the same rate, differential shrinkage can occur, which can cause strain in the body. In reality, it is very difficult to dry the total clay mass as the same time as drying occurs by removal of water from the interior and to the surface and more simply the evaporation and drying occurs from the surface, the strain is always present in a body. The amount of strain depends on the rate of drying and shrinkage in the clay mass. Excessive strain results in visible cracks in clay ware and moderate strain results in invisible cracks during drying; however, extra strain experienced during firing opens up the cracks and defective ware is produced. Additionally, high shrinkage may exhibit show irregular dry strength.

b. Warping: Warping is a drying defect that occurs when the plastic mass experiences plastic deformation. It is caused by strains during forming, and mainly occurs when one part dries faster than the other, eventually distorting the plastic part, which may be pulled out of shape.

c. Cracks: Cracks are mainly due to strain that occurs during shrinkage and strains when forming or pressure due to a large weight.

d. Checking: Checking is similar to cracking, but here the cracks are so small that they are invisible to the naked eye and the material may seem to be within the acceptable limits. It is more prevalent in the case of fire clays because the rate of diffusion of water from the interior to the exterior is slow and evaporation from the outer surface cannot be stopped. Checking may be reduced by adding grog, but grog particles must be small otherwise local checking may occur.

Despite these critical defects, it is necessary to keep tracking to avoid any other defects that may appear during drying that could lower the aesthetic look, properties, and market value of the finished product.

7.5 Firing Whiteware

In order to obtain a mature whiteware body with high mechanical strength and aesthetic value, firing is an essential and ultimate process. It provides the necessary mechanical strength, chemical resistance, and other improved physical properties to the resultant body. The glaze applied to bodies undergoes an inspection for the presence of any defect on the body during casting, glazing, or manual handling, before being placed on the kiln car. The bodies undergo firing in various types of kilns according to the time cycle required to obtain the desired properties. Before placing the bodies on the kiln cars, the glaze should be properly dried to avoid different glaze defects on the bodies. An efficient kiln consumes less fuel during firing of the components.

7.5.1 Classification of Kilns

There are various ways to categorize kilns, which are broadly divided into

1. Periodic kiln
2. Continuous kiln of chamber type
3. Continuous kiln of tunnel car type

A periodic kiln is a batch-type kiln in which the temperature in the chamber containing the glazed bodies is raised to the desired temperature and then gradually decreased to room temperature. Here, either the car containing the products moves from one end to the other through the heat zone or the flame moves through the stagnant cars. These are also divided into subcategories such as downdraft and updraft periodic kilns.

An updraft kiln is so named because the hot gas travels upward and is used for firing bricks. An example is a scove kiln, which is the most primitive type of kiln. It is used to fire clay bricks for common application. The kiln's dimensions vary according to the ability of its firing system to achieve the homogeneous heating and mechanical strength of the green brick. When bricks are set in the kiln, the outer layer is covered by clay, called "scoving," hence the name a scove kiln. In the case of a clamp kiln, the scoved region is replaced with a solid brick wall where fireboxes and permanent grates are employed for coal firing. To get rid of hot spots with low resistance to flue gas motion, these hot spot regions are made rigid to force the flue gas flow toward a cooler region. To avoid an overfired and unmatured body due to placement in the kiln, overhead kilns are used to position the batch as a whole and remove after firing.

In downdraft kiln, the hot gas pass toward its crown and bringing this back down using the product pieces as draft and release them onto the stack. It is widely used in comparison to an updraft kiln due its higher efficiency. It is further classified into "round" and "rectangular" downdraft kilns. Conventionally, the kiln height depends on several parameters including the green strength of the stacked wares and their plausible high temperature deformation.

The concept behind the use of a continuous kiln is the efficient utilization of the thermal energy generated in the kiln. Here, during cooling, the heat radiated from the hot-fired products is utilized for preheating the flue gas and extracting the heat energy from the flue gas for drying the prefired product. There are two types of continuous kilns: chamber type and car type. For industrial application, continuous tunnel kilns are generally employed in industrial practice owing to their high efficiency. The shape of the kiln may be straight, "U" shaped, oval or circular depending on the availability of the region and the thermal cycle. A kiln is a single tunnel–type structure that is divided into various segments depending on the zonal temperature, such as the preheat zone, hot zone, and cooling zone, as illustrated in Figure 7.5 [12].

7.5.2 Stacking Whiteware in Kiln

There are several ways to set whiteware in a kiln. Small wares are placed on saggers that are made of fire clay. These act as a protective casing for the wares against the kiln flame. Another purpose for employing saggers in a kiln is to provide the wares with a miniaturized muffle furnace-like surrounding in stacked loading. Large wares are placed directly on the kiln cars on a cordierite-base plate. To reduce the dead weight in the kiln car, thin silicon carbide (SiC) and fused alumina (Al_2O_3) plates are used. It is important to monitor that there is no contact between the wares placed on the car; the base on which the wares are loaded on the kiln car has to be slightly tilted toward the center of the car. The reason for this kind of arrangement is that the outer region of the furnace comes into contact with the thermal energy earlier than the interior regions, so wares start shrinking and lean toward the outer wall, where they may come into contact with the flame or fall out of the car. When the car is pushed into the kiln, the feeding rate must be moderate to avoid generating thermal shock in the body.

7.5.3 Phases in Bisque Firing

Bisque-fired or biscuit-fired bodies are unglazed fired bodies with a matte finish. The texture on the outer surface gives very good aesthetic value. This kind of body contains no or very little glassy structure on the surface. The main purpose of biscuit firing is to make the body impervious to water, resistant to damage during handling, and absorbent for glazing. The biscuit firing temperature is essential to control in the target of porosity left, formation of glassy structure and phases during firing that eventually help to final-stage glazing on the body. The employed thermal energy provides the activation energy to complete the different chemical reactions. The basic chemical reactions that occur during this process are

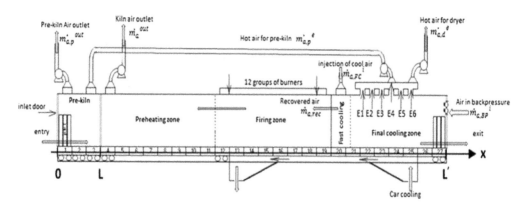

FIGURE 7.5
Schematic representation of a tunnel kiln including the heating zone profile [12].

1. Removal of physically combined H_2O
2. Structural breakdown of clay and removal of chemically combined H_2O
3. Formation of primary mullite phase
4. Growth of primary mullite crystals
5. Formation of eutectic liquid of quartz and flux
6. Dissolution of quartz into the eutectic liquid
7. Formation of acicular secondary mullite crystals
8. Growth of secondary mullite crystals
9. Excess SiO_2 remains as residual cristobalite and tridymite
10. Gradual transformation of residual SiO_2 to quartz

In addition to these reactions, other chemical reactions also occur in the body during firing, such as the chemical decomposition of carbonates ($CaCO_3$, Na_2CO_3, etc.) and sulfates ($CaSO_4$). Figure 7.6 illustrates quartz grains (large dark particles) surrounded by a glassy structure, and the presence of acicular mullite grains along with small pores (darker vacant areas) [13].

Small cracks on the surface of quartz grains are attributed to the stress generated due to thermal expansion coefficient mismatch during the thermal cycle. The porosity of the bisque (generally more than 15%) makes it an ideal medium to absorb water from the glaze suspension and hold it in place; typically, the glaze will dry sufficiently for handling in just a few seconds. Generally, biscuit firing is done at temperatures near 1000°C. However, the retained decomposed clay, residual silica, minute glassy phase, and alumina–silicate structure facilitate the formation of the mullite phase at relatively high temperatures leading to the prerequisite strength for handling during glazing as well as a defect-free glazed surface afterward.

7.5.4 Physical Changes during Firing

Firing leads to drastic changes in the physical properties of whiteware bodies from a soft, plastic clay body to a hard, shaped mass having high mechanical strength, chemical resistance, and aesthetic value. Various attempts have been made to observe high

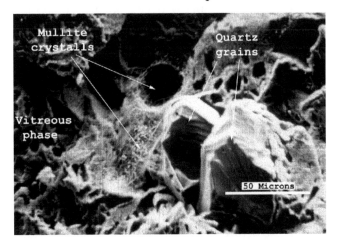

FIGURE 7.6
Typical microstructure of a whiteware body after firing.

temperature–assisted physical variations of clays used in a whiteware body. The weight of the dried mass varies due to the evaporation of the physically and chemically bound water; the decomposition of carbonaceous and sulfate content in the clay; and the volatilization of alkalis, chlorides, and fluorides. Color variation is a result of the chemical reactions in clay matter. All clays exhibit a similar trend from losing physically combined water to the individual silicate and aluminum hydroxide layer of clay to the formation of molecularly bound mullite, which eventually alters the dark gray color of an unfired clay body to the white color of a fired body, and form a soft moldable mass to a rigid hard structure.

As color parameters essentially define the aesthetic value, they should be maintained in order to obtain the ultimate goal of the industry, which is customer satisfaction. Variations in the different parameters are observed during the firing of clay bodies, such as the percentage of porosity and shrinkage, strength, weight, and color, as shown in Figure 7.7. The removal of mechanically bound water begins at approximately 450°C and ends at approximately 600°C, leading to an average weight loss of 5%–20% of the industrial level. At around this temperature, the decomposable volatile matter present in clay starts to decompose to form oxides such as SO_2 and CO_2. A characteristic difference is noticed when this composition becomes active beyond 900°C, when all the feldspar content of the body enters its molten state to fill all the pores. This eventually leads to the formation of pinholes on the clay body surface after firing. The oxide content of the structure, such as Al_2O_3 and SiO_2, remains solid in this temperature [14].

The strength of the clay body gradually increases with increasing temperature due to the formation of primary and secondary mullite from the interaction of clay and free quartz in the liquid flux medium. However, some residual quartz that remains after firing can be observed. In the initial stage of firing, shrinkage occurs positively up to 150°C, gradually becoming saturated until the inversion temperature of quartz, i.e., 573°C, when an enhancement in volume can be observed due to the polymorphic transformation of α-quartz to β-quartz. This shrinkage remains stagnant until the temperature reaches ~850°C, in this circumstance the flux being reactive to form additional phase and densification as well. The liquids from these fluxes act as a platform for the solid particles to dissolve and transform within the structure to fill the pores and form a dense solid mass. Shrinkage can be controlled by adding various clays of different CTE. The porosity in the whiteware body initially increases rapidly owing to the loss of mechanically bound water. After this stage, the rate of formation of porosity in the structure reduces, which can be attributed to the gradual decomposition and oxidation of volatile materials. The decrement in porosity is initiated when vitrification starts at ~850°C and gradually ceases when complete densification has occurred.

7.5.5 Influence of Phases on Mechanical Properties

The major crystalline phase present in the fired body is mullite ($3Al_2O_3.2SiO_2$). Mullite shows comparatively higher mechanical strength than clay. Primary mullite formed in the structure provides mechanical strength to the system, whereas acicular secondary mullite acts as a bridging element in other grains that eventually resist the crack propagation under stress. SiO_2 provides a glassy structure to the system. The left-out solid SiO_2 stays as residual crystalline quartz. Essentially, an excess amount of residual quartz has to be avoided because it is associated with volume change during cooling, which may lead to appreciable amounts of stress generation in the ware structure. This stress can lead to a drastic breakdown of the structure. The feldspar or other fluxes added to the system as liquid fills the pores present in the structure, resulting in a higher modulus of rupture value. However, it provides lower water absorption efficacy by lowering the surface pores.

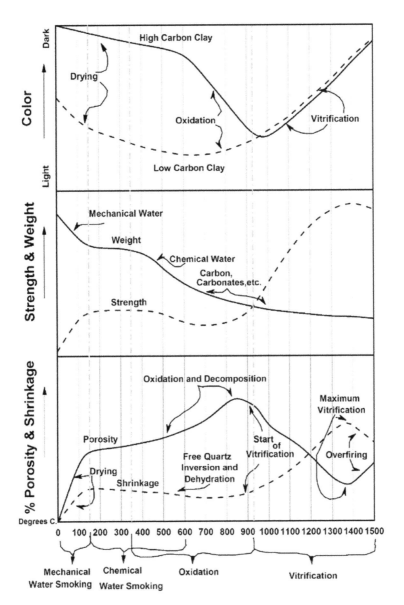

FIGURE 7.7
Variation in the physical properties of whiteware during firing.

7.6 Glazes

Glazes are traditionally glassy materials generally applied to sanitary ware and earthenware to render the ware impervious, remove surface defects, and improve the aesthetic value. In general, glazes are lustrous and glossy constituting numerous tiny crystals embedded in the glassy matrix obtained by cooling metal salts/oxides from an elevated temperature, which melts on the surface of the ceramic objects. Sometimes, an intermediate coating, known as "engobe," is applied between the glaze and body to conceal the body color and improve adhesion between the glaze and the body during firing. Glazes are traditionally prepared using principle raw materials and the oxide minerals consist of silica, boron oxide, alumina, lime, magnesia, lead oxide, zinc oxide, potash, soda and their combination thereof. However, other oxides and various pigments have also been explored to develop glaze compositions.

7.6.1 Classification of Glazes

Glazes are classified in multiple ways: (a) fusibility: easily fusible, moderately fusible, difficult to fuse etc.; (b) major constituents: raw glaze, lead glaze, lead-less glaze, fritted glaze, etc.; (c) effect on finished product: opaque, transparent, matte, satin, etc.; (d) application on ware type: porcelain glaze, majolica glaze, etc. Various classifications of ceramic glazes are given in Table 7.3.

Also, glaze composition is classified according to the chemical composition, name, and origin of the glaze (see Table 7.4).

7.6.2 Opaque Glaze

Opacity, also known as lack of transparency, develops in a glaze by suspended fine particles with the same particle size as the wavelength of visible light (0.4–0.7 μm). Opacity primarily develops when visible light is fully absorbed or light is scattered by fine crystals that exist at the surface or within the glaze. Various opacifiers are used during glaze preparation, which are generally insoluble or sparingly soluble in a liquid glaze. SnO_2,

TABLE 7.3

Glaze Classification According to Different Criteria [15]

Criterion	Classification
Fusibility	Fusible
	Hard or less fusible
Presence of an important component	Lead containing
	Non-lead containing
Further application and firing process	Single-firing covering
	Single-firing pavement
	Double conventional-firing covering
	Double fast-firing covering
Production application bases	Bases
	Airbrushing (pulverized)
	Pips
	Serigraphy
Effect on the finished product	Shining
Places of origin	Matte
	Semi-matte
	Satin
	Transparent
	Opaque
	Bristol
	Albany Slip

TABLE 7.4

Ceramic Glaze Classification According to Chemical Composition and Name

	Criterion	Name and Origin	Classification
Raw glaze	Lead containing	Porcelain, Bristol	With alumina
		Aventurine, matte	Without alumina
	Non-lead containing		With alkaline earth
			With alkaline and alkaline-earth
			With alkaline and alkaline-earth and ZnO
			With boron
			Salt glazes
Fritted glaze	Lead containing	Lead borosilicate	With alumina
	Non-lead containing	Lead-less borosilicate	Without alumina and boron
			With boron
			Without boron
			With high BaO content

ZrO_2, $ZrSiO_4$, Sb_2O_3, TiO_2, ZnO, etc., are the most popular opacifiers used throughout the sanitary ware industry while preparing opaque glaze [16].

7.6.3 Translucent Glaze

Transparent glazes are prepared without the addition of any opacifier (examples are mentioned in the previous section). The degree of transparency is observed to be comparatively less in the case of translucent glazes. Transparent glazes are visually like a transparent glassy material with a perfectly smooth surface with the underlying surface visible. However, the scattering of light may cause a transparent glaze to look like a translucent glaze. Some factors contribute toward the scattering of light from discontinuities or surfaces on or within the glass matrix, such as entrapped air bubbles inside the glassy matrix, impurities present in the raw materials, phase separation, immature particles in the glaze content, crystal growth during devitrification, etc. In simple words, the absence of a glossy smooth surface leads to an illusion of a translucent glaze. Sometimes, a translucent glaze is applied purposefully for decoration as a glossy surface. LOI materials are used for fabricating transparent glazes as these produce comparatively less gaseous materials. Other parameters such as the addition of frit, low water content, high holding time during firing, and fast cooling give rise to a transparent glaze. These kinds of glazes provide a depth for the addition of stain decoration on the surface, giving better aesthetic value. For high value brilliant transparency, optimum Na_2O and K_2O are added, which on high content may cause crazing owing to their high thermal expansion coefficient. For enhanced glossiness, a ratio of $SiO_2:Al_2O_3$ is maintained to be more than 10.

7.6.4 Maturity of Glaze and Glaze–Body Interaction

The glaze on a ceramic body is a thin coating of glaze or glass-ceramics. During firing, the objectives are (1) to obtain a good spread and surface coverage of glaze on the ceramic body; and (2) to develop a glaze with an excellent match to the CTE with the body during the heating and cooling cycle.

The viscosity of the glaze at maturity should be low to allow glass bubbles to easily escape through the surface, but it should not be low enough to run off the edges. Similarly, if the surface tension is too high, the glaze crawls away from the edges and cracks appear during cooling. If the surface of a glaze develops cracks due to shrinkage, then the molten glaze may crawl and form spheres. Adhesion between the glaze and body is mainly governed by two factors: glaze–body reaction and the thermal expansion–contraction of the glaze in respect of the body during heating and cooling. Whether the glaze is under tension or compression depends on the thermal expansion coefficient (α) of the glaze and body. If the α_{glaze} is higher than the α_{body}, the glaze experiences tension; however, lower CTE of glaze experience compressive stress on the surface. During cooling, the body is very rigid compared to the glaze and the rate of compression is different. The glaze must behave as a viscous liquid to release the strains developed on it. Eventually, a point is reached where the glaze becomes rigid and does not allow further compression. On further cooling, the strain developed by the glaze may cause the glaze to crack or pull apart and break away from the body. This phenomenon is known as "crazing" and happens if $\alpha_{glaze} > \alpha_{body}$. The opposite phenomenon is known as "peeling" or "shivering." However, shivering is less likely to occur than crazing because glazes are generally stronger in compression and weaker in tension. If the body and the glaze contract at the same rate, then no stress is generated. Thus, the α_{glaze} resemble with α_{body} expedite the excellent surface finishing. Traditionally, the development of tension in a glaze during cooling is avoided as the glaze or glass is stronger in compression but weaker in tension and the α_{glaze} is generally kept lower than the α_{body} so that the body likely contracts more and keeps the glaze in compression, preventing damage.

7.6.5 Ceramic Colors and Their Roles

Several methods are used to apply colors or stains to underglaze or overglaze for decorative purposes, including painting, stenciling, spraying, stamping, and printing. Underglaze colors are generally applied underneath the glaze. The colors are developed by calcining inorganic color oxides or pigments along with kaolin, silica, alumina, and small amounts of flux (~5 wt.%). The pigments are usually calcined at a higher temperature than the glaze maturity temperature, thereby minimizing the effect of the colorant dissolving in the liquid glaze. On the other hand, during overglaze, the color/stain is applied directly onto the body, mixing together with the glaze composition. It is a general rule of thumb that the color/stain along with flux vitrifies at a relatively lower fusion temperature before the glaze starts softening. The colors are prepared by mixing coloring oxides with glaze frits at low fusion temperatures, and grinding them into powder form. During firing, the frit melts first and dissolves the colorant in the liquid. The various types of oxide crystals used as stain are (a) spinel with the formula AB_2O_4 where the transition elements can present either at the A or B site, e.g., $(Co,Fe)(Fe,Cr)_2O_4$; (b) zirconium oxide–vanadium oxide $[(ZrV)SiO_4]$ as blue and zirconium oxide–praseodymium oxide $[(ZrPr)SiO_4]$ as yellow; (c) chromium oxide as green; (d) alumina–chrome as red; and (e) zirconium oxide–iron oxide $[(ZrFe)SiO_4]$ as peach coral, are being developed [17].

7.6.6 Glaze Defects

The presence of glaze defects after firing (glaze maturity) may render the final products rejected or discarded as scrap. Glaze defects occur for various reasons, primarily the raw materials used, glaze composition, mode of application, and firing. Sometimes, irregularity in the body composition onto which the glaze is applied may serve as an alternative reason for glaze defects. Some common glaze-related defects are blisters, pinholes, peeling, crawling, warpage, crazing, and shade variation (which is sometimes augmented due to pigments used concurrently with glaze).

Blisters, as the name suggests, are swellings or depressions on the glaze surface originating mainly from the release or out-gassing of carbonaceous materials/gaseous species that exist in the glaze/body composition during firing. During firing, gaseous species vaporize out of the clay and the molten glaze. However, sometimes, highly viscous glazes interfere with the release of gaseous species within the firing cycle and slow down glaze devitrification. In such cases, blistering is likely to occur. In order to avoid blisters appearing on the final glaze surface, the following precautions must be taken: a thin layer of glaze must be applied; entrapped air in the body must be released during body preparation; the ceramic body must be free from any dust particles, carbon, or organic residue; the glaze slurry/slip must be free from entrapped air; and the body must be allowed to fire slowly and a quick firing cycle must be avoided so that the gaseous species can escape before the glaze matures on the body.

Pinholes are small openings or pits on the glaze surface and are mainly due to the escape of volatile matter and gaseous species during firing or are due to bubbles forming in the molten glaze that do not completely heal. The bubbles arise mainly from the entrapped air originating in the powdered glaze particles as the glaze starts to mature. Sometimes, under firing a glaze composition and carbon deposition in the early stages of firing may give rise to pinholes at the later stages of firing when it escapes as carbon dioxide. To reduce the formation of pinholes, the following precautions should be taken: an increase in the firing cycle; avoiding the kiln from entering the reduction cycle during the early stages of firing; proper glaze composition formulation without leaving any ungrounded particles or dust; adequate clay beneficiation preventing the presence of volatile organic matter.

Peeling or shivering is a glaze defect in which the glaze breaks away from the body/ceramic ware mainly due to a mismatch in the thermal expansion coefficient between the glaze and the body. Peeling is the opposite phenomenon of crazing. To avoid peeling, the feldspar (sodium feldspar or nepheline syenite) content must be increased in the glaze composition, and the silica content must be decreased proportionately in either the body or the glaze.

Crawling appears as patches on the fired body through the top glaze surface mainly due to improper adherence of the glaze layer onto the underneath clay body during firing. The leading cause of crawling is a weak bond between the glaze and body. It may also occur if the glaze composition contains excessive powder materials that do not adhere to the body during firing. Crawling is common in matt glazes and can be minimized by the addition of extra flux. However, fluxes such as zinc and magnesium increase the chances of crawling, and their quantity in a glaze must be restricted.

Waviness or surface undulation is a defect that mainly originates from the imperfect application of a thick glaze on the ceramic body that, upon firing, cannot evenly smooth out, leading to a wavy appearance on the glaze surface. However, the warpage is a surface distortion mainly due to differential shrinkage between the glaze and body for complex geometry.

Crazing appears as a spider web pattern or network of cracks developed on the glaze surface after firing. The cracks are often fine and can only be visible when soaked in a color solution (methylene blue). The significant reason for crazing is the mismatch of the thermal expansion coefficient between the body and glaze, which creates tensile stress during which glaze cannot withstand cooling. Despite this incidence, coefficient of thermal expansion (CTE) difference in between body and glaze composition expedite crazing as well. Thus, crazing can be minimized by altering the glaze and body composition so that a mismatch between the CTE can be completely avoided. This can be achieved by increasing the silica content in the body or glaze, decreasing the feldspar (sodium and potassium in general) content in the body or glaze, and increasing the boron and alumina content in clay.

7.7 Fabrication Protocols

7.7.1 Sanitary Ware

Sanitary ware consists of vitreous/semivitreous ceramic products such as closet bowls, flushing tanks, urinals, sinks, bathtubs, lavatories, and allied products. Vitreous and semivitreous sanitary ware is generally produced by an amenable slip casting method because intricate and complex shapes can be easily prepared.

The sanitary ware produced by the ceramic industry is mainly composed of bodies made with clay, quartz, and feldspar as the primary constituents with hard, translucent, porcelain glaze coverage rendering it resistant to abrasion, ware, and chemical species. The body composition of traditional sanitary ware varies from place to place but mostly consists of 25% quartz, 25% feldspar, and 50% clay from which ball clay and china clay remain in the 50:50 or 40:60 ratio.

Slip casting using plaster of paris molds is mainly used for preparing the body and to impart a smooth texture on the outside surface. A typical schematic of a product control flow chart including casting, drying, firing, sorting, and mode of reusing of a defective product is shown in Figure 7.8. In addition, pressure casting with plastic and metal molds is also frequently used to reduce the casting time and increase the production capacity. Slip properties such as viscosity, density, thixotropy, and other rheological properties are essential parameters, which are controlled for each batch composition and optimized in the laboratory to obtain sound casting. Nowadays, industries are adopting sophisticated casting techniques in which certain intricate parts are first cast in a plaster of paris mold by the "over-casting" technique and then in a larger mold with further slip addition to join the precast pieces together into a single large component.

"Drain-casting" is another technique that is used to produce hollow components with certain controlled thickness developed in the mold. The extra slip is drained from the center leaving a hollow cavity between the thick cast. Drying a green sanitary ware body must be carefully performed to avoid the buildup of drying-related defects (cracks, surface bulging, warping, etc.). Generally, the entire product must be dried in a humidity-controlled drying chamber at a uniform rate so that moisture can be evaporated uniformly across the surface. Dried sanitary ware is fired in a tunnel kiln using a single-fired technique where the glaze and body mature at the same time. Using the double-fired technique, the body is initially vitrified by firing before the glaze is applied using a spraying, painting, or dipping method and then fired again to mature the glaze slightly at a lower temperature than the first firing temperature. Nowadays, porcelain glaze with a typical composition of $0.1–0.2$ ZnO, $0.2–0.3$ K_2O, $0.1–0.2$ Na_2O, $0.25–0.4$ CaO, $0.35–0.5$ Al_2O_3, $0.3–0.4$ B_2O_3, and $4.0–6.0$ SiO_2 along with opacifiers (zircon) and stains mature along with the body using the single-firing technique. However, the composition of glaze is a trade secret and varies for various vitreous/semivitreous bodies from industry to industry [18].

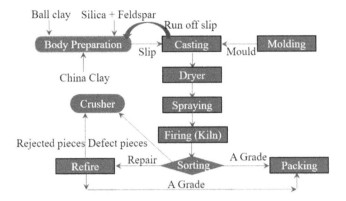

FIGURE 7.8
Schematic presentation of sanitary ware fabrication by the wet process.

7.7.2 Tiles

Tiles are generally classified into two categories: wall tiles with a water retention capacity of ~8%–15% and floor tiles with a water absorption capacity of ~6%–10%. Both categories of tiles are generally covered with a decorative glaze that is sometimes stained and develops a vitrified body on top upon firing. Tiles are mainly used for imparting aesthetic looks to kitchens, bathrooms, floors, hallways, residential buildings, hotels, subways, and public places. Therefore, the tile's outer surface is covered with a hard porcelain–type glaze that must withstand scratch, wear, and heat, and must possess good abrasion resistance and active protection from corrosive media.

The body composition of tiles, which is crucial for the development of strength, has undergone several changes in the last few decades. Initially, a triaxial body with 50% clay (50:50 ratio of china clay and ball clay), 25% flint, and 25% feldspar was used throughout the industry. However, the body composition was further modified with the addition of small amounts of wollastonite, talc, pyrophyllite, and their combination in place of kaolin to improve the modulus of rupture and avoid delayed crazing. The incorporation of talc (3%–5%) and pyrophyllite (10%–15%) greatly improves the thermal expansion coefficient mismatch between the body and glaze and avoids the formation of crazing.

Soda and potash feldspars are also being partially replaced with limestone and dolomite as a flux in the body of tiles. Also, several auxiliary materials including fire clay, laterite, and scrap tiles are being explored for commercial purposes and are greatly improving the manufacturing cost. The body of tiles with a clay-based composition is generally fired in a roller kiln with a peak temperature reaching 1050°C–1150°C, forming a bisque body with mullite and cristobalite, as the primary phase(s). A low-melting glaze is generally applied to the body by spraying consisting of optimum amount of frits and other constituents of lime, soda, potash, boron oxide, alumina, silica, and opacifiers with or without stains. However, this optimum amount depends on the type of glaze and their mode of applications. The "dry process" in the manufacture of tiles, which is relatively cheap and improves the manufacturing time, is being adopted by the industry throughout the world in preference to the "wet process." During the dry process, the clay and other raw materials are grounded, graded, screened, and stored in different silos before being fed into a hopper for blending with water at a certain stoichiometric ratio of clay compositions (see Figure 7.9).

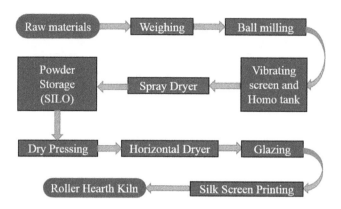

FIGURE 7.9
Schematic presentation tiles fabrication by the dry process.

The mixed clay composition is further pulverized in a hammer or roller mill and screened once again before pressing in a hydraulic automatic/semiautomatic press. The moisture content of a clay batch is generally maintained between ~8% and 12%, which is an extremely critical parameter to enable plastic deformation during pressing and to concurrently avoid clay particles sticking with the die and punches. The pressing operation is carried out in a fully automatic proportional-integral-differential (PID) controlled manner where the clay powder is first fed into the die cavities followed by compaction of the clay within the die to produce green tiles. Generally, four to six tiles are produced per stroke at a rate of ~20–30 strokes per minute producing 80–120 tiles. The die punches are mainly made with high strength alloy steel and are occasionally preheated to ~50°C–60°C to avoid sticking. Generally, the die punches used throughout the industry produce tiles with dimension of 8″ × 8″, 12″ × 12″, and 16″ × 16″ for floor tiles and 6″ × 8″ for wall tiles. The pressed tiles are passed through a tunnel drier so that the body becomes leather-hard with the moisture content reduced to ~3%–4%. Occasionally, the edges of the tiles coming out of the press machine are cleaned manually or with a fettling machine and excessively broken edges are discarded. The leather-hard dried tiles are passed through a conveyor belt where engobe and glaze spraying is performed in an engobe and glaze booth fitted with an atomizer, respectively. The glaze booth performs a uniform coating of engobe and glaze on the tile surface and the excess liquid glaze dripping from the sides is scraped using a fettling machine. The green tiles covered with glaze are passed through a roller-hearth tunnel kiln with three to four decks. The tiles are evenly distributed into the decks and passed through the kiln with the peak temperature reaching 1150°C for a single-firing cycle ranging from 30 minutes to 1 hour. The fired tiles are sorted into different categories depending on the presence of minor defects such as pinholes and blisters, and shade variations, and are sent to the warehouse for dispatch. The tiles are also laboratory tested on a periodical basis for crazing, modulus of rupture, ware, and abrasion resistance for certification and quality standard purposes. Wall tiles are made from a semivitreous body containing ball clay ~20%–25%, kaolinite ~5%–10%, talc ~40%–45%, and flint ~30%–35%. The fired strength of wall tiles is very low because they are mainly used for decorative purposes and are covered with transparent or colored glaze with water absorption reaching ~10%–15%. Wall tiles are prepared by the dry process akin to semivitreous china and porcelain bodies. The water from the slip is removed using a filter press, pugged, de-aired, and dried to the desired moisture content before being aged for a couple of days. The dry pressed bodies are stacked in saggers before bisque firing in a tunnel kiln.

7.7.3 Table Ware

Tableware is mainly earthenware to semivitreous porcelain bodies prepared for crockery and dinnerware sets. It is mainly triaxial bodies that are used, which are generally single fired between 1200°C and 1250°C, displaying moderate strength. Triaxial bodies contain plastic clay (preferably ball clay) ~25%–30%, china clay ~30%–35%, flint ~15%–20%, and feldspar ~10%–15%. The bisque-fired body is overglazed and matured at ~1050°C–1100°C. Due to the presence of a significant amount of china clay, the fired body shows complete vitrification with a porosity of <0.3%. A porcelain body is generally fired at a slightly lower temperature than vitreous/semivitreous triaxial/china clay bodies. The fired porcelain body constitutes mullite crystals suspended in a siliceous glassy matrix. The mullite crystals are needle shaped and form a continuous network throughout the glassy matrix. Mullite crystals rapidly appear when the temperature reaches ~1250°C and the weight percentage is enhanced as the temperature increases and remains stable up to 1545°C. During cooling, the interlocking mullite crystals with glassy matrix provide strength to the body.

On the contrary, bone china bodies with their distinct translucency have a different composition with an appreciable amount of bone ash. Bone ash is a raw material typically consisting of tricalcium phosphate $[Ca_3(PO_4)_2]$ and small impurities. The composition of bone china bodies varies with china clay ~25%–30%, bone ash ~35%–45%, and feldspar ~25%–30%. Bisque firing of bone china bodies is generally performed at ~1225°C–1300°C. Since bone china bodies suffer from high shrinkage, extra precaution must be taken during the drying and firing cycles. Bone ash that contains P_2O_5 has a highly fluxing nature, increasing the fluidity of the melt and providing a short firing range.

A fritted glaze that has matured at 1100°C–1150°C is generally used. The major difference between porcelain and bone china ware is that porcelain ware is made when the body and glaze have matured during the same firing, whereas bone china bodies are made first by vitrification of the body followed by maturing the applied glaze at a slightly lower temperature. Hotel chinaware, primarily used in hotels and restaurants, possess properties akin to porcelain and bone chinaware. Hotel chinaware is generally fired to complete vitrification and a glaze is applied to the ware, which is matured at a somewhat lower temperature. The typical composition of a hotel china body is kaolin ~30%–35%, ball clay ~8%–12%, flint ~35%–45%, potash feldspar ~12%–15%, soda feldspar ~2%–5%, and dolomite ~1%–1.5%. Due to presence of high alumina in the body composition a large amount of mullite formation is observed in the fired body, which provides high strength. The vitrified body also contains siliceous glass generally derived from undissolved silica particles of flint. The fluxes melt early and form liquid phases that help to dissolve the alumina and silica from the clay composition and allow the crystallization of mullite.

7.7.4 Electrical Insulators

Electrical porcelain insulators are mainly used for low tension and high tension insulation purposes for power transmission and distribution to residential areas and factories. An increase in the power transmission capability over long distances with minimum power loss increases the demand for quality insulation in terms of dielectric properties, thermal shock resistance, low thermal expansion coefficient, and excellent mechanical strength. The body composition of triaxial-type, vitrified, hard porcelains is generally preferred due to its combination of high strength and thermal shock resistance. The body composition of porcelain insulators is mainly composed of ball clay (~25%–30%), china clay (~15%–25%), quartz (~10%–20%), whiting (~0%–2%), pyrophyllite (~0%–20%), feldspar/nepheline syenite

(~25%–30%), and talc (~0%–2%), which are generally single fired to cone 10–12 to render a completely vitrified body with zero porosity [19]. The fired body consists of a glassy matrix anchored with crystalline mullite and undissolved quartz. The glassy matrix usually contains residual silica, which contributes to the low thermal expansion. Sometimes, alumina (~15%–20%) is added to increase the mechanical strength by replacing the equivalent amount of china clay. Traditionally, wet processing is followed for parts of the fabrication where grounded individual components are weighed and loaded into a hopper. Feldspar and quartz are crushed and grounded in a ball mill before being put into a blunger along with clay, water, and deflocculant. The slip thus produced is filter pressed and the cake is aged for a couple of days. The aged cake is fed into a pug mill ensuring compositional homogeneity and trapped air is removed. The pugged mass is extruded into cylindrical forms to the desired shape. Generally, porcelain glaze is used along with pigments to impart a brown color to the insulator. The glaze is generally applied by a dipping or spraying method. Grogs are used in the glazed area where metallic parts are to be cemented. The metallic parts are cemented using a Portland cement, plaster, or epoxy resin.

7.8 Conclusions

This chapter provides a brief insight into classical and ancient materials and whiteware, looking into the background of this primitive field of ceramic science and incorporating various advanced technologies including material behavior variation with the addition of additives during the phase transition. Raw material selection, compositional variation, fabrication routes, and the effect of corresponding process parameters on various whiteware bodies have been highlighted. The typical composition of whiteware bodies along with their thermal (phase transition and in situ phase formations), physical (surface modification by glaze application), and electrical (insulating property) behavioral changes have been systematically discussed. Essential topics such as the science behind the rheological behavior of slip, the suspension of colloidal clay particles, and the effect of surfactants on surface potential have been covered in depth. The microstructure after densification of whiteware bodies has a great impact on their behavior, and this has also been discussed. In-depth knowledge about various kinds of whiteware bodies and their application has been provided. Whiteware body and glaze defects along with fabrication protocols and mode of prevention have been emphasized. Factors such as differential temperature and humidity have been found to be essential parameters in the prevention of drying and firing defects. In summary, an attempt has been made to discuss the different classes of whiteware and glazes including their basic understanding to stimulate the research and development of a new class of whiteware.

References

1. Miller, G. L. 1980. Classification and economic scaling of 19th century ceramics. *Historical Archaeology* 14(1):1–40.
2. Salmang, H. and Scholze, H. 1968–2007. *Keramik*, 5th–7th Edition. Springer Verlag, Berlin-Heidelberg.

3. Holdridge, D. A. 1959. Composition variation in ball clays. *Transactions of the British Ceramic Society* 11:645–659.
4. Wachtman, J. B. 1991. *Materials & Equipment/Whitewares*. John Wiley & Sons: New York.
5. Ryan, W. 1978. *Properties of Ceramic Raw Materials*. Pergamon Press: Toronto, Canada.
6. Darcy, H. 1856. *Les fontaines publiques de la ville de Dijon*. Dalmont: Paris.
7. Molina, F. V. 2014. *Soil Colloids: Properties and Ion Binding*. CRC Press: Boca Raton, FL.
8. Sarkar, D. 2019. *Nanostructured Ceramics: Characterization and Analysis*. CRC Press: Boca Raton, FL.
9. Worrall, W. E. 1986. *Clay and Ceramics Raw Materials*, 2nd Edition. Elsevier Applied Science: London.
10. Carty, W. M. and Senapati, U. 1998. Porcelain—Raw materials, processing, phase evolution, and mechanical behavior. *Journal of the American Ceramic Society* 81(1):3–20.
11. Hewitt, W. 1927. *Ceramics-Clay Technology*. McGraw-Hill Company: New York.
12. Soussi, N., Kriaa, W., Mhiri, H., and Bournot, P. 2017. Reduction of the energy consumption of a tunnel kiln by optimization of the recovered air mass flow from the cooling zone to the firing zone. *Applied Thermal Engineering* 124:1382–1391.
13. Gorea, M., Kristály, F., and Pop, D. 2005. Characterisation of some kaolins in relationship with electric insulator ceramics microstructure. *Acta Mineralogica-Petrographica* 46:9–14.
14. McNamara, E. P. 1949. *Ceramics – Clay Products & Whitewares*, volume III. Pennsylvania State University: University Park, PA.
15. Casasola, R., Rincón, J. M., and Romero, M. 2012. Glass–ceramic glazes for ceramic tiles: A review. *Journal of Materials Science* 47(2):553–582.
16. Parmelee, C., and Harman, C. 1973. *Ceramic Glazes*, 3rd Edition. CBLS: Ohio.
17. Taylor, J. R., and Bull, A. C. 1986. *Ceramics Glaze Technology*. Pergamon Press: Oxford.
18. Blonski, R. P. 1994. The effect of zircon dissolution on the color stability of glazes. *Ceramic Engineering & Science Proceedings* 15:249–265.
19. Chowdhury, T. 2010. *Ceramic Insulator*. VDM Verlag: Germany.

8

Glass Processing and Properties

Susanta Basu and Debasish Sarkar

CONTENTS

8.1 Introduction

The earliest known manmade glasses date back to around 3500BC, in Mesopotamia. When a mixture of a large quantity of beach sand (SiO_2), sea salt (NaCl), and mammal bones (CaO) was fired together, and a crude soda-lime glass was obtained without in-depth analysis of the basic need for network formers, intermediates, and modifiers. Discovery of glassblowing around the 1st Century BC was a very important step in glass making. Glasses were conventionally formed by cooling from a melt. The homogeneity and quality of the glass are of major concern in the past due to use of crude processes.

Glass can be defined in many ways:

- An inorganic product of fusion which has been cooled to a rigid condition without crystallization.
- An amorphous solid completely lacking in long-range and periodic atomic structure, and exhibiting a region of glass transformation behavior near to melting point.
- A liquid with a very high viscosity like a supercooled liquid.

With technological advancements, a wide range of products namely flat glass, containers, thermo-chemically resistant laboratory ware, colored decorative glass, vacuum tubes, glass sculptures, optical fibers, medical products, and more could be produced. Several modern techniques are also in use to make glasses, for example, vapor deposition, sol-gel processing, etc.

In time, extensive research resulted in the development of several classes of glass for different applications, containing ceramic oxide particles, metals, or polymers; thus, glass is no longer restricted to ceramics, nor is silica an essential constituent to make glass. Any material, whether inorganic, metallic, or organic, developed by any technique that exhibits glass transformation behavior is considered to be a glass. Consequently, there is always scope for the invention of new classes of glass and applications with the help of science and technology.

Recently, metallic glass has come to the forefront of research and development, as it exhibits both high strength and low stiffness, resulting in very high resilience and thus, the ability to store and release elastic energy [1]. Metallic glass or amorphous metals are novel engineering alloys; metallic glass comprises a disordered and random atomic arrangement, similar to transparent soda-lime glass, but it is opaque, and its electrical conductivity is generally two orders of magnitude lower than that of crystalline components. It also experiences glass transition into a super-cooled liquid state and controlled viscosity, creating the possibility of a high degree of flexibility in the shaping of metallic glass. Several classes of amorphous metal alloys have been developed using a controlled high cooling rate (100 to 1000 K/s). Interestingly, due to its amorphous atomic structure, metallic glass does not have crystalline defects, and it possesses high mechanical properties compared with its crystalline counterparts. Recently, an excellent video demonstrated how a metallic glass facilitates the effective and continuous bouncing of a bearing ball compared with a stainless steel surface; this is attributed to the low strength and plastic deformation of steel causing faster damping of the kinetic energy of the bearing ball [2]. Metallic glasses can be classified according to the base material used into metal–metal glasses (e.g., $Zr_{41.2}Ti_{13.8}Cu_{12.5}Ni_{10}Be_{22.5}$, $Pd_{77.5}Cu_6Si_{16.5}$, etc.) and metal–metalloid glasses, in which transition metals such as Fe, Co, and Ni and metalloids such as B, Si, C, and P are used (e.g., $Fe_{40}Ni_{38}Mo_4B_{18}$, $Co_{66}Fe_4Ni$, $B_{14}Si_{15}$, etc.). This class of material is used to make accurate standard resistors and in computer memories, magnetic sensors, surgical instruments, prosthetic materials, and magnets for fusion reactors, for nuclear waste disposal, for reinforcing particulates to enhance their mechanical properties, etc.

By convention, polymers either melt or decompose at much lower temperatures compared with ceramic oxides and metals. Polymer glass can also be made through faster rate of cooling from the polymer melt to below the glass transition temperature (T_g). Below T_g, polymer becomes hard and brittle, like glass. Some polymers, such as polystyrene and poly-methyl-methacrylate, are preferentially used below their T_g, whereas rubber elastomers such as poly-isoprene and polybutylene are used above their T_g. It is to remember that the glass transition temperature is not the same as the melting temperature. Melting is a phenomenon of crystalline solids, but glass transition refers to when materials are in an amorphous and non-arranged state. However, a mixture of ordered crystalline and amorphous states may exhibit both a melting and a glass transition temperature, respectively. Glassy polymers are technologically important across a wide horizon of applications: structural (hyperbaric windows), electronic (ionic conductors, surface coatings for printed circuit boards), environmental (membranes for industrial gas separation), and for the packaging of luxury cosmetics and lifestyle products [3]. Nonetheless, the glass transition temperature (the temperature below which a super-cooled liquid behaves as a glass) is a common feature for any quality of glass; it is where the transformation behavior begins. What, then, is required to define glass transformation behavior from a thermodynamic perspective? This will be discussed in the next section.

8.2 Enthalpy–Temperature Diagram

The glass transformation behavior of any material is usually discussed on the basis of variation of either enthalpy or volume with respect to temperature, as shown in Figure 8.1 [4]. Enthalpy is the heat content of a system, or the amount of energy within a substance, existing in the form of both kinetic and potential energy. Prior to discussing the glass transformation behavior, recall the first law of thermodynamics; it states that the energy change **ΔE = q − w**, where w = work done by the system and q = heat flow = enthalpy change for the process at constant pressure (ΔH).

Let us consider a small volume of melt above the melting temperature T_m. This melt gradually changes its atomic structure as the temperature decreases. The final change of characteristics occurs in one of three ways:

I. The liquid readily crystallizes at its melting (freezing) point to form the equilibrium crystalline phase and then cools down to room temperature.

II. The liquid by-passes the equilibrium crystallization and super-cools to the glassy state at a relatively slower rate of cooling.

III. The liquid by-passes the equilibrium crystallization and super-cools to the glassy state at a relatively faster rate of cooling.

What would be the consequences of different cooling rates and the resulting atomic structure? To answer this question, first consider the common crystallization process, in which

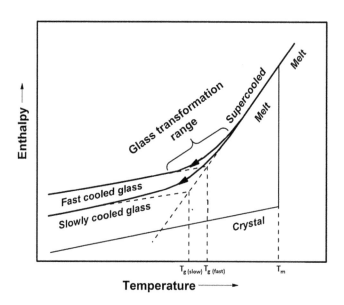

FIGURE 8.1
Enthalpy of a liquid cooled from above its melting point to room temperature.

cooling below T_m facilitates the nucleation of the crystallite phase and thus, an abrupt decrease in enthalpy (or volume) to form the long-range periodic atomic arrangement of a crystal. Further cooling results in a decrease of enthalpy or volume because of the specific heat or thermal contraction of the crystal. Crystallization can be avoided by fast cooling, which means that the system does not have enough time for nucleus formation and atomic rearrangement in a systematic fashion as occurs in a crystal. Thus, without crystallization, a super-cooled melt is obtained, and it obviously behaves as a metastable state compared with the crystallization state. In this circumstance, there is no abrupt decrement of enthalpy due to discontinuous structural rearrangement. However, further cooling enhances the viscosity and restricts the long-range periodicity. This causes the enthalpy (volume) equilibrium line to begin to deviate, following a curve of gradually decreasing slope until it becomes independent of temperature. The temperature region where the enthalpy (or volume) is either that of the equilibrium liquid or that of a glass is denoted as the *glass transformation temperature*. Now, the resultant non-periodic atomic arrangement of the frozen liquid mass becomes a rigid glass.

8.3 Theory of Glass Formation

The formation of a glass from the liquid state, in contrast to crystallization, is a continuous path from the liquid, through the viscous (super-cooled), to the brittle state. A typical difference in atomic arrangement between crystalline and non-crystalline vitreous silica is illustrated in Figure 8.2. Lebedev identified the connection between glass formation and polymorphism in 1921 and defined glass structures as being comprised of submicroscopic crystals connected by structurally disordered regions, as estimated from the broadening

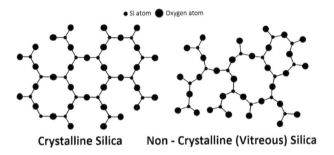

FIGURE 8.2
Structure of crystalline (long-range periodicity) and vitreous silica.

of diffraction patterns [5]. Following this, Zachariasen (1932) [6] established the connotation of a continuous random network (CRN), which should satisfy all the four rules for forming glass, as listed in the following:

- An oxygen atom is linked to a maximum of two cations and does not form additional bonds with other cations.
- The coordination number of the glass-forming atom is small.
- The oxygen polyhedra share the corners but not the edges or faces.
- The polyhedra are linked in a three-dimensional network.

The single bond (SB) strength approach coined by Sun (1947) critically identified the basic materials for making glass and the consecutive influence of glass network formers, intermediates, and glass network modifiers on the strength of oxides. These are summarized in Table 8.1 [7].

In this context, Stanworth classified a group of materials based on the mixed cation–anion bond concept: cations in *network formers* have fractional, near to 50%, ionic character; these are known as Group I and produce good glasses. Cations with slightly lower electronegativity, known as Group II, form more ionic bonds with oxygen and cannot form glasses independently but are capable of partially replacing cations from Group I; these are referred to as *intermediates*. Finally, cations with very low electronegativity, known as Group III, can form highly ionic bonds with oxygen; these never act as network formers but only serve to modify the network and are termed *modifiers* [8]. In consideration of

TABLE 8.1

Influence of Glass-forming Oxides on the Single Bond Strength

SB Strength of Oxides	kcal/Bond	Type	Property
High	>80	Glass network formers	Difficult to break/reform into ordered lattice on cooling; high viscosity, so good glass formers; e.g., B_2O_3, SiO_2, GeO_2, P_2O_5, Sb_2O_5, V_2O_5, As_2O_5, etc.
Medium	60–80	Intermediates	Cannot form glasses on their own, but aid other oxides to form glass; e.g., TiO_2, ZnO, PbO, Al_2O_3, ThO_2, BeO, CdO, etc.
Low	<60	Glass network modifiers	Easy to break/reform into preferred crystal; e.g., Na_2O, K_2O, MgO, CaO, BaO, Li_2O, SrO, etc.

material characteristics and glass formation efficiency, the American Society for Testing and Materials (ASTM) International defines that glass can form when molten liquids are cooled to below the T_m sufficiently fast to avoid the nucleation of a periodic arrangement of crystal seeds and the growth of crystals; however, control of viscosity facilitates working to the strain point of glass.

8.4 Viscosity of Glass

Viscosity is a phenomenon that describes the resistance of a fluid to flow. It is a measure of liquid fluidity and varies inversely with temperature. As glass cools, it becomes more viscous. The unit of absolute viscosity η is the poise = dyne-second/cm². The higher the poise number, the more resistant to flow the material will be. Note: 10 poise = 10 P = 1 Pa·s; so, log (viscosity in poise) = 1 + log(viscosity in Pa·s). By convention, the crystallinity affects the range of viscosity with respect to the processing temperature of the substance; for example, comparative data on borosilicate glass, 96% silica glass, and fused glass are represented in Figure 8.3. From these data, one can see that the melting and working temperatures for 96% silica glass and fused glass do not exist in the limited temperature region; interestingly, at the same time, the softening, glass transition, annealing, and strain point temperatures increase with the crystallinity of the system. Thus, the selection of the composition and its effective working range is a critical consideration during the manufacture of defect-free glass. Details of fabrication and probable imperfections are discussed in the next section.

   The change in viscosity is a dynamic phenomenon during the manufacture of glass from a melt. A typical example of the changing viscosity characteristics of float glass is shown in Table 8.2; it describes different zones to select the temperature for a particular process of interest.

8.5 Economic Importance

To understand the importance of glass and the market scenario, we can check the status of the glass business in the EU. Recent data demonstrate that this sector produced near to *8.5 million tons of float glass* from the *55 float lines* operating in the EU alone [10]. An extensive number of people are directly employed in the manufacture of flat glass, but the entire value chain (glass processing, transformation, window assembly, installation, recycling, etc.) generates almost 1 million jobs in the EU. A float plant is highly capital intensive, typically costing around €70 million to €200 million depending on size, location, and product complexity. Once operational, a float glass furnace is *designed to operate continuously, 365 days per year, throughout its lifetime of between 15 and 18 years.* Float lines are normally capable of several "lifetimes" following major repair or upgrade programs (€30 million to €50 million). The economics of the continuous flow–based float operation require a high capacity *utilization rate, typically above 70%,* before a plant becomes profitable. Energy and raw material costs are significant, representing almost two-thirds of the production costs. The handling of glass is a critical issue, as it is brittle, and it is necessary to take great care during transportation; thus, the distribution cost is relatively high, sometimes nearly 15% of the total cost of the glass [11].

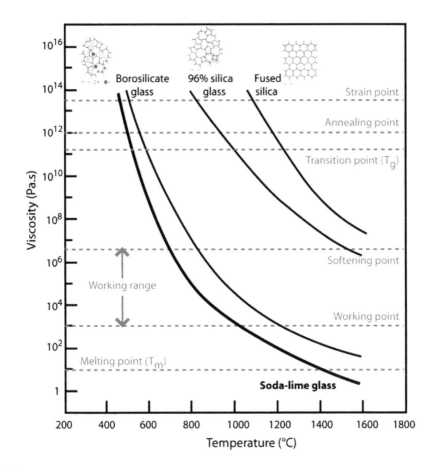

FIGURE 8.3

Comparison of the viscosity of several kinds of glass versus temperature [9].

TABLE 8.2

Characteristic Change in Viscosity with Respect to Temperature during Manufacturing of Float Glass

log10 (η, Pa·s)	Temperature (°C)	Description
1	1438	Melting point: melt is fluid enough for fining and/or homogenization.
3	1020	Working point: glass is formed into ware.
4	905	Flow point: at this point, glass begins to flow freely if unrestrained.
6.6	723	Littleton softening point: glass deforms under its own weight. Standard fiber elongation test (1 mm/min).
9	626	Dilatometric softening point: glass flattened by dilatometer; depends on the load. Measured by TMA (thermo-mechanical analyzer).
10.5	584	Deformation point.
11.3	560	Transition point: glass stress dissipates quickly (T_g).
12	545	Annealing point: temperature at which "stress is substantially relieved" in a few minutes. It is measured by a standard fiber elongation test.
13.5	520	Strain point: stress is substantially relieved in several hours without significant structural rearrangements, or no permanent flow.

The manufacturing protocol of float glass is discussed in the text.

8.6 Float Glass

By convention, flat glasses are mostly manufactured using a float process that operates continuously for around 12–15 years without interruption. In this concise review, emphasis has to be placed, first, on the manufacturing of flat glass followed by the technology involved in the processing of modern annealed glass from the perspective of different civil constructions, including windows, doors, etc. The *annealing* of glass is a slow cooling process at a predetermined temperature to relieve residual internal stresses introduced during the manufacturing of hot glass. By virtue of this, the entire float glass processing protocol is divided into five major steps: batch preparation, melting, forming, annealing in a lehr, and final-stage processing. A typical flow diagram is illustrated in Figure 8.4. The details of each step, including starting raw materials and batch preparation, are described systematically in the following.

> **Step I—Batch Preparation:** Prior to glass manufacturing, the selection of a defined composition provides a wide range of properties for a targeted application. As ceramic oxides are prime resources, the percentage content and quality of their base raw materials are important parameters for consistency of the resulting properties. However, at the same time, one has to monitor the production cost, and thus, raw materials together with cullet are mixed in the batch plant. In this context, a typical soda-lime-based window glass composition and the percentage resources from raw materials are described in Tables 8.3 and 8.4, respectively. To meet the target glass composition, calculated quantities of the raw materials, such as silica sand, soda ash, dolomite, limestone, sodium sulfate, carbon, colorants, and cullet (broken glass for recycling), are weighed, mixed, and fed into a glass melting furnace. Sodium sulfate or salt cake is used for the refining of glass; that is, the removal of bubbles.

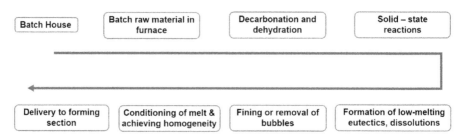

FIGURE 8.4
Flow diagram of the process from batch preparation, followed by several melting stages, to the final stage of processing.

TABLE 8.3

Typical Example of Window Glass Composition

	Constituents								
	SiO_2	Na_2O	CaO	MgO	Al_2O_3	Fe_2O_3	TiO_2	K_2O	SO_3
Weight (%)	72.3	13.9	9.7	3.3	0.4	0.08	0.06	0.01	0.25

TABLE 8.4

Typical Example of Contribution of Oxides from Raw Materials

	Constituents (%)							
	SiO$_2$	Al$_2$O$_3$	Fe$_2$O$_3$	Na$_2$O	K$_2$O	CaO	MgO	TiO$_2$
Silica sand	99.4	0.5	0.04					0.08
Soda ash				58.3				
Sodium nitrate				35.9				
Salt cake				43.5				
Dolomite	0.6	0.05	0.03			31.5	20.3	
Limestone	0.6	0.07	0.03			55.1	0.7	
Rouge			97.5					

Step II—Melting: Raw materials, consisting of the predefined batch composition, including cullet mix, are fed into a glass tank furnace, where a gradient temperature profile is monitored through six side-fired ports connected by left and right regenerators. Figure 8.5 demonstrates a typical glass melting furnace, which comprises three major zones maintaining different temperatures for glass melting and refining: the batch entry zone (1200°C), the melting zone (1550°C), and the refining zone (1400°C). The continuous raw material disintegration and simultaneous reaction during melting facilitate the combination of physical and chemical transformation, gradually transforming into molten glass with substantial bubbling in the homogenization process. The molten glass passes through an immediate narrow zone, designated the "waist zone," maintained at a relatively low temperature of 1300°C, and then enters the working end or conditioning zone in the temperature range of 1100–1300°C. Finally, the molten glass enters into the forming zone through the "canal zone," maintained at a temperature of ~1125°C.

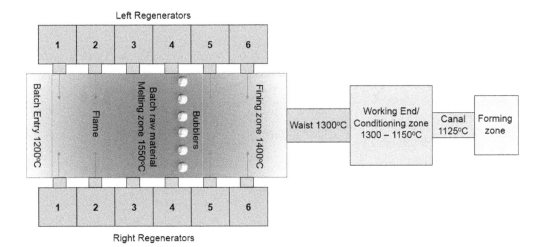

FIGURE 8.5
Six side-fired ports glass melting furnace zones and temperatures. Arrow indicates the flame in different zones. Temperature regions are shown in different shades, and the continuous bubbling phenomenon is shown by spheres.

Step III—Forming: In this zone, the molten glass at a temperature around 1100°C is poured on to the mirror-like surface of molten tin (Sn) through a refractory spout. Preferably, the glass spreads onto the molten tin and attains a targeted equilibrium thickness (T) of 6.6 ± 0.1 mm, as shown in Figure 8.6a. By convention, the float bath is made of tin, as tin does not react with glass components, and glass is able to float because of its lower density (2.4 g/ml) compared with tin (6.5 g/ml). It is worth mentioning that continuously measure and maintain an optimum oxygen partial pressure in order to prevent the compositional change of metallic tin; otherwise, the disturbed atmosphere in the tin bath promotes unwanted chemical deposition on the annealing lehr rolls. To achieve a uniform glass thickness on a molten tin bath, a typical surface tension for the glass/air (S_g), glass/tin (S_{gt}), and tin/air (S_t) interfaces in the range of 0.32, 0.53, and 0.50N/m, respectively, is desired.

Despite differences in specific gravity and surface tension, the glass experiences very high viscosity (102–109 Pa.s) compared with molten tin (0.001 Pa.s); at the same time, the metallic tin experiences high thermal conductivity (46 W/m.K) compared with glass (1 W/m.K). However, consistent thickness is a foremost factor from the perspective of application; thus, the desired thickness and width of the glass ribbon are achieved with the help of the assisted direct stretch (ADS) machine's penetration, speed, and angle of rolling wheels, as shown in Figure 8.6b. The glass finally leaves the float bath as a solid ribbon on the lift out roll (LOR) at 600–610°C and enters the annealing lehr.

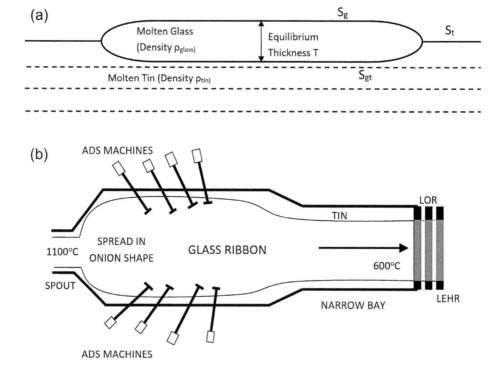

FIGURE 8.6
(a) Float glass process equilibrium thickness and (b) top view of the float glass on molten tin (Sn) bath. S_g: surface tension of molten glass; S_{gt}: surface tension at the interface; S_t: surface tension of molten tin.

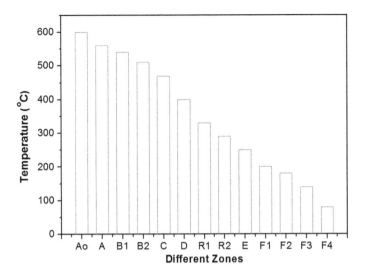

FIGURE 8.7
Typical float glass annealing lehr zone and temperature profile for a typical glass processing unit.

Step IV—Annealing Lehr: The glass ribbon then enters the *annealing lehr*, where the internal stresses are released, ensuring the perfect flatness and solidity of the final product. The upper end of the annealing range, or the annealing point relieve the internal stress through internal viscous flow. Followed by, the lower end of the annealing range, or beyond the strain point strain-free glass can be cooled quickly without introducing permanent strains For typical soda-lime glass, annealing points average around 545°C and strain points 520°C. A typical zone-temperature profile practice in glass industryis shown in Figure 8.7.

Step V—Inspection, Cutting, Packaging, and Warehousing: After sufficient reduction of stress and temperature, the glass ribbon passes through the online thickness measuring machine (TMM) and laser scanner for defect monitoring and isolation. Successful and precise inspection ensures a high-quality and consistent product that boosts customer confidence and results in the existence of product on the market for a long time. After this, the glass sheet is cut into pieces of different sizes based on customer requirements using automated equipment, stacked, and packed for storage in the warehouse. As well as glass density (2,500 kg/m³), different properties are measured to ensure quality control, some of which will be discussed later.

8.7 Window Glasses

8.7.1 Insulated Glass Units (IGUs)

Despite the direct use of float glasses in construction for either glass windows or doors, a multi-unit assembly of annealed float glass is an effective and popular choice for a comfortable home. This unit consists of two or more sheets of glass with a space between and sealed around the edges. IGUs, also called *double-glazed units* (DGUs), *triple-glazed units*

(TGU), or *sealed insulating/insulated glass units* (SIGU), are designed to improve the thermal insulation of buildings. IGUs can retain much more heat during winter, which in the long run, minimizes heat loss and saves energy. Furthermore, they help to reduce noise and frost condensation on glass windows. In summer, with the use of solar control glass, it is possible to reduce heat gain and increase the efficiency of air conditioning [12]. Prior to the fabrication of such IGUs, cut-to-size glasses are thoroughly washed and inspected to ensure defect-free, clean glass surfaces. A hollow aluminum spacer and a strip of polyisobutylene (PIB) are placed between the glass sheets as a primary seal. The spacer is filled with desiccant (molecular sieve) to prevent condensation after sealing. Silicone is then applied along the perimeter as a secondary seal; this provides good tensile strength, low vapor and gas diffusion, quick adhesion between the glass and the metal spacer, and finally, better structural bonding. Typical double- and triple-glazed IGUs made of window glass are shown in Figure 8.8. IGUs minimize heat exchange, increase soundproofing, and reduce condensation and frost, and it is obvious that triple-pane windows offer a higher degree of performance compared with double panes. Despite the high cost of IGUs, there is a probability of heat and sound conduction through metal spacers and that may reduce the performance of IGU. Eventually, the IGU performance depends on the U-value and it is essential to analysis before establish such glass window, as it quantifies the heat transmission through a wall or a glass window. It is measured as W/m^2K in metric, and the imperial unit is $BTU/(hrFf^2)$.

The value indicates how much heat per unit of time flows through an area of 1 m^2 at a constant temperature difference of 1 K (1°C) between the warm and cold sides of the structure. The lower the U-value, the better will be the insulating glass unit. Thus, the U-factor imperial decreases with an increasing number of glass panes, as depicted in Figure 8.9a. However, deliberate modification is required to improve the performance of IGUs; thus, an optimum content of non-reactive and non-toxic argon (Ar) gas between the glass panes and proper sealing can enhance the efficiency of IGUs. In this technology, three-paneled argon-filled windows provide effectively two layers of insulation, as argon has a thermal conductivity approximately 30% lower than that of air. At the same time, sealing is a major concern to maintain the consistent performance of argon gas–entrapped window panes; the constraint is either expansion or contraction, which eventually promotes slow leakage and dissipation that is actually difficult to identify.

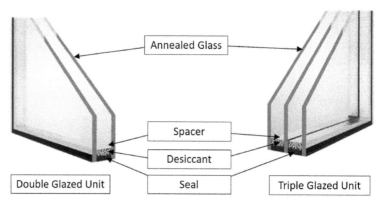

FIGURE 8.8
Insulated glass unit (IGU) made from window glass; two classic examples of double- and triple-glazed units consisting of spacer, desiccant, and sealing material.

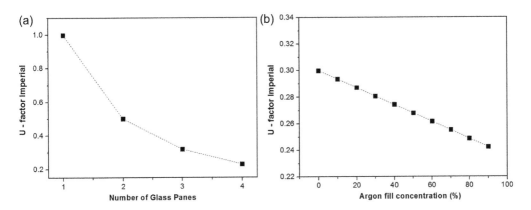

FIGURE 8.9
(a) U-factor variation with number of glass panes and (b) variation of U-factor with respect to percentage argon fill concentration within two glass panes.

The content of argon gas reduces the thermal conductivity and is responsible for reducing the U-factor imperial, as illustrated in Figure 8.9b. An alternative to argon gas is krypton (Kr), which can be used for the effective insulation of IGUs, but these are more expensive than argon-filled IGUs. Thus, U-value minimization results in high performance of IGUs, and several academic and industrial research institutes are at the forefront of research to develop high-performance IGUs. In modern practice, several other processes have also been adopted to make single-pane or multi-pane window glass for construction purposes. For example, in the presence of low-emissivity (low-e) glass, the U-factor is further reduced from 0.24 to 0.20 when 90% argon gas is introduced into the IGUs, whereas krypton in the same multi-pane reduces the U-factor to 0.18. Apart from the use of highly dense noble gas, several techniques, including low-e glass, high-absorbing tinted glass, reflective glass, and impact-resistant glass, are employed to enhance the effect further. A multiple low-e coating may provide analogous results to gas-filled glass windows.

8.7.2 Low-emissivity (Low-e) Glass

The *emissivity* of the material surface describes how effective it is at radiating heat or light energy. The specialty of this glass sought in low-e glass windows is protection of the temperature inside the construction, resulting in pleasant room conditions by reflecting heat in the summer and retaining it in the winter, and lighting efficiency inside the building construction. A special coating (e.g., silver nanoparticles) is applied to the glass surface that reflects infrared (IR) light but allows visible light through. More details and a fabrication protocol can be found elsewhere [13]. The atmospheric heat or light energy typically absorbed by glass is either shifted away by air movement or re-radiated by the glass surface. The understanding of some basic features makes low-e glass effective. These include

- Dull, darker-colored glass has high emissivity.
- Highly reflective material provides low emissivity.
- The quantity of radiation is dependent on the emissivity and the surface temperature.
- A low-e window glass surface promotes insulating capacity.

- A low-e coating minimizes ultraviolet (UV) and IR light transmission without much reducing visible light transmission.
- A low-e-coated window *costs about 10 to 15% more* than a regular window; at the same time, it can reduce energy loss by 30 to 50%.

8.7.3 Heat-absorbing Tinted Glass

This concept can be implemented in single-pane windows as well as IGUs. Tinted heat-absorbing glass is a glass body having different colors that are capable of absorbing solar energy radiation. Different colorants are used during the melting process of glass. Colorants used for making tinted glass include chromium oxide, cobalt oxide, nickel oxide, iron oxide, selenium metal, etc. The most popular glass of this type is gray and bronze tinted, in which selenium, iron oxide, and cobalt oxides are used as the basic colorants.

Depending on thickness and colorants, tinted glass absorbs the solar heat incident on the glass surface and thus reduces the heat transmission. The light transmission behavior of differently tinted glass with identical thickness at different visible and IR wavelengths is illustrated in Figure 8.10. As well as energy efficiency, heat-absorbing glass also offers security, safety, privacy, and design versatility.

8.7.4 Reflective Glass

Here, we have to start with a special metallic coating on float glass that cuts off solar heat energy transmission and provides a one-way mirror effect, preventing visibility from the outside and thus preserving privacy. Reflective glass is manufactured by either a pyrolytic process or a vacuum (magnetron) process. During the pyrolytic (online, hard-coat) process, semiconductive metal oxides (e.g., TiO_2) are directly employed. This coating, called a *hard coating*, is fused to the glass while the glass is still hot and is relatively less harmful to the environment. Pyrolytic-coated glass is durable and easy to cut, and it can

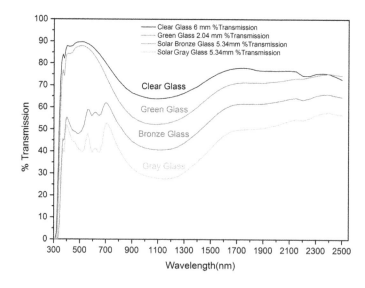

FIGURE 8.10
Light transmission behavior of colored glasses at different wavelengths. Gray glass has the lowest transmission in the range of 800–1200 nm.

be heat-strengthened or toughened. It is found to be cost effective. The vacuum (magnetron) sputtering process (offline) offers a vacuum process used to deposit very thin films on glass by applying a high voltage across a low gas pressure (usually argon at about 5 millitorr) to create a "plasma," which consists of electrons and gas ions in a high-energy state. In the course of sputtering, energized plasma ions strike a "target" composed of the desired coating material and cause atoms from that target to be ejected with enough energy to travel and bond with glass. These coatings are soft and applied on the inner side of glass panes to protect them from the external environment. Their low resistance makes them better when used in a double-glazing system. Vacuum-coated products have better shading coefficient values than pyrolytic products; however, this glass is relatively expensive. Both online and offline coating methods will be discussed in Section 8.9.1.

8.7.5 Impact-resistant Glass

In hurricane-prone places, glass windows experience premature damage due to high-speed winds and consequent impact by erosive debris. Thus, an alternate choice to improve the glass quality is needed that can resist impact and extend the service life. By convention, this quality of glass is a laminated product, and lamination is the preferred answer to the demand for impact-resistance characteristics. For example, multiple layers of polyvinyl butyral (PVB) sandwiched between two glass sheets followed by autoclaving under high pressure enable the glass to be tempered and enhance its strength, resulting in resistance to small missile (projectile) impacts. At the other extreme, polyethylene terephthalate (PET)–laminated glass and glass-clad polycarbonate are designed for relatively larger missile impacts. Windows of these qualities are finally attached to the frame with silicone to make them perfect.

8.7.6 Selection Criteria for Glass Windows

One has to know the preliminary selection criteria of a particular window glass quality for a particular climate and mode of use. A few classic specifications to select glass quality for specific zones of interest are as follows:

- **A lower U-factor**, which measures heat transmission through a wall or glass window. Lower numbers (between 0.2 and 1.2) indicates a product that is better at keeping heat in.
- **A lower solar heat gain coefficient** (SHGC), between 0 and 1, indicating a product that is better at blocking unwanted heat gain.
- **Visible transmittance** (VT) measures (between 0 and 1) the amount of light coming through; a higher VT indicates a higher potential for day lighting.
- **Air leakage** (AL) measures (between 0.1 and 0.3) the rate of the amount of outside air coming into a home or building through the window. A lower AL is better at keeping air out.
- **Condensation resistance** (CR) measures (between 1 and 100) resistance to the formation of condensation. A higher CR is better able to resist condensation.

Window glass selection criteria for a particular climate:

- A lower SHGC is used in warmer climates.

- A lower U-factor is used in cooler climates.
- Low-e coatings on exterior window panes prevent heat gains from exterior radiation.
- Low-e coatings on interior windows prevent heat loss.

8.8 Toughened Windshield Glass

Toughened windshield glass is a safety glass, processed using controlled thermal or chemical treatment with the target of mechanical strength enhancement. The windshield or windscreen is the glass used in the front windows of aircraft, cars, buses, and more for protection from wind and flying debris such as dust, insects, and other flying objects. Glass and ceramics both experience the early stages of fracture under tension compared with compression. When glass accommodates an impact because of wind speed or is hit by flying objects, the impacted side of the glass surface experiences compression, while the other side is under tension. Normal annealed glass has very low tensile strength, around 30–40 MPa, which is very low for windshield application, but experiences compressive strength of around 1,000 MPa; put another way, this indicates that to shatter a 1 cm cube of glass requires a load of some 10 tonnes. However, after the toughening process, the tensile strength of glass becomes 120–200 MPa (depending on thickness, edgework, holes, flaws, etc.), which satisfies the windshield application requirements. The glass fails beyond its maximum tensile stress, and thus, toughened glass is in the forefront compared with annealed glass (four to five times higher tensile strength) for windshield purposes. During the toughening process, a greater contraction of the inner layer induces compressive stresses in the surface of the glass, balanced by tensile stresses in the body of the glass. A successful toughened glass and a safety glass should possess compressive strength higher than 69 and 100 MPa, respectively [14]. However, excessive impact breaks the toughened glass, which crumbles into small granular chunks instead of splintering into jagged shards as annealed glass would do; the latter may cause serious injury during driving (Figure 8.11).

8.8.1 Manufacturing Toughened Glass

Prior to manufacturing toughened glass, one has to pick up the annealed float glass, as has already been discussed in Section 8.6. Chemical toughening is another process to toughen the glass; however, in this chapter, we will discuss the temperature-assisted toughening process only. As well as annealing, the following five steps are usually adopted to manufacture toughened glass.

Step I: It is mandatory to cut the annealed glass sheet to the targeted size before toughening, as annealed glass cannot be cut after toughening. This is followed by grinding the edges to avoid any uneven appearance on the edge, which is responsible for enhancing the chance of failure during toughening. At this stage, holes are drilled if required in the final product.

Step II: The processed glass sheet is heated above the glass transition temperature but below the softening point. A schematic representation of the manufacturing protocol is illustrated in Figure 8.12. The bottom surface or tin-side of float glass may contain micro-flaws due to contact with lift off rolls and annealing lehr rolls

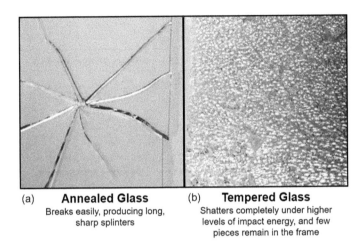

(a) **Annealed Glass**
Breaks easily, producing long, sharp splinters

(b) **Tempered Glass**
Shatters completely under higher levels of impact energy, and few pieces remain in the frame

FIGURE 8.11
Patterns of glass breakage: (a) annealed glass consisting of sharp pieces and (b) tempered glass shattered into small pieces and partially remaining in the frame [15].

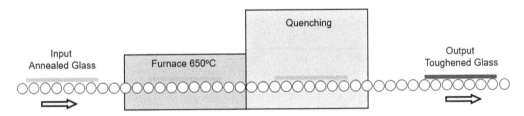

FIGURE 8.12
Schematic representation of toughened glass manufacturing protocol.

during manufacturing. This flaw can be identified by fluorescence when illuminated with UV rays. Thus, it is preferable to place the tin-side annealed glass in contact with rolls that enable the closing of micro-cracks under compression and the manufacture of flawless toughened glass.

Step III: The glass sheet may either be maintained optically flat or formed into a definite curvature or any other desired shape. By preference, the deforming temperature for flat and curved glass is maintained at 625 ± 15 and $650 \pm 10°C$, respectively. A typical history of temperature-dependent stress and temperature profile distribution is shown in Table 8.5 and Figure 8.13. The compressive and tensile stresses develop due to the variation between the center and surface temperatures of a glass object. Considering the data set, it can be noticed that heating zone A does not have any temperature difference between the center and the surface and is not responsible for generating any stress in the system. However, enhancing the surface-to-core temperature difference influences the high tensile stress on the glass surface (zones B and C), whereas the same system experiences more compressive stress during rapid quenching in the temperature difference of $-140°C$ (zone D).

Step IV: The cooling has to be performed in a uniform manner to maintain the top and bottom surface temperatures equal to each other and lower than the center part temperature of the glass.

TABLE 8.5

Samples with Different Toughening Parameters

	Time (s)	Center Temperature (°C)	Surface Temperature (°C)	Difference (°C)
A	00 (Heating)	25	25	0
B	20 (Heating)	140	210	70
C	90 (Heating)	450	510	30
D	05 (Quenching)	435	295	−140

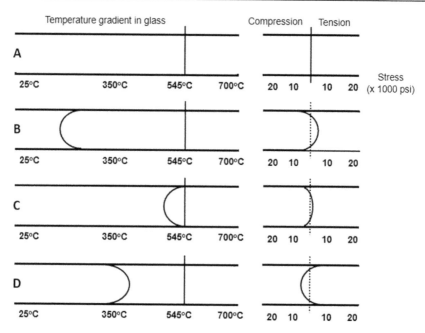

FIGURE 8.13

Variation of stress during temperature treatment from the perspective of the development of toughened glass.

Step V: After attaining a temperature below the strain point, the toughened glass needs to cool down to ambient temperature, and it will then be ready for use.

8.8.2 Probable Reasons for Failures of Toughened Glass

Discussion of the additional toughening process to improve the impact resistance of annealed glass also requires the basic reasons for the probable failure of such toughened glass. A few of them are highlighted to help in understanding their effects:

- **Installation issues:** improperly installed windshields are more prone to vibration and stress, resulting in the initiation of flaws, and premature cracks and failure.
- **Temperature changes:** Excessive use of the car heater or defroster may expedite glass expansion or shrinkage, so that uneven thermal stress around the edge leads to cracking and failure of the glass.
- **External objects:** Beyond the compressive strength limit, any flying debris such as dust, insects, other flying objects, or hail hitting the windshield glass repeatedly at unexpected speed can cause failure.

- **Glass imperfections:** In general, very high-quality standards are maintained while manufacturing automotive-quality glass, although common glass defects, primarily optical imperfections, are more responsible for the early stage of cracking under impact. Apart from these defects, a non-identifiable nickel sulfide (NiS) inclusion is detrimental and may initiate spontaneous breakage.

- **Spontaneous failure due to NiS inclusion:** Small inclusions of NiS are often present in float glass, and thus, absolutely NiS-free float glass seems to be impossible. NiS is crystalline and does not dissolve in molten soda-lime silicate glass.

Prior to discuss the failure mechanism, it is necessary to highlight the resource of NiS. In actual, several components including heating elements in float glass plant are made of stainless steel, where contain certain amount of nickel. In the same time, extensive amount of fuel feeding is the prime resource of Sulfur. Thus, the probable contact between two elements (Ni and S) expedite formation of small inclusion of NiS in the size range of 0.05mm within glass matrix. Followed by, the annealing process leads to the transformation from the high-temperature α-phase to the low-temperature β-phase, which results in a 2–4% volume expansion of NiS stone. When the same float glass is used to make toughened glass, the β-phase further changes to the α-phase above the transition temperature (379°C) of the stone. A rapid cooling process does not allow switching to the low-temperature β-phase; rather, it slowly transforms to a stable β-phase accompanied by the 2–4% volumetric expansion. This unexpected inclusion can initiate cracking, and damage can become severe if (a) the inclusion is located in the tensile zone of the glass; (b) the diameter of the inclusion is >50 μm (approx.); (c) the temperature and duration enable allotropic transformation. An alternative destructive heat soak test (HST) can determine the influence of NiS; in this test, the NiS inclusions are forced to transform at a higher temperature, which encourages the glass to break inside the furnace at the factory and minimizes the chances of breakages in use.

8.9 Surface-Coated Glass

Surface engineering of metal and ceramic materials is quite often employed to fulfill different objectives, and thus, the surface engineering of glass is in the forefront of accommodating different properties by using various coatings. The ion exchange protocol is an excellent tool to modify chemical, mechanical, electrical, optical, and other properties. In fact, these properties cannot be introduced while producing the basic glass during the melting process, since it is not possible in large-scale production. The most common glass-coated products are self-cleaning, anti-reflective, IR reflective, and low-e coatings for automotive, architectural, and photovoltaic solar panel applications. Based on the customer's requirements, the properties are modified after forming in either online (hard coating) or offline mode (soft coating). The coatings are applied via complex online pyrolytic spray coating or chemical vapor deposition (CVD) and offline sputtering or physical vapor deposition (PVD) technologies. Apart from these coating techniques, powder spray, sol-gel coatings, and thermal evaporation techniques are also frequently used in the glass industry. In the following, some basics on commercial-scale online and offline coating technologies will be discussed, which can be adopted after the fabrication of annealed float glass or toughened glass.

8.9.1 Online Coating

Online surface modification is done while the glass is hot, just after it is formed or while it is still in the annealing lehr. Vaporized chemical coating compounds react with the glass in situ during production before the glass is cooled to a temperature of 600–700°C. This technique is analogous to tinted float glass, as the size and tolerance limitations are similar for both processes. This process encompasses several advantages, including low cost of production; higher productivity, around 70–85% of the non-coated basic glass production capacity (300–600 MT per day); excellent adhesive bonding, sufficiently hard to remove; and long lasting compared with offline soft coating. However, in industrial practice, there are major challenges in making a consistent product:

- The coating is being applied to glass moving at a speed of 7 to 12 meters/minute.
- The deposition rate is fast (60–100 nm/s).
- The temperature of the glass during coating is more than 600°C.
- Continuous coating demands an appreciable input of glass to make an economical product.
- The coating is less uniform compared with the offline process.
- The chemicals coming out of the exhaust cannot be reused, increasing product cost.

Pyrolytic spray coating is an online method for depositing a thin optical coating onto hot glass. A brief online coating protocol is given in Figure 8.14. A metal organic precursor is applied to hot glass at a temperature of 600°C; the high temperature of the glass causes the in situ decomposition of the precursor, which facilitates the oxidation of metal in the precursor, causing it to bond with the glass surface. This method is also often called the *chemical vapor deposition* (CVD) process. In online coating, organometallic compounds, predominately organic precursors of tin, cobalt, nickel, chrome, iron, titanium, etc., dissolved in organic solvent are used to develop different colors. Usually, pyrolytic coatings are in demand for light and heat reflection (LHR) glass.

In the early days, a reciprocating apparatus was used for spraying. This was restricted to a glass moving speed less than 6 m/min, and its high vibration resulted in a defective coating. Later, a rotary apparatus was developed, including an exhaust system, and used to overcome the previous problems. Two different online arrangements are shown in Figure 8.14a and b. Online coatings have advantages of hardness and durability compared with offline coatings and are also competitive in terms of bending and toughening.

8.9.2 Offline Coating

Unlike online coating, offline coating involves a two-step process. In the former process, coatings are applied separately to the manufactured glass, and the coating later develops in situ during glass manufacturing. Among the offline coating application methods, the most widely used is *sputtering*, also called *physical vapor deposition* (PVD). In this process, atoms of coating materials are ejected and condensed on the glass substrate in a high-vacuum environment. Sputtering can be done with virtually any non-magnetic metal or alloy. A metallic coating can be deposited in the presence of an argon atmosphere; however, a metal oxide or nitride results from the presence of oxygen and nitrogen, respectively, in the vacuum chamber. The resultant color depends on the coating thickness, the

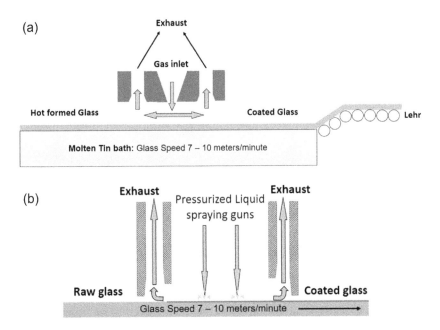

FIGURE 8.14
Pyrolytic coating equipment located (a) inside tin bath before exit and (b) between bath exit and annealing lehr entrance.

material, and the object configuration. The advantages of this type of coating include the following:

- One or more coatings can be effectively applied to achieve the desired properties of finished glass.
- Fine tuning enables uniform coatings throughout the surface.
- A 3,000 × 2,000 mm flat glass can be sputter coated in minutes.
- It is possible to deposit pure metal, metal oxides, metal nitrides, etc.
- The obtained coating is a better reflector of UV and IR rays.

However, there are some disadvantages: primarily, it is an expensive batch process, although it results weak coating adherence and softer surface, and the life of the coating is shorter.

The basic direct current (dc) sputtering system is a vacuum chamber operated at a very low pressure. In this system, a dc voltage is set up between the target and the substrate to achieve effective coating; electrically neutral argon atoms are introduced, which eventually ionize and create a plasma-like hot gas phase consisting of ions and electrons. These ions move toward the target, and continuous collisions are generated with the target atoms, which flow toward the substrate and settle as a thin film. Electrons released during argon ionization move toward the anode substrate, collide with additional argon atoms, and create more argon ions and free electrons, and the cycle continues. A brief schematic is illustrated in Figure 8.15a. However, the control of coating characteristics is difficult using this process, and therefore, an advance magnetron sputtering system has been introduced to obtain a high-end product; such a setup is shown in Figure 8.15b. In a magnetron sputtering system, a strong magnetic field is

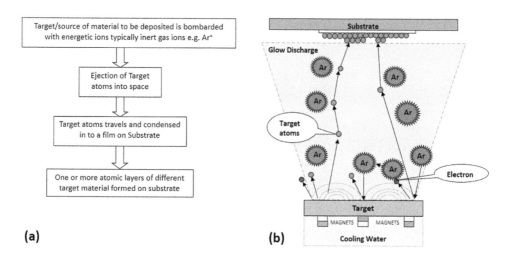

FIGURE 8.15

Process flow diagram of (a) dc sputtering and (b) magnetron sputtering system.

added to the basic dc sputtering system near the target area to facilitate and control the flow of ionic species. Thus, it has the following special features in addition to those of the conventional dc sputtering system:

- Travelling electrons spiral along the magnetic field lines near the target.
- The confined plasma region facilitates uniform coating.
- The probability of ionization of argon atoms increases as electrons travel a longer distance.
- The plasma remains more stable with higher ion density.
- More ion generation increases the efficiency of the sputtering process.

From the application viewpoint, different-colored dielectric coatings are employed through this process to enhance interference effects, and thus, it allows higher light transmission with increased selectivity [16]. It is worth remembering that toughening or deforming cannot be done after offline coating; however, this is an effective process to develop combined coatings in double-glazed and laminated products. Another class of coating, known as *dichroic coating*, is usually composed of multi-layered coatings and exhibits different coloring phenomena by reflection and transmission as a function of the viewing angle.

8.10 Mirror Glass

A mirror is a surface that reflects light and produces a clear image of objects in front of it. Mirror silvering is a chemical process that deposits a coating of metal, usually silver, onto the surface of glass. This deposit is usually protected by a layer of copper, which in turn, is protected by a paint backing. Generally, glass sheets of different sizes and thicknesses are used as the substrate onto which coatings are applied to produce mirrors.

8.10.1 Manufacturing Process of Mirrored Glass

To manufacture mirrored glass, we need defect-free float glass or a definite shape of glass to develop the mirror. The entire process protocol can be subdivided into five major steps:

Step I—Preparing the Glass: Cut the glass sheet to the required size and grind the edges. The creation of mirrors begins with a single sheet of glass, commonly referred to as a *lite*. The lite is placed onto the silver line and proceeds through a physical washing process. Wash and polish with Cerirouge, a very fine abrasive powder, and finally, clean with hot distilled water. The objective is to remove all contaminants from the glass surface, and it is an essential step to ensure the proper bonding of metal to the lite.

Step II—Tinning and Silver Coating: Prior to the silver coating, an intermediate coating of an ultra-thin layer of tin on the clean and polished glass surface is essential to enable the strong adhesion of silver; tin has the ability to form a strong bond with both glass and silver. In this process, first, a solution of stannous chloride in distilled water is sprayed onto the glass followed by washing with distilled water. After the tinning process, the lite is preferably rinsed with hot deionized water to maintain the optimum lite temperature in the range of 32–38°C for the silvering process. The silver coating on the glass adds the reflective properties, and finally, a mirror is formed. The silver is produced chemically, as it is applied to the glass by the reduction of a silver salt solution. The silver hardens almost immediately on contact with the tin layer. The layer of silver is applied at a temperature of ~38°C. After the silver application, the lite is passed under a high-velocity air knife to remove excess silver and water before entering the coppering process. The composition of the silver solution and the reducing solution is as follows:

Silver solution: silver nitrate + sodium hydroxide + ammonia + distilled water.

Reducing solution: sucrose (sugar) + formaldehyde + conc. sulfuric acid + distilled water.

The solutions are mixed in a multiple-jet spray gun containing silver and reducing solutions separately and sprayed onto the surface of the glass to give a thin coating layer of silver up to 3 nm (3×10^{-9} m) thick. The silver may also be applied to the glass using a "rocking table," on which the solutions are mixed by rocking back and forth, or by pouring the solutions over the glass as they mix. The chemistry of this process is also used in the laboratory in Tollen's test for an aldehyde (the silver mirror test). Aldehydes are easily oxidized to the corresponding carboxylic acids, and when this happens, they reduce silver ions to metallic silver in alkaline conditions (Equation 8.1). Ammonia is used to form the diammine silver complex ion and thus prevents the precipitation of insoluble silver hydroxide or silver oxide [17].

$$Ag(NH_3)_2^+ + e^- \rightarrow Ag + 2NH_3$$

$$\frac{RCHO + H_2O \rightarrow RCOO^- + 2e^- + 3H^+}{2Ag(NH_3)_2^+ + RCHO + H_2O \rightarrow 2Ag + RCOO^- + 3NH_4^+ + NH_3} \tag{8.1}$$

Note that the sucrose is actually a "non-reducing" sugar and so cannot itself be oxidized to a carboxylic acid. However, it is hydrolyzed to glucose and fructose in the acidic solution, and thus, eventually, glucose is considered as a reducing sugar.

Step III—Copper Protection: Sometimes, a copper backing over the silver surface is also used to protect the precious silver coating from corrosion and enhance the mirror's performance. This is usually done by reducing cupric salt with a zinc suspension.

> Copper solution: copper sulfate + conc. sulfuric acid + distilled water.
>
> Zinc suspension: zinc dust + distilled water.

The copper solution is mixed in a spray gun with the zinc suspension and applied to the lite, where the zinc reduces the copper ions to copper according to the equation $Cu^{2+}(aq) + Zn(s) \rightarrow Cu(s) + Zn^{2+}(aq)$. The lite is then carefully rinsed, dried in pressurized heated air ovens, and sent to the painting process.

Step IV—Protecting the Backing: The mirror is dried in a hot chamber after the application of the metals to evaporate all moisture. The back is then painted with a special paint to protect the plated metals. The paint is applied using a double-roll coat system, which applies two separate coats of low-lead paint to the back of the mirror. After the protective paint layers are applied, the mirror enters the final heating process.

Step V—Curing, Cleaning, Inspection, and Packaging: The mirror cures in the oven as it continues down the silver line. The mirror is heated to a temperature of 130–140°C for approximately 6 minutes for proper curing. After exiting the oven, the mirror is cooled and run through a final cleaning process. Once the mirror reaches the end of the silvering line, the quality of the mirror is inspected, and it is removed from the line and packed onto a rack for custom fabrication or packaging. If the mirror is scheduled for custom fabrication, it is moved to the fabrication station for cutting to pre-specified sizes by a computer numerical controlled (CNC) cutter, beveling, edge polishing, grooving, and safety tape backing.

Sometimes, it is necessary to perform the re-silvering of old mirrors. For this, the mirror is soaked in dilute hydrochloric acid for about 2–3 days to remove all mirroring materials. The glass is then re-silvered using the previously described process.

8.10.2 Imperfections in Glass Mirrors

Imperfections are a common occurrence during the manufacture of any product, but their identification is essential to minimize the defects. From this perspective, a brief summary of common glass (mirror) defects follows:

- **Chip:** an imperfect edge due to the breakage of a small fragment out of the regular surface. There are three dimensions: chip width, the distance between the edge of the mirror and the inner edge of the chip; chip length, the distance between the edges of a chip measured parallel to the edge of the mirror; and chip depth, the distance from the face of the mirror into the thickness.
- **Flare:** a protrusion on the edge of the mirror.

- **Blemishes:** imperfections in the body or on the surface of the mirror. These can be further distinguished into several types, such as point, linear, cluster, and crush blemishes. Point blemishes comprise inhomogeneity in the form of a vitreous lump in the mirror (knot); a small foreign particle imbedded in the glass surface (dirt); a crystalline inclusion in the mirror (stone); or a round or elongated bubble causing a cavity in the mirror (gaseous inclusion). Linear blemishes highlight the damage on the glass surface in the form of a line (scratch); abrasion of the mirror surface producing a frosted appearance (rub); or a deep and short scratch in the glass surface (dig). Cluster generation is found when more than two point blemishes are separated by less than 50 mm. Another defect, a lightly pitted area in the glass surface, causes a dull gray or white appearance, known as *crush*.

- Apart from the aforementioned defects, other defects, such as a small area from which the silver coating is absent (spot silver fault), corrosion changing the color or reflectance along the mirror edge (edge corrosion), or a frosted appearance in the reflected image from a silvered mirror (visible clouding), are common in practice.

8.11 Bottle Glass

A glass bottle is a container made of glass. The volume varies over a wide range; say, 10 ml to 5 liters. Common uses for glass bottles include food condiments, soda, liquor, cosmetics, pickling, and preservatives. Two major types of glass bottles are manufactured: (i) borosilicate glass and (ii) soda-lime glass. The composition of bottle glass is analogous to that of float glass; however, the former has a relatively lower percentage MgO content (see Table 8.3). A typical bottle glass composition is given in Table 8.6.

8.11.1 Manufacturing of Bottle Glass

Manufacturing bottles is a highly automated process with enhanced productivity. Continuous research and development successfully reduced the weight of glass bottles by more than 50% between 1970 and 2000. A typical glass bottle consists of different parts, as shown in Figure 8.16. The entire manufacturing protocol will be discussed in seven steps, followed by their testing modules to ensure product reliability.

Step I—**Batch Preparation:** The correct mix of different raw materials is calculated to achieve the chemical composition of the final glass. The calculated amounts of individual raw materials are weighed and mixed as a batch, and later, a calculated amount of cullet is added before moving to the batch hopper. The batch-making

TABLE 8.6

Typical Composition of Soda-lime Bottle Glass

	SiO_2	Al_2O_3	Fe_2O_3	Na_2O	K_2O	CaO	MgO	TiO_2	SO_3 (Highly Reduced Glass)	SO_3 (Highly Oxidized Glass)
Weight (%)	74	1	0.10	12.9	0.2	11	0.3	0.01	0.025	0.30

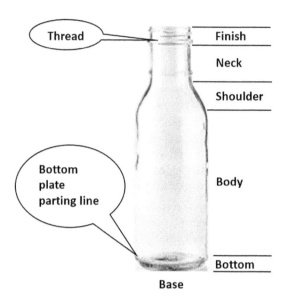

FIGURE 8.16
Different parts highlighted in glass bottle.

process is continued to meet the furnace melting and maintain the productivity. It is worth remembering that borosilicate glass cullet is not recycled, because it is a heat-resistant glass and does not melt at the same temperature as soda-lime glass; thus, it would alter the viscosity of the fluid in the furnace during the re-melt stage, eventually affecting the production process and product quality.

Step II—Glass Melting: The batch is inserted inside the melting furnace using a suitable feeding system; for example, a piston batch charger, a screw batch charger, a pusher type charger, a vibratory feeder, etc. The furnace melts cullet, silica sand, soda ash, dolomite, limestone, and other raw materials together. During glass melting, the maximum temperature of silica refractory crown is maintained below 1630°C. Other than transparent glass, amber glass is the most common color for bottle glass; for this, it is essential to add together iron, sulfur, and carbon in the batch, and the color is generated in a reducing environment. Amber-colored glass absorbs nearly all radiation consisting of wavelengths shorter than 450 nm and prevents the transmission of UV radiation, which is a mandatory requirement for bottles containing beer, few acids and organic compounds, and some drugs.

Step III—Gob Distribution: The molten glass then leaves the furnace and is distributed to different fore-hearths for the respective production lines. Molten glass is conditioned in the forehearth and maintains its homogenized temperature gradient. The stream of molten glass is intercepted with the help of plungers and mechanical shears to make gobs and is distributed into blank molds by individual section machines.

Step IV—Forming: Gobs of glass at its high deformable plastic temperature (1050–1200°C) are fed into blank molds to form parison. This is then transferred to a mold, where compressed air (pressure 2.8–3.2 bar) and vacuum are applied for stretching and cooling to the exact inner cavity of the mold, forming the desired bottle shape. The glass mold temperature varies between 400 and 500°C.

Step V—Annealing Lehr: Once formation is complete, bottles may suffer from stress as a result of unequal cooling rates. The glass bottles are then passed through a long heated oven called a *lehr* for annealing glass. Typically, the temperature of a glass bottle entering the lehr is 455–480°C. Bottles are then heated to the annealing point (averaging around 538°C) and slowly cooled to room temperature to remove internal stresses, making the glass more durable. The typical average strain point of soda-lime silica glass is 504.4°C. General guidelines for a typical setup of lehr parameters for the manufacture of glass bottles are as follows:

- Lehr entrance temperature = 482.2°C
- Heating rate entrance temperature to annealing point = 27.8°C/minute
- Annealing point, soda-lime glass (η = 1013 poise) = 537.7°C
- Strain point, soda-lime glass (η = 1014.5 poise) = 504.4°C
- Maximum thickness = 12.7 mm
- Maximum cooling rate, annealing through strain point = 4.4°C/minute
- Maximum cooling rate, strain point through lehr exit temperature = 14°C/minute
- Proposed exit temperature = 121°C

Step VI—Cold End: After annealing, a layer of polyethylene wax is applied using a water-based emulsion technique, which makes the glass surface slippery and protects it from scratching. It also prevents the bottles from sticking together in the annealing lehr.

Step VII—Inspection and Packaging: All the parameters are monitored continuously during the production process. At the cold end, the bottles are checked using a computer and a laser scanner and finally, in a manual inspection booth. It is most important to meet health and safety standard requirements. Typical imperfections in glass bottles are noticed, including small cracks in the glass called *checks*, foreign inclusions (stones, cords and knots, bubbles, and blisters), thickness differences, and forming defects. The last stage is packaging, which may vary from factory to factory depending on the specific type and size of bottles. Rejected bottles are recycled back into the furnace.

8.11.2 Testing of Glass Bottles

Continuous quality checking ensures consistent productivity, and laboratories perform different tests to ensure reliability. A few important testing protocols are listed here:

- The pendulum impact test (PIT) assesses impact resistance.
- Vertical load resistance measures the resistance of glass containers to external force in the direction of the vertical axis.
- Coating assessment implies the longevity of surface protection coating and is usually tested using slip tables.
- Internal pressure strength is usually measured by filling an article with water and applying continuous pressure in an incremental manner until it breaks.

- Capacity and headspace measure the volume of liquid filled and the space left.
- The thermal shock test is performed by taking a glass bottle from a hot oven and placing it on a cooled worktop, and measuring the number of cycles it can resist.

8.11.3 Important Terminology

A novice graduate trainee always faces difficulty in understanding the terminology that is often used on the shop floor. Therefore, a cluster of common terms and their technical meaning related to glass bottle manufacturing are listed here:

- **ACL (applied ceramic labeling):** colored ceramic paints fused on glass bottles as lettering or design.
- **Amber:** a brown color of glass that absorbs wavelengths lower than 450 nm, thus protecting the contents from UV radiation. This is essential for beer and certain drugs.
- **Batch:** a mixture of raw materials that is heated and melted to form different shapes.
- **Blanks:** the mold used for preliminary formation of glass, called *parison*.
- **Blow and blow:** the molding process in which parison is formed by compressed air for narrow-finish bottles.
- **Bottom plate:** a piece of the mold used to shape the bottom of the glass bottle with identification marks.
- **Chemical durability:** the resistance of glass to attack by solvents or products.
- **Cullet:** broken or waste glass used as glass raw material for re-melting.
- **Density:** a unit measured as weight per unit volume. The density of bottle glass ranges from 2.48 to 2.53 g/ml.
- **Finish:** a special shape surrounding the bottle opening, which accepts the cap.
- **Forming or finish mold:** the mold in which the bottle is blown into its final shape.
- **Melting furnace:** a large enclosed container made of refractories for melting raw materials.
- **Gob:** a lump of molten glass with a specific weight, shape, temperature, and viscosity.
- **Headspace:** the space between the level of contents in the neck and the closure, kept to accommodate expansion due to heat after packing.
- **IS machine:** individual section machine used for the formation of glass bottles.
- **Lehr:** a long hot oven through which the glass bottles are passed for annealing to remove stresses after formation.
- **Mold:** a set of iron forms for shaping the glass bottle. The part names are bottom plate, blank molds, finish molds, neck rings, and tip.
- **Narrow mouth:** a bottle in which the diameter of the finish is considerably smaller than the body.
- **Neck:** The portion of the bottle where the shoulder cross-section reduces to join the finish.

- **Parison:** the initial shaped hot glass that hangs from the neck rings when the blank mold opens. It is also called *pattern* or *blank*.
- **Parting line:** the horizontal ridge formed from the excess glass due to mold part joints. The vertical line formed due to two halves of the same mold is called the *seam*.
- **Press and blow:** the molding process in which parison is formed by pressing the glass against a blank mold with a metal plunger for large-diameter-finish bottles.
- **Refractories:** materials that retain their strength at very high temperatures.
- **Shoulder:** the portion of the bottle where the body area or maximum cross-section reduces to join the neck.
- **Weathering:** attack on the glass surface by atmospheric elements.

8.12 Bioactive Glass

Bioactive glasses are a group of surface-reactive biomaterials, encompassing biocompatibility. Synthetic bone grafts are used to cure diseased or damaged bones in the human body. First-generation biomaterials were close to bio-inert so as to minimize the formation of scar tissue at the interface with host tissues. The second generation is bioactive glass that undergoes surface dissolution in physiological environments and facilitates cell proliferation. After implantation in the human body, it reacts with body fluids and produces a surface hydroxyapatite (HA) layer that forms a stable chemical bond with the adjacent living bone tissue.

Bioglass 45S5 is commonly referred to by its commercial name *bioglass*. Commercial-grade dental bioglass and bioglass ceramics have distinct compositions and exist in different phases, which results in different degrees of in vivo bonding; classical compositions are discussed in Table 8.7. Class A bioactive materials bond to both bone and soft connective tissues. The surface reactions rapidly form a hydroxycarbonate apatite (HCA) layer that binds collagen fibers of soft and hard tissues. Class B bioactive materials bond to bone via osteoconduction but do not bond to soft tissues. The surface reactions to form an HCA

TABLE 8.7

Composition and Properties of Bioactive Glasses and Glass-ceramics Used Clinically for Medical and Dental Applications [18–20]

Composition (wt.%)	45S5 Bioglass	S53P4	A-W Glass-ceramic
	(NovaBone, Perioglas) (NovaMin)	(AbminDent1) (BonAlive)	(Cerabone)
Na_2O	24.5	23	0
CaO	24.5	20	44.7
CaF_2	0	0	0.5
MgO	0	0	4.6
P_2O_5	6	4	16.2
SiO_2	45	53	34
Phases	Glass	Glass	Apatite, beta-wollatonite, glass
Class of Bioactivity	A	A	B

layer are too slow to bind the collagen fibers of soft tissues. Medical professionals can choose a suitable composition according to the patient's health and preferences.

The sequence of surface reactions occurring on the glass surface to form a bioactive bond is as follows:

- Ion exchange of sodium ions with protons and hydronium ions takes place.
- Silanols formed by the ion exchange undergo a condensation reaction to form a high–surface area silica gel.
- Calcium and phosphate ions from body fluids, along with ions released from the glass, form an amorphous calcium phosphate phase on top of the silica gel.
- Crystallization of the calcium phosphate–rich phase occurs, incorporating carbonate ions from solution to form polycrystalline HCA. The HCA crystals bind to collagen from the host tissue to form the bond.

The bioactive bond is the interfacial bond between a bioactive glass and bone. It is composed of HCA crystals bonded to collagen fibers. The rate of bonding depends on the species of animal and the location of the implant. Bonding to rat femora occurs in days; for humans, it takes weeks. The bond to bone is above 100 micrometers within a few weeks and stabilizes at 200–300 micrometers by nearly 6 months. Due to the microporosity of the thick silica gel and calcium phosphate bi-layer on the glass surface, the interfacial bond has a gradient of elastic modulus from the glass to the tissue that mimics the gradient of elastic compliance between tendons and ligaments, and bone implants have interfacial strength equal to or greater than that of host bone. Several synthesis protocols and bioactivity are discussed in the current literature [21–23].

8.13 Glass Properties

Glass is produced by cooling molten material at a rapid rate, so that it does not get enough time to form a crystal lattice, and thus, it is transparent to visible light. Some ingredients are added to glass to change its properties based on requirements. Lead (Pb)-containing glass has a higher refractive index and appears to be more brilliant. The thermal and electrical properties of glass can be manipulated through the boron content; however, optical properties in the form of preferred color are simulated by different classes of metal and metal oxides. Thus, a wide range of properties are required to monitor and fulfill the reliability of commercial-grade glass products. In this context, a series of glass properties are discussed.

8.13.1 Optical Properties (Visible Light)

While we are discussing the optical properties of glass, some critical parameters are important to understand from the perspective of market demand and mode of use. In this domain, light transmittance (LT) is the proportion of visible light at near normal incidence that is transmitted through the glass. Light reflectance (LR) is the proportion of visible light at near normal incidence that is reflected by the glass.

Quantitative estimation can predict the physical significance of such optical features, and therefore, several parameters for different-colored glass having a consistent thickness of

5.66 mm (as an example) are tabulated in Table 8.8. The refractive index or index of refraction is a measure of the bending of a ray of light when it passes from one medium into another. The refractive index n is defined as the ratio of the sine of the angle of incidence to the sine of the angle of refraction; i.e., $n = \sin i / \sin r$, where i, the angle of incidence of a ray in vacuum, equals the angle between the incoming ray and the normal (the perpendicular to the surface of a medium), and r, the angle of refraction, equals the angle between the ray in the medium and the normal. The refractive index is also equal to the velocity of light c of a given wavelength in empty space divided by its velocity v in a substance, or $n = c/v$. Some important standard refractive index data at 22°C are given in Table 8.9.

The thermal coefficient of refractive indices (dn/dT) or the refractive index is affected by alteration of the glass temperature. This can be ascertained from the temperature coefficient of refractive index (for light of a given wavelength), which changes with wavelength and temperature. The temperature coefficients of the glasses help to calculate the change of refractive index for any wavelength from near UV to near IR in a given temperature range. Some special glasses with negative temperature coefficients in an optical system can help to keep wave front deformations caused by temperature changes to a minimum level. Such glasses are called *athermal glasses*.

8.13.2 Thermal Properties

Several thermal properties, including annealing, heat treatment, and coating, and the resultant performance of the manufactured glass depend on the composition and process parameters. The linear expansion of float glass is effectively measured as the coefficient of linear expansion; within a range of 20–300°C, this is $8–9 \times 10^{-6}$ m/m/k. According to this definition, if the temperature increases by 100°C, 1 meter glass will expand by 0.8–0.9 mm. Thermal conductivity (k) refers to the amount/speed of heat transmitted throughout a material. The reciprocal of thermal conductivity is called *thermal resistivity*. Materials of high thermal conductivity are used in heat sink applications, and materials of low thermal conductivity are used as thermal insulation. Some indicative values of k at 293 K are

TABLE 8.8

Light Transmittance and Reflectance Data for a Typical Glass Thickness of 5.66 mm

Characteristic Features	Clear	Green	Bronze	Gray
Light transmission	88	75	54	44
Total solar heat transmission	79	59	51	44
Reflection (outside–in)	8	7	6	5
Reflection (inside–out)	8	7	6	5

TABLE 8.9

Refractive Indexes of Some Regular Substances

Material	Refractive Index
Vacuum	1.0000
Air at STP	1.0003
Ice	1.31
Fused quartz	1.46
Clear flat glass	1.52
Diamond	2.42

aluminum: 237 W/m·K, copper: 401 W/m·K, and glass: 0.9 W/m·K. Usually, the thermal conductivity of materials is temperature dependent. The conductivity of soda-lime silicate float glass increases gradually with increasing temperature, and its indicative values are k = 0.90 W/m·K at 293 K and 1.3 W/m·K at 773 K. It is worth mentioning that different qualities of glass will possess different thermal shock resistance; for example, annealed, heat-strengthened, and fully tempered glasses are capable of resisting thermal shock at a temperature of 38, 121, and 204°C, respectively. Emissivity is defined as the ratio of the energy radiated from a material's surface to that radiated from a blackbody (a perfect emitter) at the same temperature and wavelength and under the same viewing conditions. The normal emissivity of float glass, $\varepsilon = 0.89$, means that 89% of absorbed heat is re-radiated [24].

The incidence of sunlight on a glass panel is a common phenomenon, involving reflection, transmission, and absorption, and thus, the following important parameters must be taken into account to control the solar radiation:

- Direct solar energy transmittance (ET): this defines the proportion of solar radiation at near normal incidence that is transmitted directly through the glass.

- Solar energy reflectance (ER): this defines the proportion of solar radiation at near normal incidence that is reflected by the glass back into the atmosphere.

- Solar energy absorptance (EA): this defines the proportion of solar radiation at near normal incidence that is absorbed by the glass.

- Total solar energy transmittance (TET): this is referred to as the 'g' value or solar factor, the fraction of solar radiation at near normal incidence that is transferred through the glazing by all means. It is composed of direct transmittance (short-wave component) in addition to the part of the absorptance dissipated inwards by long-wave radiation and convection (long-wave component). The proportions of the absorbed energy that are dissipated either inside or outside depend on the glazing configuration and the external exposure conditions.

- Selectivity index (S): this is the ratio of light to heat, i.e., LT/TET, where LT = Light Transmission

The solar radiant heat admission in glasses can be compared in terms of their shading coefficients, which can be illustrated as follows:

- The total shading coefficient (TSC) is derived by comparing the properties of any glass with a clear float glass having a total energy transmittance of 0.87 (such a glass would have a thickness of about 3 mm). It comprises a short-wavelength and a long-wavelength shading coefficient.

- The short-wavelength shading coefficient (SWSC) is direct energy transmittance divided by 0.87.

- The long-wavelength shading coefficient (LWSC) is the fraction of the absorptance released inwards, again divided by 0.87.

- Thermal insulation heat loss: this defines the thermal transmittance or U-value (U). The U-value, usually expressed in S.I. units of W/m²K, is the heat flux density through a given structure divided by the difference in environmental temperatures on either side of the structure in steady-state conditions. It is the rate of loss of heat per square meter, under steady state conditions, for 1 Kelvin (or degree Celsius) of temperature difference between the inner and outer environments separated by the glass or other building element.

The temperature has a noteworthy influence on the electric conductivity as well. For example, the conductivity of metals decreases with temperature, while the conductivity of semiconductors increases with temperature. However, a typical soda-lime silicate glass has an electrical conductivity of about 2×10^{14} Ω at normal temperature; in other words, it is a very bad conductor.

8.13.3 Mechanical Properties

The elasticity behavior of glass shows that it is a perfectly elastic material, as it cannot be permanently deformed until breakage. Young modulus is a coefficient of elasticity of a material and is expressed as the ratio between a stress that acts to change the length of a body and the fractional change in length caused by this force. For float glass, it is 72 GPa [25]. The more a material resists deformation, the higher is the value of the elastic modulus. Young modulus is dependent on the glass composition, which affects the final stress level of tempered glass. A small change in tempering parameters can change the elastic modulus of an identical composition of glass, resulting in a different degree of elastic modulus from manufacture to manufacture. Compressive strength is the resistance of a material to compressive stress. Glass is extremely resilient to pressure up to 700–900 MPa. Flat glass can withstand a compressive load 10 times greater than the tensile load.

8.13.4 Interaction with Water

In the presence of prolonged contact with water, scale and a rough surface are noticed, and this occurrence is attributed to corrosion or stain. This corrosion reaction mechanism (Equation 8.2) is initiated by a diffusion-controlled ion exchange process involving sodium ions in the glass and hydrogen ions from the water.

$$SiONa\,(glass) + H_2O\,(solution) \rightarrow SiOH\,(glass) + NaOH\,(solution) \tag{8.2}$$

If this corrosion is allowed to continue uninterrupted, pH levels gradually increase due to an accumulation of hydroxide ions (OH^-) in solution, promoting other leaching processes that further damage the surface. Once solution pH levels reach 9.0 or more, the second important reaction in the glass corrosion process begins, as described in Equation 8.3. At this point, the hydroxide ion concentration is sufficient to begin attacking the silicate network. The main reaction is the severing of silicon–oxygen bonds, and the glass itself is slowly dissolved:

$$Si\text{-}O\text{-}Si\,(glass) + OH^-\,(solution) \rightarrow Si\text{-}OH\,(glass)$$
$$+O\text{-}Si\,(dissolved\,glass; sodium, calcium\,silicates) \tag{8.3}$$

During the beginning stages of this reaction, microscopic pitting of the surface occurs. If the reaction is allowed to continue, the resulting surface damage to the glass may appear as widespread iridescence or a dense and translucent haze. Finally, the optical quality of the glass deteriorates, even though the overall mechanical strength of the glass is maintained. The removal of a few microns of the surface by grinding and polishing may restore the optical characteristics of corroded glass, but this remedy is neither practical nor economical, and sometimes, it is necessary to discard all heavily corroded glass.

8.14 Conclusions

Glass is an ancient material and a subject of interest from the research laboratory to civil construction. As well as the versatile use of float glass in daily life, advanced processing protocols facilitate its use, such as impact resistance, reflecting mirrors, insulating glass, etc. It is worth recalling that the glass fabrication environment and performance expectancy depend on several factors, including composition selection, control over viscosity at adequate temperature, processing temperature and atmosphere, working point, annealing point, and defect control. As well as float glass, recently, bottle glass has been in high demand for the beverage and food packing industries, and it is obvious that each and every manufacturer has its own processing protocol to make defect-free components. The market in medical practice highlights the enormous scope of bioglass to make different scaffolds, and it is obvious that for health reasons, this process demands more precise control for consistent composition and purity. The evaluation of critical properties provides the level of confidence in the product, and therefore, a cluster of properties, primarily optical, thermal, mechanical, electrical conductivity, and plausible interaction with water, have been discussed. The precise incorporation or introduction of crystalline phases results in what are conventionally known as *glass-ceramics*, which experience different degrees of properties compared with only amorphous glass. However, this concept and their manufacturing process have not been covered in this book.

References

1. Hufnagel, T. C., Schuh, C. A., and Falk, M. L. 2016. *Acta Materialia* 109:375–393. doi:10.1016/j. actamat.2016.01.049.
2. Johns Hopkins University, Material Science. and Engineering, Metallic Glasses https:// engineering.jhu.Edu/Materials/Research-Projects/Metallic-Glasses/#.WquUI1RuYdV.
3. Hill, A. J., and Tant, M. R. The structure and properties of glassy polymers: An overview. https://pubs.acs.org/doi/pdf/10.1021/bk-1998-0710.ch001.
4. Shelby, J. E. 2005. *Introduction to Glass Science and Technology*. RSC Paperbacks: London.
5. Lebedev, A. A. 1921. On polymorphism and annealing of glass. *Trans Opt Inst Petrograd* 2:10.
6. Zachariasen, W. H. 1932. The atomic arrangement in glass. *Journal of the American Chemical Society* 54(10):3841–3851.
7. Sun, K. H. 1947. Fundamental condition of glass formation. *Journal of the American Chemical Society* 30(9):277–281.
8. Stanforth, J. E. 1948. The ionic structure of glass. *Journal of the Society of Glass Technology* 32:366–372.

9. Rossi, M., Vidal, O., Wunder, B., and Renard, F. 2007. Influence of time, temperature, confining pressure and fluid content on the experimental compaction of spherical grains. *Tectonophysics* 441(1–4):47–65.

10. Glass for Europe, Flat glass value chain, http://www.glassforeurope.com/en/industry/index. php.

11. Industrial reports, https://www.mordorintelligence.com/industry-reports/global-glass-bottles-containers-market-industry.

12. Sarkar, D. 2019. *Nanostructured Ceramics: Characterization and Analysis*. CRC Press: Boca Raton, FL. ISBN: 9781138086807.

13. Glass Education Centre, http://glassed.vitroglazings.com/glasstopics/how:lowe_works.aspx.

14. The benefits of glass tempering and strengthening. *Swift Glass*. Sheila Reynolds on July 16, 2015. https://www.swiftglass.com/blog/the-benefits-of-glass-tempering-and-strengthening/.

15. Remodel connection, Window Experts, https://www.proreplacementwindows.com/window-glass-company/.

16. Knapp, B. J., Kimock, F. M., Petrmichl, R. H., and Daniels, B. K. 1994. Highly durable and abrasion-resistant dielectric coatings for lenses, US5846649A.

17. Curtis, H. D. 1911. Methods of silvering mirrors. *Publications of the Astronomical Society of the Pacific* 23(135):15–19.

18. Bioactive Synthetic Bone Graft, Novabone, http://www.novabone.com/perioglas.html.

19. Hench, L. L., Splinter, R. J., Allen, W. C., and Greenlee, T. K. 1971. Bonding mechanisms at the interface of ceramic prosthetic materials. *Journal of Biomedical Materials Research* 5(6):117–141.

20. Hench, L. L. 2013. *An Introduction to Bioceramics*. World Scientific.

21. Kokubo, T., Kushitani, H., Sakka, S., Kitsugi, T., and Yamamuro, T. 1990. Solutions able to reproduce *in vivo* surface-structure changes in bioactive glass-ceramic A-W3. *Journal of the American Ceramic Society* 24(6):721–734.

22. Pirayesh, H., and Nychka, J. A. 2013. Sol–gel synthesis of bioactive glass-ceramic 45S5 and its *in vitro* dissolution and mineralization behavior. *Journal of the American Ceramic Society* 96(5):1643–1650.

23. Hong, Z., Liu, A., Chen, L., Chen, X., and Jing, X. 2009. Preparation of bioactive glass ceramic nanoparticles by combination of sol–gel and coprecipitation method. *Journal of Non-Crystalline Solids* 355(6):368–372.

24. Bansal, N. P., and Doremus, R. H. 1986. *Handbook of Glass Properties*. Academic Press: Cambridge.

9

Miniaturization of Complex Ceramics

Partha Saha and Debasish Sarkar

CONTENTS

9.1 Introduction

Questions have been raised if ceramic components need to be miniaturized and if it is necessary when the turnover of major traditional ceramic industries is already worth billions of dollars. Both questions are answered in the affirmative. It is obvious that materials and prototypes need to be improved to be applicable to more sophisticated and portable devices. Prototypes are required to be included in a wide range of performance assessments in different environments, and such modules are illustrated in the next chapter. In this chapter, a number of high-tech ceramic processing techniques including additive manufacturing (AM), atomic layer deposition (ALD), microinjection molding, and precious micromachining to make tiny, complex-shaped components are discussed.

9.2 Additive Manufacturing

A daunting task for today's engineer is the evolution and fabrication of complex geometry ceramics for high-end structural, electronic/electrical, biomedical, and environmental applications. The demand for advanced engineering ceramics is primarily governed by the end-user and thus requires fabrication of 3-D complex shapes with an ordered porous body or in dense monolithic form for critical applications. Product innovations have always pushed the emergence of new technologies with improved productivity catapulting the use of advanced ceramic processing routes for engineered ceramics in major areas. There is no denying that if advanced ceramics need to find wide industrial use in the future, then the technology should be ready to fabricate the near-net shape components with complex geometry. Employing different machining techniques to achieve a high degree of complex geometry is a common phenomenon, but it is an expensive and time-consuming process. Thus, an alternative cost-effective manufacturing protocol targeting the complex geometry of ceramics is required. Also, the limitations of technologies to produce complex 3-D parts with the desired microstructures and properties has led to the emergence of novel methods such as additive manufacturing (AM), solid free forming (SFF), rapid prototyping (RP), etc., which are becoming increasingly favorable techniques where 3-D objects can be assembled by a point-by-point, layer-by-layer operation. However, consistency in the preparation of AM input material is a critical and challenging task as well [1]. The predefined contours are stacked upon each other and a 3-D solid model is developed by curing, binding, and sintering mechanisms. Figure 9.1 provides a brief classification of the AM process in order to fabricate the complex geometry of ceramics with consideration of the critical dimension, probability of aggregation, mode of layer formation, and commercial availability.

9.2.1 Brief Sketch of AM Techniques

Addition is the reverse phenomenon of subtraction. Thus, AM encompasses layer-upon-layer addition of precursors to complete a predefined 3-D computerized design (computer-aided design [CAD] model, STL file) without the use of additional fixtures and cutting tools, in opposition to the common subtractive manufacturing protocol known as machining. AM technology started to develop in the 1980s. It is an innovative approach that is capable of making any complicated shape, including holes, grooves, threads, and helical

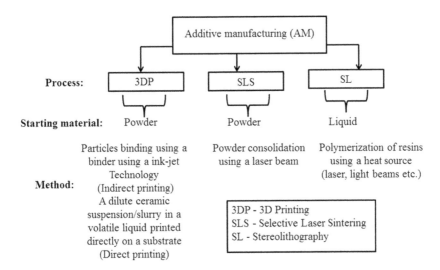

FIGURE 9.1
Major classification of additive manufacturing in view of process protocols [2].

shapes, in a very short time, and this protocol has been coined "rapid manufacturing" or rapid prototyping (RP). Machining, stamping, and other subtractive processes create waste; however, there is minimal waste of raw material when manufacturing complex miniaturized ceramics using the AM process [2]. This new flexible approach overcomes traditional fabrication protocol constraints, allowing effective and optimal design for lean, on-demand, tailor-made production without waste that will eventually revolutionize the manufacturing industry, providing tremendous benefits to society at large. In discussing AM, it is worth mentioning that it is capable of producing functional intricate parts made of a wide range of materials including polymers, metals, ceramics, and their combinations in various textures such as composites, hybrids, and functionally graded materials (FGMs). First-generation AM was designed for widely used polymers and was coined "rapid prototyping"; however, an extensive range of materials in the nanoscale dimension have been developed for the aerospace and pharmaceutical industries, and with the increasing awareness of this process, several complicated 3-D shapes have been made [3, 4]. The International Organization for Standardization (ISO) as well as the American Society for Testing and Materials (ASTM) 52900:2015 standard classify AM processes into seven categories: (1) binder jetting (BJ); (2) directed energy deposition (DED); (3) material extrusion (ME); (4) material jetting (MJ); (5) powder bed fusion (PBF); (6) sheet lamination (SL); and (7) vat photopolymerization (VP). Thus far, the various technologies developed to fabricate 3-D parts include 3-D printing (3DP), selective laser sintering (SLS), stereolithography (SLA), fused deposition modeling (FDM), laminated objective manufacturing (LOM), laser metal deposition (LMD), and laser engineered net shaping (LENS) [5–7]. In this section, 3DP, SLS, and SLA are considered to explore the basic advantages of AM techniques; however, concise text of different processing protocols are tabulated in Table 9.1.

9.2.2 Three-Dimensional Printing

The process of joining and solidifying materials to form a 3-D object, using a computer to control the process, is called 3DP, and it may include the fusing together of liquid molecules or powder grains. Basic use of 3DP is found in AM and RP. Components consist of

TABLE 9.1

Summary of Techniques, Principles, Materials, Advantages, Disadvantages, and Applications of Various Additive Manufacturing [22,23,24]

Categories	Techniques	Principles	Materials	Advantages	Disadvantages	Applications	Manufacturer/Country
BJ	3D Inkjet Technology	Liquid binder(s) jet printed onto thin layers of powder. The part is built up layer by layer by gluing the particles together	Polymers Ceramics Composites Metals Hybrid	Ability to print large structures Quick printing Design freedom Relatively low cost	Fragile parts with limited mechanical properties May require post processing	Biomedical Functional parts	ExOne, USA PolyPico, Ireland
DED	LENS	Focused thermal energy melts materials during deposition	Metals Hybrids	Reduced manufacturing time and cost Excellent mechanical properties Controlled microstructure Accurate composition control Excellent for repair	Low accuracy Low surface quality Need for a dense support structure Limitation in printing complex shapes with fine detail Limited to metals/ metal-based hybrids	Aerospace Retrofitting Repair Cladding Biomedical	Optomec, InssTek, Sciaky, USA Irepa Laser, France Trumpf, Germany
ME	FDM	Material is selectively pushed out through a nozzle or orifice	Polymers Composites	Widespread use Inexpensive Scalable Can build fully functional parts	Weak mechanical properties Limited materials (only thermoplastics) Layer-by-layer finish	Rapid prototyping of advanced composite parts	Stratasys, USA
MJ	3DP	Droplets of build materials are deposited	Polymers Ceramics Composites Hybrid Biologicals	Fine resolution High quality Low waste Multiple material parts Multicolor	Slow printing High porosity in the binder method Support material is often required	Biomedical Large structures Buildings Prototyping	Stratasys, 3D Systems, 3Dinks, USA PolyPico, Ireland WASP, Italy

(Continued)

TABLE 9.1 (CONTINUED)

Summary of Techniques, Principles, Materials, Advantages, Disadvantages, and Applications of Various Additive Manufacturing [22,23,24]

Categories	Techniques	Principles	Materials	Advantages	Disadvantages	Applications	Manufacturer/Country
PBF	SLS	Thermal energy fuses a small region of the powder bed of the build material	Metals Ceramics Polymers Composites Hybrids	Relatively inexpensive Powder bed acts as an integrated support structure Large range of material options Fully dense parts	Relatively slow Lack of structural integrity Size limitations High power required Finish depends on precursor powder size	Biomedical Electronics Aerospace Lightweight structures Metal/ceramics preforms Heat exchangers	ARCAM, Sweden EOS, Concept Laser Cusing, MTT, Realizer, Germany Phoenix System Group, France Renishaw, UK Matsuura, Japan Voxeljet, 3Dsystems, USA
SL	LOM	Sheets/foils of materials are bonded	Polymers Metals Ceramics Liquid materials Inks	High surface finish Low materials, machine, process cost Reduced tooling A vast range of materials	Inferior surface quality and dimensional accuracy Limitation in manufacturing of complex shapes	Metallic sheet Ceramic tape Electronics Smart structures	3D Systems, USA MCor, Ireland
VP	SLA	Liquid polymer in a vat is light-cured	Polymers Ceramics	Large parts Excellent accuracy Excellent surface finish and detail	Very limited materials Slow printing Expensive Limited to photopolymers only Low shelf life, poor mechanical properties of photopolymers Expensive precursors/slow build process	Biomedical Prototyping	Lithoz, Austria 3D Ceram, France

BJ, binder jetting; DED, directed energy deposition; ME, material extrusion; MJ, material jetting; PBF, powder bed fusion; SL, sheet lamination; VP, vat photopolymerization.

different shapes, geometries and critical hidden features are produced in extensive horizon in consideration of digital data such as CAD and their systematic conversion to a 3-D model or by another electronic data source such as an AM file (AMF) in case of requirement of sequential layering.

Commercially available 3DP technology allows either direct part printing or binder-assisted printing (indirect) analogous to the inkjet printer heads layer by layer. While direct printing is the process whereby the print head dispenses all parts of the component, binder printing refers to the spectrum of processes whereby additives and binders are printed onto a powder bed to form the bulk of the body. To make a ceramic component by direct printing, maintaining tailor-made properties including starting particle size, shape and distribution, choice of binders, viscosity, suspension stability, and more is a challenging task. However, direct printing is a popular choice to fabricate completed shape of several class of polymers. Binder-assisted or indirect printing has many limitations including the accuracy and the mechanical strength of the desired microstructure, particularly for large components compared to subtracting machining. However, continuous improvement in both the direct and indirect 3DP of ceramic components across research laboratories and industries is ongoing to fulfill the targeted object geometry—strength and cost-effectiveness. Figure 9.2 provides a schematic of binder printing. The systematic binder addition on a powder bed facilitates printing and results in a 3-D object. Herein, a very small portion of binder is being delivered results immediate powder – binder spherical agglomeration and binder leads bonding to the existence matrix or previously printed. After printing one layer, the bed of powder is lowered and another layer of powder is spread on to it, a mechanism that is referred to as "counter-rotating rolling." The counter-rotating rolling mechanism is very similar to the recoating methods of the PBF processes. The addition of printing binder to the bed and recoating the bed with another layer of powder is repeated several times until a group of parts is produced. Herein, 3DP patterning in a one-dimensional avenue is synchronized through adjusting the binder ejection from attached multiple nozzles in the printer head. The process can be economically scaled by increasing or decreasing the number of nozzles, hence the method is measured as a linewise patterning process.

This rapid printing technique and its low energy consumption enable low-cost production. Once the printing is completed, the printed part usually remains on the powder bed so that the binder can fully set and the green body achieve appreciable strength for

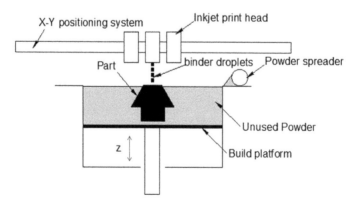

FIGURE 9.2
Schematic of indirect binder-assisted 3D printing process.

further handling. A post-printing process comprises removing the body from the bed and eliminating unbound powder using pressurized air to give it good strength and other mechanical properties. In brief, the 3DP process has many advantages similar to powder bed processes [8].

1. Parts have the ability to stand on their own, called "self-supporting nature," thereby eliminating the requirements for support structures.

2. Parts can be arrayed in one layer or stacked in multiple layers in the bed in order to multiply or increase the number of parts that can be built at once.

3. The fabrication of assemblies of parts and kinematic joints is possible because loose powder can be easily removed between parts.

Figure 9.3 illustrates direct and indirect 3-D printed alumina and carbon-based components, respectively. In the former case, the alumina suspension with a solid loading up to 40% was prepared and printed from a micro-dimensional nozzle (60–80 μm) to fabricate a green impeller.

In this process, the deposited materials were spread on the substrate and merged with neighboring drops to form independent, large-shaped, low density green ceramics. Extensive shrinkage and near to 80% density were achieved, as shown in Figure 9.3a [9]. A commercial-grade carbon component made through indirect printing is illustrated in Figure 9.3b. It is worth mentioning that sometimes a printed, complex-shaped component experiences low strength, so post-processing methods such as silicon or metal infiltration and polymer impregnation followed by restricted sintering assist to produce carbon–ceramic or carbon–graphite composites [10].

9.2.3 Selective Laser Sintering

Selective laser sintering (SLS) is an excellent process for making in situ complicated shapes with a higher density compared to 3DP. In addition to ceramics [7, 11], SLS is at the forefront in the manufacture of metal [12] and polymer [13] components for a wide range of applications. Selective laser melting (SLM) is solely used for certain metals such as steel and aluminum. Laser scanning in SLS leads to sintering but does not fully melt the powders; rather it elevates the localized temperature on the surface of the particles resulting in

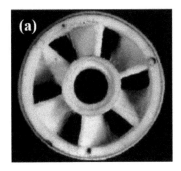

FIGURE 9.3
(a) Sintered alumina impeller made through direct 3DP green component followed by sintering [9] and (b) binder jet printed component made of a carbon–ceramic composite [10].

fusion between the powders. On the other hand, the powders fully melt and fuse together during SLM, resulting in superior mechanical properties.

The major advantage of the SLS technique is that it can process powders directly without using a sacrificial binder, unlike other solid free forming (SFF) methods that demand special materials such as photopolymers in stereolithography (SLA. A thin layer of powder is spread out and levelled over the top surface of a building platform, as shown Figure 9.4. Continuous feeding of the powder below the laser scanner expedites selective powder layer scanning to fuse the particles together and form a geometrical shape predefined by a CAD model, leaving the adjacent non-fused particles as supports for subsequent building. No support structure is needed as the non-fused powder remains in place to act as a support itself. The laser energy also helps to fuse the consecutive layers together. For specialty particles that are prone to oxidation, the building chamber is backfilled with an inert atmosphere. After depositing the first layer, an elevator platform is used to lower the parts for subsequent layer deposition. After successfully building a 3-D shape, the loose powder is separated and further recycled. Powder size distribution and packing, which determine the density of the printed 3-D part, are the most crucial factors during the SLS process. 3-D parts fabricated by SLS are porous in nature and post-processing is often necessary to impart strength to the body of ceramics [7, 14]. However, during indirect SLS, a low-temperature binder is used to assemble the powder and subsequent pyrolysis and sintering steps are needed. This can be an organic polymer coating or a low-melting inorganic salt. The indirect SLS technique is mainly used for high melting point ceramics that require a high level of laser energy, which can adversely create thermal gradients and difficulty in forming a contiguous shape. The low-melting phase binder helps to overcome the problem when melted with the ceramic powders. This second phase can be an organic binder coated on the powder, which is an inorganic binder [2]. The SLS technique is widely used in various industries for advanced applications, such as scaffolds for tissue engineering, and the aerospace and electronics fields. Starting with a 3-D CAD model, Figure 9.4 presents classic tissue engineering prototypes made of hydroxyapatite, Ti6Al4V, and biodegradable poly-epsilon-caprolactone (PCL). The powder bed is used as support

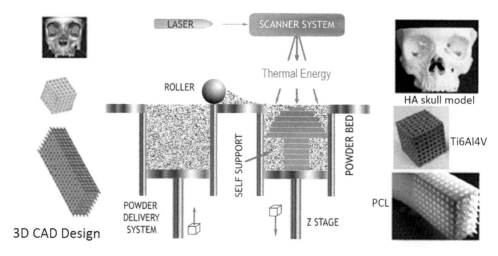

FIGURE 9.4
Schematic of SLS from a 3-D CAD design to the laser sintering process. Both the 3-D CAD model and SLS prototype are made of hydroxyapatite skull [11], synchronized porous Ti6Al4V [12], and biodegradable poly-epsilon-caprolactone (PCL) [13].

and thus the troublesome removal of the SLS component from the base material can be avoided; however, PBF is a slow process and promotes a porous structure when the powder is fused with a binder that eventually demands post processing, resulting in a high cost component.

9.2.4 Stereolithography

Ceramic SLA or photopolymerization is a 3DP process composed of ceramic particles including ultraviolet (UV) curable resin and a photoinitiator, where an ultraviolet laser beam is used to initiate a chain reaction of resin or monomer that instantly converts to polymer chains after radicalization and creates a sequential layer of a solidified object consisting of ceramic particles and organics, analogous to a CAD model [6, 15]. Owing to the ease of this process, complex-shaped ceramic precision objects fabricated by SLA can be expected to be used in both structural and functional applications. A typical schematic of SLA used to make complex threaded and porous alumina shapes is shown in Figure 9.5 [16].

Initially, a mixture of methylenebisacrylamide (MBAM), acrylamide (AM), and glycerol organic mixture is prepared, and then an alumina powder (d_{50} = 180 nm) is mixed and deagglomerated using a 0.3 wt.% sodium polyacrylate (dry weight basis of alumina powder) dispersant by high-energy milling. In order to prepare for SLA, a 2 wt.% photoinitiator of the monomer solution is added and the UV-curable alumina (Al_2O_3) slurry is dispersed into the reservoir (or vat) and spread out in a thin layer using a wiper. It is worth mentioning that a rheological study of slurry is essential prior to initiating the fabrication of ceramics using the SLA process. Subsequently, a 3-D CAD model is converted to a 2-D sliced image and a controlled UV light is selectively scanned on the slurry surface based on the 2-D image. The same phenomenology follows for each 2-D section, producing layer by layer through photopolymerization a photocurable resin in the slurry that eventually makes a 2-D network of ceramic particles embedded in a cross-lined polymer. Once a 2-D single layer is formed, the platform moves upward to separate from the slurry; the same ceramic slurry is recoated in the reservoir and a new layer is cured as previously outlined, finally developing a 3-D prototype. Post-processing organic removal and sintering are done to avoid delamination and cracking and to obtain a highly dense component.

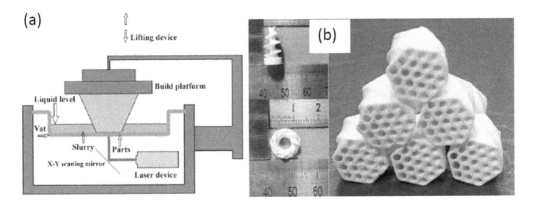

FIGURE 9.5
(a) Schematic representation of SLA equipment and (b) complex Al_2O_3 green parts fabricated by rapid prototyping techniques [16].

The unreacted slurry is further removed after completion of the process. Sometimes, a post-process treatment such as heating or photocuring may be used on parts to achieve the desired mechanical integrity. Interestingly, the 3-D printed alumina part exhibits excellent mechanical properties compared to tape casting, gel casting, and slip casting. SLA can fabricate high-quality parts at a fine resolution as low as 10 μm [17]. On the other hand, it is relatively slow and expensive and the range of materials for printing is very limited. Also, the reaction kinetics and the curing process are complex phenomena. The energy of the light source and exposure are the main factors controlling the thickness of each layer. Using SLA, the building speed is relatively high (1–3 cm per hour) and the minimum layer thickness is dependent on the curing depth. The downsides of SLA are errors due to overcuring and the exorbitant costs of the materials. However, SLA can be effectively used for the AM of complex nanocomposites and functionally graded ceramic composites. Several researchers have attempted to make ceramic components using a suspension of silica, zirconia, and alumina particles in UV-curable resin following a similar protocol to SLA [18–21].

9.2.5 Competitive AM for Metals and Ceramics

The AM of metals facilitates the fabrication of 3-D complex parts made with titanium and its alloys, nickel-based alloys, magnesium alloys, and stainless steel, which are important in the aerospace, automotive, energy, and biomedical industries. The AM of metals is a rapidly growing field where complex parts that cannot be fabricated by traditional methods are manufactured with significant quality improvements and reduced manufacturing times. It has the potential for mass production in combination with conventional manufacturing processes. Typically, the 3DP process of metals consists of melting metallic feedstock (generally powders) using a laser source or an electron beam. Upon solidification, the melted parts are transformed into a complex 3-D geometry. Well-known techniques for printing metals are PBF and direct energy deposition (DED). These processes can achieve higher accuracy and speed. In terms of PBF, their widespread use for metals derives from their ease of weldability of various metal powders. During SLS/SLM processes, the powders are melted using a laser source and completely fused, resulting in a theoretically dense component. The energy for fusing the powder is provided by two different energy sources, laser or electron beam; however, the latter is not applicable to ceramics, as most ceramic materials do not possess sufficient electron conductivity. The success of AM of metals is also based on existing knowledge concerning the handling of metal powders and laser material interactions for the main alloy systems used in AM. The scientific knowledge base on laser–metal interactions, such as laser cladding, is well established. Additionally, the commercial availability of high-quality metal powders in the appropriate size ranges renders the AM of metals a feasible process. With the wide knowledge of powder metallurgy and the laser cladding phenomena, many vendors are providing high-quality metal powders in various sizes at competitive prices. This wide availability of feedstocks makes the powder-based AM for specific metal/alloy systems more economical and feasible than ceramics, which are not as widely used [22].

In the current market, the availability of ceramic-based AM components with adequate mechanical properties is limited, although there is extensive market demand for complex ceramics made by AM. Despite the large number of AM technologies suitable for processing ceramics, the main hurdle for the AM of ceramics lies in the coupling of a specific AM technology with the formulation of a suitable feed that would enable the fabrication of dense ceramic parts with optimal properties. In addition, the post-processing of sintered ceramic parts for binder burnout and forming the desired shape is a time-consuming and costly process. However, 3DP of porous ceramics offers numerous benefits for developing structures that are tailored to different applications. Ceramic scaffolds used in tissue engineering have become a more convenient and faster process compared to traditional methods of casting and sintering [23]. The main methods of 3DP ceramics are inkjet printing, PBF, and SLA. Inkjet is appropriate for making dense ceramics that may not need posttreatment [2]. The SLS of powder is also a common method; however, thermal shock may lead to the generation of cracks in ceramic parts. For ceramic powders that do not easily fuse or melt, a low-melting binder is generally used. The binder melts during laser heating and temporarily holds the ceramic powders in the desired shape, after the removal of excess powder; the green body is sintered at an elevated temperature. Direct ceramic SLA uses a photocuring binder along with ceramic fillers that, upon curing, achieve the desired shape of green body. To improve the light curing of the binder, smaller ceramic particles with better light scattering properties are preferred in this method [2, 24]. A competitive study of different techniques, principles, type of materials, advantages, disadvantages, applications, and probable manufacturers is shown in Table 9.1.

9.3 Atomic Layer Deposition

ALD has recently received much interest as a potential deposition method for advanced thin film structures. Originally, it was known as atomic layer epitaxy (ALE), and introduced to produce extremely high-quality, large-area flat panels composed of thin film electroluminescence (TFEL) displays [25]. Today, this advanced technology is also extensively used in the industrial production of TFEL displays. By definition, ALD is a chemical vapor deposition (CVD) technique that is suitable for manufacturing inorganic material layers with a thickness in the nanometer range. ALD is based on a series of surface-controlled reactions between solid and gaseous species to produce thin film materials and overlayers with perfect conformity. It has emerged as a versatile technique to produce extremely complex, conformal, atomic layers of high-quality thin film for the microelectronics industry. In addition, ALD-derived materials have a wide range of functional applications, from catalysts to electroluminescent displays and beyond. This technique has advantages over other thin film deposition protocols (see Chapter 5) because it can produce pinhole-free uniform film in the nanometer range based on the sequential use of self-terminating gas–solid reactions.

9.3.1 Principle of ALD and Growth Characteristics

ALD is a modified version of the CVD technique with distinctly different film growth occurring in a cyclic manner. The ALD technique is based on a sequence of self-terminating gas–solid reactions, as schematically shown in Figure 9.6. Thin film formation during ALD consists of repeating the following steps [26]:

Step 1: Self-terminating reaction of the first reactant or a chemisorption reaction of a metal reactant (Reactant A).

Step 2: A purge or evacuation to remove the unreacted reactants and the gaseous species formed as byproducts.

Step 3: Self-terminating reaction, i.e., chemisorption reaction, of the second reactant typically a non-metal reactant (Reactant B).

Step 4: Further purge or evacuation.

Steps 1–4 constitute a reaction cycle that is sometimes referred to as half reactions of an ALD reaction cycle. During each reaction cycle, a material is added to the surface, referred to as the growth per cycle (GPC). To grow a material layer, the growth cycles are repeated until the required thickness of the film is formed. Depending on the process parameter(s) and the reactants used, the cycle time can vary from half a second to a few seconds. The precursor molecules chemisorb or react with the surface groups of a preheated substrate,

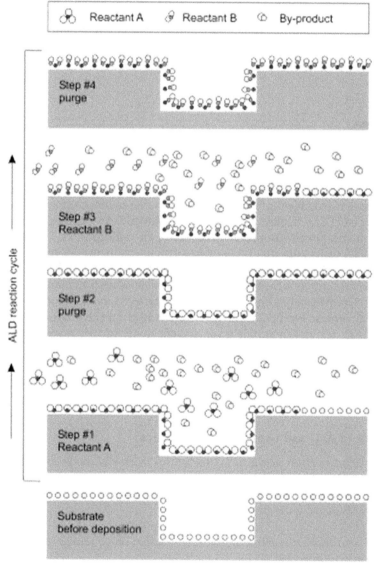

FIGURE 9.6
Schematic illustration of the ALD reaction cycle [26].

and after the formation of the chemisorbed layer, no further adsorption takes place. The use of self-terminating reactions leads to the conclusion that ALD is a surface-controlled process, where process parameters other than the reactant, substrate, flow rate, and temperature have little or no effect. Owing to the surface-controlled phenomenon, ALD-grown films are extremely conformal in nature and uniform in thickness. Moreover, the ALD processing window is often wide, enabling the thin film growth phenomena unperturbed by a small change in the process parameter(s), and allowing the processing of varied reactants to form multilayer structures in a continuous process (see Figure 9.7) [27].

The most essential requirements for the ALD processes are the irreversible and saturating reactions desired for a uniform thin film on large-area substrates. Figure 9.8a–e compares the saturating and irreversible gas–solid reactions (often called "self-terminating" or "self-limiting") with various types of adsorption-based reactions. It is clear that only irreversible and saturating reactions lead to the growth of a conformal and uniform coating on the substrates, irrespective of the amount of reactants available, the exposure, and the purge times. At the same time, the reactant partial pressure does not influence the amount of material adsorbed in the saturating and irreversible reaction. This automatic control of the amount of material deposited is a key feature of ALD [26, 28].

9.3.2 III–IV Semiconductor Film Growth by ALD Process

Over the last decades, ALD (or ALE) of III–V semiconductors including AlAs, GaAs, GaAlAs, GaP, InP, and InAs has been extensively studied and synthesized from traditional precursors, such as trimethylaluminum (TMA), *trimethylgallium* (TMG), *trimethylindium*

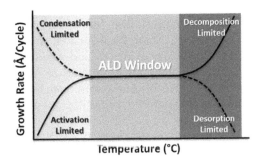

FIGURE 9.7
Schematic diagram showing the possible ALD growth behavior per cycle versus temperature and ALD window [27].

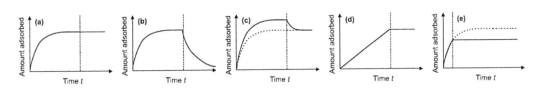

FIGURE 9.8
Schematic illustration and examples of how the amount of material adsorbed can vary with time: (a) irreversible saturating adsorption (preferred for ALD), (b) reversible saturating adsorption, (c) combined irreversible and reversible saturating adsorption, (d) irreversible non-saturating adsorption (i.e., deposition), and (e) irreversible saturating adsorption not allowed to saturate. The vertical dashed line marks the end of the reactant supply and the beginning of a purge or evacuation [26].

(TMI), PH$_3$, and AsH$_3$ [26]. The utility of GaAs in the ALD technique as well as that of other III–V compounds has been limited due to the relatively narrow operating temperature window, low growth rate, and high levels of carbon impurity. Several groups have studied the monolayer growth mechanisms and three different mechanisms have been outlined elsewhere [29]. The metalorganic vapor-phase epitaxy (MOCVD) technique provides a guideline to selecting the precursors for III-IV ALD, for example, trimethyl alkyl compounds and chlorides are successfully used precursors for ALD [30]. The ALD of other compounds (AlAs, GaP, InP, InAs) is very similar to that found for GaAs with a variation in the process temperature. Moreover, the polycrystalline III–V films are of growing interest, outclassing the need to deposit III–V compounds as epitaxial films on any substrate. In addition, substrates parameter(s) do play an important role in the growth, structure, and morphology of the III–V film. This is a viable coating technique on a 3-D substrate; therefore, a recent study has revealed the possibility of growing GaAs quantum wires embedded in a GaAs$_{1-x}$P$_x$ layer on a V-shaped, grooved GaAs through a selective isotropic/anisotropic deposition method on the surface. However, the group III-IV compound synthesis by ALD is not a commercially popular protocol because of inadequate precursor resources.

9.3.3 Oxide films

Metal oxides are difficult to deposit using CVD techniques at moderate temperatures. ALD allows the separation of reactants until the reaction occurs on the substrate surface, thus highly reactive species can be used at lower temperatures. However, the precursors, the substrate morphology, and the temperature determine the beginning of the growth of the film. Depending on the surface energy of the substrate or the already grown film, the film can grow as amorphous or in crystalline form. The crystallization temperature of Al$_2$O$_3$ is high; therefore, Al$_2$O$_3$ grows in amorphous form during ALD at low temperatures. There is a growing demand to synthesize oxides for industrial applications, such as diffusion barriers on polymer substrates using low-temperature ALD processes. Many oxides can be deposited by the thermal ALD technique at moderate temperatures. The precursors used for oxide ALD must have high reactivity, volatility, and good thermal stability over a wide temperature range.

Oxides form the most studied material group in ALD and detailed processes exist for Al$_2$O$_3$, TiO$_2$, and ZnO because of their highly reactive precursors that interact easily with water and oxygen. These favorable reaction kinetics render the low temperature, thin film deposition of oxides amenable using the ALD technique. Owing to the need for high dielectric constant oxide films in the electronics industry, the ALD of ZrO$_2$ and HfO$_2$ has been extensively studied and numerous precursors and processes have been developed. The ALD of niobium, tantalum, transition metal, and rare earth oxides has also been documented [26]. Among alkaline earth oxides, MgO is a widely studied material, and is prepared from commonly considered cyclopentadienyl compounds and oxygen as precursors [31]. It was observed that CaO, BaO, and SrO grow easily on a TiO$_2$ surface as SrTiO$_3$ and BaTiO$_3$ [32]. However, following the use of cyclopentadienyls and water as precursors, limitations and the problem of the formation of Sr(OH)$_2$ and Ba(OH)$_2$ as a passivating layer were witnessed. SiO$_2$ is one of the important oxides used in microelectronics. A major advancement in SiO$_2$ ALD was made with the discovery of a catalytic process where the reaction between tris(*tert*-butoxy) silanol and water was catalyzed by an Al$_2$O$_3$ layer [33]. The growth rate in this process is exceptionally high, too high for very thin films. Al_2O_3 ALD is performed using TM and H_2O, and repeating the surface reaction leads to a 300 nm ultrathin coating with an extremely linear growth rate of 1.1–1.2 Å on an Si wafer [34]. It was observed that alkylamino silanes even react with H_2O_2 and form relatively pure SiO_2 with little carbon and nitrogen residue. Lately, the ALD of tin oxide films has attracted much interest and precursors such as chlorides, iodides to alkyl, alkyl amine, and alkoxide compounds have been used [26]. Antimony is used as a dopant in SnO_2. Also, a working process for SnO_2:Sb synthesis is based on metal chlorides and water [35]. Fluorine is also used as a common dopant for conducting tin oxides.

9.3.4 Non-Oxide Films: Nitrides and Other Compounds

Metal nitrides have been widely used in the electronics industry for many years. Silicon nitride is known as an excellent electrical insulator. Similarly, titanium nitride also finds application as an excellent diffusion barrier for oxygen, and an electrical conductor. Owing to the high conductivity of molybdenum nitride, it has found application in microelectronics. Therefore, metal nitrides are useful materials for various niche applications, and the synthesis of thin films of metal nitrides using the ALD technique has been practiced for decades. Transition metal nitride films have been successfully fabricated by ALD using metal chlorides and ammonia gas at the preferred processing temperature of 500°C [36]. TiN has been successfully fabricated without any impurity phase at 500°C because titanium does not form any other nitrides. However, of the metal nitrides that are formed at higher oxidation states, ammonia is not suitable, that is, non-conducting Ta_3N_5 is formed instead of TaN [37]. Using ALD, NbN and MoN films were successfully prepared by the reaction between metal chloride and ammonia; however, non-stoichiometric nitride compounds also formed [38]. Non-metal nitrides such as Si_3N_4 and BN are challenging for the fabrication trough in ALD owing to the fact that covalent bonds present in the silicon precursors are immune to ammonia. However, $SiCl_4$ reacts with ammonia at ~425°C producing Si_3N_4 with a high growth rate per cycle [39]. It was found that hydrazine is a better choice as a nitrogen precursor than ammonia. The challenge in Si_3N_4 ALD is still the need for a fast thermal process, preferably employing ammonia as the nitrogen source. The silicon precursors so far examined have been limited to silane, dichlorosilane, and tetrachlorosilane [26]. BN depositions have been studied using reactions between boron chloride/bromide with ammonia. Films deposited at 750°C were crystalline and purer than the amorphous films made at 400°C [40]. BN films deposited at low temperatures contain significant amounts of hydrogen. Although BN ALD looks rather straightforward, there is still a need for process development to obtain pure crystalline film at reasonable temperatures.

The synthesis of metal chalcogenides has a long history since the inception of the ALD technique. The pioneering work of Suntola on ZnS deposition using the ALD technique was the first published scientific work and it gained much attention [41]. Elemental Zn and S were used as precursors that were evaporated and deposited on a rotating substrate. However, later precursors such as $ZnCl_2$ and H_2S proved useful for ZnS growth. The synthesis of ZnS-based visible region luminescent material doped with rare earth metals and alkaline earth metals using ALD has also been reported and is well documented in the literature [42]. The deposition of other metal sulfides has also been carried out for use as buffer layers in photovoltaic cells, particularly CdS [43] and InS [44].

9.4 Microinjection Molding

"Micro" implies a very low dimension compared to a centimeter dimension. Prior to highlighting this topic, a brief discussion of common injection molding is necessary. Powder injection molding is a unique fabrication protocol covering device dimensions from a few centimeters to the micron scale. Thus, we essentially need a prerequisite quality of powder or a binder-added feed to flow through a screw mechanism and obtain the desired shape, which is followed by debinding and sintering as required. In the case of a polymer, additional binder can be avoided, but single or multiple binder systems are mandatory for metal and ceramics. This technology is one of the prime techniques to produce a net-shaped and high volume complex geometry of polymer, metal and ceramic parts, which have small dimensions as well. Owing to the processing features, this effective and economic technology has different advantages including improved mechanical response, corrosion resistance, reduced surface roughness, and quality finishing of complicated shapes and tiny parts. Hence, this processing technology, particularly in the case of micro-component fabrication, is in high demand in industries such as medical and healthcare, consumer electronics, aerospace, and optical fiber communication. To comprehend the manufacturing protocol, this section is divided into four major parts: working principle and complete fluxogram; important processing factors and difficulties; conventional components; and microcomponents for both metal and ceramic systems.

9.4.1 Design and Principle of Injection Molding

Prior to highlighting injection molding, a brief discussion on the origin and different types of molding protocol will provide a better insight on the subject. Molding is a processing technique to manufacture a rigid-shaped component or structure using liquid, pliable, or solid material. A mold is a hollow cavity that consists of two halves depending on the complexity of the output. During molding, the flowable mass hardens after cooling and adopts the required shape followed by release through dissembling the movable mold parts. This method is cost-effective, consumes less time, and enables complex geometries to be drawn. Rapid and accurate geometry in the manufacturing of household to aircraft components are the major advantages of this process. Conventionally, different classes of molding processes are available, such as injection molding, extrusion molding, blow molding, matrix molding, compressive molding, spin-casting, centrifuge casting, transfer molding, powder metallurgy, and sintering. Among these, injection molding is an effective process to fabricate a wide range of components imparting flexibility in materials such as polymers, metals, ceramics, and their composites as well. The machine is primarily divided into three units, namely, an injection unit, a mold, and a clamping unit, as illustrated in Figure 9.9.

The injection unit contains a moving plunger or injector that is reciprocally screwed inside a barrel covered by a heating coil. At the end of the barrel, a hopper is connected through which the feeds are poured. The feed may be granules (e.g., polymer) or binder-added powder (e.g., metal or ceramics). A hopper opens inside the barrel to melt or maintain the flowability of the feed, which is injected into the mold by a hydraulically powered high-pressure plunger or injector. The material moves axially with the grooves of the injector and is finally injected into the molds through a nozzle to obtain the desired shape. The molding unit consists of molds that split equally into two halves and are made of steel or aluminum. Of the two halves, one half is movable while the other half is attached to the

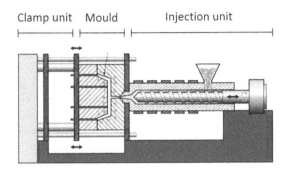

FIGURE 9.9
A typical injection molding consisting of an injection unit, a particular mold of interest, and a clamping unit.

machine that eventually forms the definite geometry of the component during injection of the material. The clamping unit helps to secure the closure of the two halves of the mold. The front half of the mold, called the "mold cavity," is mounted onto a stationary platen and aligned with the feeding nozzle, while the movable rear half of the mold is clamped with a movable platen that passages toward the stationary platen to keep the mold closed and secured. After adequate insertion of the material of interest, it is subjected to cooling down and careful ejection from the mold is carried out. The polymer component requires no further processing steps apart from surface finishing as required, whereas several steps are mandated for making metal and ceramic parts, as discussed in next section.

9.4.2 Powder Injection Molding

The market for powder injection molding is forecasted to increase up to $2.6 billion due to increasing demand for high-performance materials and the miniaturization of components for different sectors [45]. Thus, powder injection molding is one of the most widely used, highly efficient molding techniques for producing shaped components from metal or ceramic powders, where polymer cannot withstand load or temperature. This process overcomes the lacuna to make complex miniature components in comparison to other conventional processes such as production limits in isostatic pressing, premature defects in slip casting, tolerance limits in investment casting, shape and geometry limits in traditional powder compacts, and the mechanical strength of die-cast products. The components are prepared in the as analogous to the polymer injection technique, first mixed with a metal or ceramic particles with polymer binder, followed by the removal of polymer and the sintering process. Hence, the steps involved in powder injection molding are polymer–powder mixing, molding, removal of binder, and densification, depicted in Figure 9.10. This classic fabrication protocol allows the complex geometry and integrated system of different multiple parts to be made [46].

The sequential processing steps in powder injection molding to fabricate metal or ceramic components are [47]

> **Step 1:** Mixing and granulation in feedstock preparation is the first step. Mixing and granulation involve the mixing of metal or ceramics and an organic multicomponent binder. The material is mixed in an extruder or grinder followed by the formation of granules or pellets using a nodulizer or pelletization machine for moving the feedstock into the molding machine. The binder works as a medium

for the powder and is later removed during the debinding process. The binders used for binding the powder are ethylene vinyl acetate (EVA), polyethylene (PE), polypropylene (PP), polystyrene (PS), polyethylene glycol (PEG), polymethyl methacrylate (PMMA), and others.

Step 2: Injection molding is the second step. This technique is similar to the conventional molding technique except that the hardware of the molding machine accords with the specificity and type of feedstock material depending on its compressibility and viscosity. Control of the molding process is very important for maintaining high tolerance in the subsequent steps.

Step 3: Debinding is the next step in injection molding. It is necessary to remove the binder from the molded specimen. This process is expensive and time-consuming and can be performed by three methods, namely, thermal, solvent, and catalytic processes. Thermal debinding is the removal of binder by degradation, evaporation, or liquid extraction in a temperature range of 60°C–600°C. On the other hand, solvent debinding is relatively fast using an organic solvent or water. Catalytic debinding involves the degradation of a solid to vapor through catalysis, and is a much faster debinding method compared to thermal or solvent binding.

Step 4: This is the drying step. After debinding, the molded part is dried using an air dryer or oven at 30°C–60°C prior to sintering.

Step 5: Sintering is the final stage of densification and is one of the most important processes in powder injection molding. The molded specimen is fired in the temperature range 1,200°C–1,600°C depending on the material of interest. It provides inter-particle bonding and densification resulting in a greater improvement in properties compared to the loose powder mass. Depending on the types of material, different sintering temperatures are used in powder injection molding, which results in high sintered density ranging from 95% to 99.5% of the theoretical density, thus superior mechanical and corrosive material properties can be achieved.

Despite fabrication complex components of a few centimeters, advanced technology demands high precision and miniaturization of components. In this context, new invention come to the forefront to fabricate a few micron sized critical geometry, their details are discussed in Section 9.4.4.

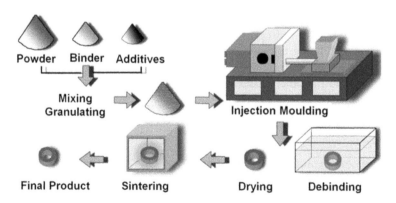

FIGURE 9.10
Overview of powder injection molding.

9.4.3 Controlling Parameters and Difficulties

The typical parameters that need to be controlled depend on each of the processing steps and the management of materials. However, some of the governing parameters are the injection speed, injection pressure, melt temperature, packing pressure, packing time, mold temperature, and cooling time. The rate of flow affects the physical appearance of the layered structure as high/low flow rates can cause unstable filling. Controlling the temperature of the mold and the cavity is of utmost importance in the molding process. Most cases adopt the variotherm method for microparts, where the feedstock is raised to a temperature near the melting point in the mold, followed by feedstock injection and cooling to achieve sufficient strength in the green compact after demolding. The cavity should be evacuated prior to injection to avoid damage such as thermal degradation of the feedstock and also hot spots generated from compressed air. A low injection pressure can lead to a short shot but higher pressure can cause flash because of high average pressure in the mold cavity [48]. The initial packing pressure determines the flow rate of the melt. If the pressure is low, the melt front cools causing surface defects; on the other hand, a high initial packing pressure causes high residual stress due to a high melt flow rate. The packing time could be determined/controlled by stabilizing the part weight in time and by reducing the flow of melt through the gate. The cooling time should be sufficient; however, it should be taken into account that while an increase in the cooling time can reduce parts warpage, it also affects the performance of the equipment through longer cycles and contributes to costs thereafter.

9.4.3.1 Polymer System

In the polymer system, the thermoplastic behavior of the materials during injection molding is important. It is characterized by the temperature conditions and also low viscosity materials are preferred. The temperature for amorphous and semi-crystalline materials should be below the decomposition temperature and above the glass transition and melting temperatures, respectively. When using fiber-filled polymers in a micropatterned mold, friction between the mold inserts should be avoided. The thermoplastic materials commonly used are polymethylmethacrylate (PMMA), polymethacrylmethylimid (PMMI), polycarbonate (PC), acrylonitrile-butadiene-styrene (ABS), carbon black, carbon-fiber-filled plastics such as polyoxymethylene (POM-C), and Nylon 6 (N6), and ceramic-filled polymers [49].

9.4.3.2 Metal System

Metal-based injection molding requires a very high injection pressure (up to 200 MPa) relative to polymers. The injection units must be wear resistant, which can be provided by hard-plate inline layers or having the whole unit made up of hard metals. Superior wear protection not only reduces the chances of contamination but also achieves high-purity sintered parts. However, this increases the cost of investment. Plunging effects should be countered by adjusting the compression ratios for the defined residence time of the feedstock in the hot zone, which helps to avoid unwanted decomposition after a few cycles. Unexpected freezing and over-molding of feedstock can reduce the product reliability and thus it is essential to maintain the optimum clamping force. Most used are alloys of tungsten and ferrous (stainless steel, iron-nickel magnetic alloys, tool steels, tungsten-copper); hard materials such as cermets (Fe-TiC), cemented carbides (WC-Co), and cobalt-chromium;

and materials such as titanium alloys and aluminum. The fine powder size has greater influence on the surface quality and replication compared to the molding process only [50].

9.4.3.3 Ceramic System

The major requirement in developing ceramic microparts is to develop highly flowable feedstocks to achieve sufficient green strength. To achieve high viscous powders with high flowability, the particle size must be adjusted and an optimum moldable powder–polymer mixture is required. As the particle size distribution narrows, the moldability of the powder slows. Parameters such as homogeneous green density and easy demolding with no defects are crucial for microparts production. Considering the machinery, there is little difference between the metal- and the ceramic-based systems. Compared to metals, ceramic feedstocks require a longer cooling time due to the low thermal conductivity of the ceramic–polymer mixture. Polymer removal during the sintering process can cause swelling and surface blisters and can also form large pores reducing the resultant density and compromising the mechanical properties. To avoid such phenomena, solvents are used to remove polymers or they are thermally degraded, known as the "thermal debinding process." During debinding, the temperature profile must be controlled to avoid crack formation and distortion of the product [51]. Depending on the types of binders such as polymers, paraffin, or wax, a high or low-pressure injection is carried out due to differences in viscosity. Ceramic materials such as Si_3N_4, Al_2O_3, ZrO_2-Al_2O_3, Al_2O_3-TiN, and others have been commercially used.

9.4.4 Micro-Powder Injection Molding

The complex and micro-dimensional polymeric component manufacturing industry is adopting microinjection molding technology, and thus micro-dimensional metal and ceramic components are extensively using the micro-powder injection molding protocol. Micro-powder injection molding of metals and ceramics is an excellent innovation, making it more efficient to fabricate in dimensions of a few microns when the polymeric material is no longer sustainable at high temperatures or in abrasive environments and hazardous fluids. It is a highly sophisticated, precise injection molding technique used to produce complex geometry and micro-miniature components analogous to powder injection molding. Excessive care is needed to maintain the accuracy of the dimensions as the size of the components becomes smaller for modern technical applications. Thus, it requires a specialized molding machine with optimal injection speed and pressure, precise and uniform control over the feed temperature, and ultra-fine resolution with an accurate program to produce microcomponents of complex geometries. The mold cavity is in the micrometer range, so mold fabrication technology is another arena that can critically adjust to precise dimensions. The principal differences between microinjection molding and conventional molding are the evacuation of the cavity and the variotherm process. Furthermore, very fine powder is expected during fabrication of micro-dimensional metal or ceramic components, thus debinding and sintering without deformation or cracks is an extreme challenge. In consideration of Laplace's equation, very fine particles lead to very high sintering stress (σ) during densification as shown in Equation 9.1 [52]:

$$\sigma = \gamma \left(\frac{1}{R_1} + \frac{1}{R_2} \right) \tag{9.1}$$

For example, a commercial-grade metal powder injection molded component prefers 20 μm feed particle size, but its reduction and starting particle size of 0.2 μm encompasses substantial sintering stress increment. Typical particle sizes in ceramic injection molding are 1–2 μm, but much finer particles down to the submicron or nano region are also being used in advanced microinjection molding. To avoid distortion of the microcomponents during densification, sintering cycles contain long holding times at critical densification temperatures to accommodate the induced stresses and allow enough time to relax before the temperature is ramped up again when new stresses are induced. Despite several critical considerations including mold fabrication, processing parameters, the details of densification behavior are discussed elsewhere [53]. Here, some classical, real-life metal and ceramics components are summarized.

9.4.4.1 Micro-devices by Metals

Microcomponents are preferentially divided into three categories, namely, micropecision parts, micropatterned parts, and microparts. Micropecision parts are in the dimensional range of a few millimeters up to decimeters with special features. Micropatterned parts have a dimensional range from a few micrometers to centimeters fabricated on a surface. The geometric feature of advanced micropatterned parts is a larger length compared to their width, so they have a high aspect ratio. Microparts have a size range much smaller than 5 mm. Microcomponents experience a large surface to volume ratio that leads to faster cooling or even freezing of the injected melt in cooled tools. Conventionally, polymers have low thermal conductivity and show a "self-isolating effect," with the injected material freezing on the molding tool and microcavities not filling completely within time. As a consequence of the thin walls and large surfaces of micropatterns compared with their volume, both the mold and molding material temperature attain rapid equilibrium in milliseconds. Therefore, the micropatterned mold inserts with high aspect ratios are heated to temperatures commonly above the melting point or even above the no-flow temperatures of the molding materials, which are usually about 30 K above the melting point. The design and fabrication of microsystem products have remarkable potential in manufacturing industries and are leading the research and development of this technology. Fundamental research on microinjection molding was started through LIGA (*Lithographie Galvanoformung and Abformung*) for fabricating and manufacturing high aspect ratio micropatterns or microparts. The unique technique allows the manufacture of microcomponents of critical and arbitrary lateral geometry with a micron range resolution, but with structure heights in the millimeter range. Despite its silicon material, this lithographic technique is well adapted for non-silicon candidates including polymers, metals, and ceramics. Using the LIGA technique, a mold insert was prepared to fabricate 316L SS microgear with a diameter of 480 μm and a minimum tooth width of about 50 μm, as depicted in Figure 9.11. The quality of a featured replication or molding greatly depends on several factors such as size, aspect ratio, and overall geometry [54].

While considering different microcomponents made of metal powders, it has been noticed that the processing parameters, including the mold temperature and the holding pressure, have the most significant effect in controlling the surface to volume ratio, precise dimension, and geometrical features. In a recent article, different sets of classic microcomponents such as planetary gear sets, connectors, impellers, cams, and microreactors were fabricated through micro metal powder injection molding and their dimensional tolerance with respect to product size is illustrated in Figure 9.12 [55].

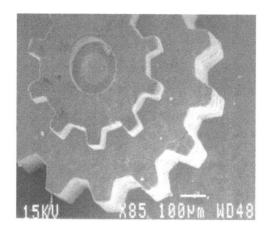

FIGURE 9.11
Stepped LIGA gear wheels made of 316L stainless steel with a structural diameter of about 480 μm and a minimum tooth width of about 50 μm; both values were measured after sintering [54].

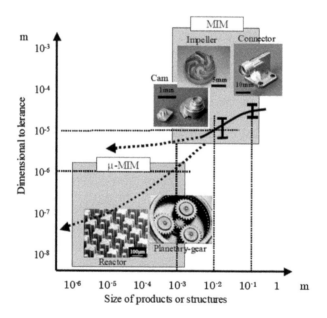

FIGURE 9.12
Dimensional tolerance versus size of MIM products [55].

9.4.4.2 *Micro-devices by Ceramic Oxides*

Apart from metals, ceramic materials such as alumina and zirconia powder feedstock are used in the engineering of new micro-dimensional components through microinjection molding, which eventually exhibit high wear and corrosive resistance, biocompatibility, insulation properties, thermal stability, and a high modulus in low weight where metal and polymer components fail. To fully exploit the extraordinary properties of these materials, however, there is a comprehensive need for the entire processing technique to include the synthesis of raw materials, the design and manufacture of molding dies through the molding process itself, and the sintering and finishing of the desired product. Ceramic microscale processing provides great dimensional freedom with multivariate geometrical features such as internal and external threads, undercuts, inclined drill holes, and freely formed falls without any error or rewriting effort. The process of feedstock preparation is analogous to the processing steps in ceramic powder injection molding, such as feedstock preparation, molding, debinding, drying, and sintering. High green density is desirable and several intermediate steps such as cold isostatic pressing can be adapted before debinding and sintering. Several sintering protocols including pressureless sintering, microwave sintering, and hot isostatic pressing are smart choices to obtain high density and different degrees of opaqueness to transparency.

Some classic and reliable commercial-grade microcomponents made of ceramic oxides are shown in Figure 9.13a. The electronic, health, and thermal sectors demand high-precision capillaries, receptacle guides, and grippers; thus, ceramic is one of the best choices to make such stringent tolerance, thin walls and tiny holes of ZrO_2 ceramics, as illustrated in Figure 9.13b and c [56]. However, several difficulties have been encountered during the fabrication of such components, for example, incomplete feedstock filling in such a small cavity, fragile tendency during demolding, and the density measurement and estimation of mechanical properties of real prototypes. The use of fine powder is required but it enhances the viscosity of the feed flow because of its high surface area to volume ratio, and thus excess binder may reduce the viscosity but it produces low green density followed by sintered density. Thus, during the fabrication of the critical geometry of such micro-dimensional objects, it is necessary to avoid sharp edges in the mold, material accumulation, a lengthy and free-standing core, unnecessary variation in wall thickness, and abrupt changes in cross sections.

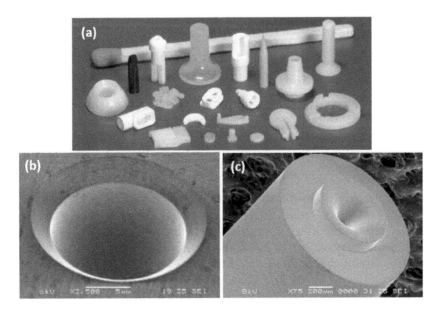

FIGURE 9.13
(a) Various microcomponents of ceramic oxide. (b, c) Exceptional freedom of design for component shapes in ZrO_2-sintered drill hole geometry with outlet in final form [56].

9.5 Micromachining

Machining is a basic approach to completing a job by changing its shape, size, and surface through a controlled material removal process for a specific purpose, be it a few meters to a few microns, the opposite of the AM phenomenon. Professional machining provides very good dimensional tolerance with internal and external geometry. With the advances in technology, a machining job is becoming more precise and efficient, working with a wide range of materials of different geometries. Conventional machining is not sufficient when a complex miniaturization component in the micron dimension accuracy is at the forefront. Thus, the manufacturing sector has introduced precise machining protocols to overcome the existing problem in conventional machining, and such novel processes are called "micromachining." Micromachining is a precise and sophisticated process that works on a very low dimensional range up to several micrometers, and is in great demand for the miniaturization of components for advanced microsystem technologies. In the industrial world, microscopic-scale manufacturing was first introduced as a result of the rapid growth in silicon-based microelectromechanical systems (MEMS) research and development, and their high aspect ratio patterns or 3-D workpieces are in progress. In discussing such advanced technologies, the use of polymer, metal, and ceramic has great prospects in electronics, machineries, diesel engines, biotechnologies, diagnoses, medicines, and surgery. A basic understanding and different case studies on the fabrication of micro-dimensional ceramics through computer numerical control (CNC) micromachining, photochemical machining (PCM), laser machining (LM), LIGA, microelectrical discharge machining (EDM), micro ultrasonic micromachining (USM), electrochemical machining (ECM), electron beam machining, and mechanical micromachining are discussed. Finally, a brief discussion is provided on grinding and polishing to obtain the desired surface finish in the interest of a particular application.

9.5.1 CNC Micromachining

A professional blueprint and product specification are the starting point prior to initiating the machining process. Expert programming and manpower achieve excellent dimensional tolerance end finishing. However, several common errors such as dull tools, incorrect clamping, and vibration provide an inappropriate surface finish, coined "chatter." Several well-established machining processes make a wide range of components, and it is obvious that a very precise multi-axes CNC machine enables high accuracy compared to a manual lathe or milling machine. CNC machining executes a wide range of manufacturing tasks controlled by programming language and computerized devices that eventually enable the development of the analogous geometrical and dimensional features of a CAD model. Since the digital revolution in the early 1950s, there has been a demand for new material design and a number of technological advancements have occurred. In this context, CNC micromachining is an innovative approach to the precise milling and manufacturing of extremely small parts and products. This process has been streamlined step by step, reducing manufacturing to one thousand of a millimeter, helping make the production of tiny parts more efficient and realistic. Such an effective process is also popular in micro/meso-scale mechanical manufacturing, in brief M4 process [57]. Despite machining protocols and their difficulties, geometrical features, setting, and wear resistance properties are of great concern for single edge and multipoint cutting tools. It is also essential to remember that the optimum relative motion of a workpiece in contact with the cutting tool and the depth of the cut determine the degree of machining and the surface roughness parameters. This multimode machining practice can effectively machine tiny components for different sectors including miniature gears for the ophthalmic industry; sonic nozzles with precision venturis for flow measurement; miniature components for nuclear armament timing devices; components for respiration monitoring equipment; electron microscope accessories; fiber-optic components including MEMs; optical switch components such as collimators; and impellers for ventricular assist devices (VAD) such as heart pumps. A typical schematic of a five-axis CNC machine and micromachined components made from polymer, metal, and ceramic is illustrated in Figure 9.14 [58–61].

CNC micromachining is used to successfully manufacture different classic components made from different class of materials for industries such as aircraft, aerospace, defense, energy, fiber optics, food manufacturing, medical, oil and gas, and tools and dies.

9.5.2 Photochemical Machining

PCM is also referred to as photochemical etching or the photofabrication process. In this process, a photoresist is used and material is selectively removed by corrosive etchants that eventually enable the production of highly complex parts having microscale details and without geometrical distortion or stress. This process is relatively economical and technically superior compared to punching, stamping, laser, water jet cutting, or wire EDM for ultrathin precision parts. This method effectively manufactures high precision micro-dimensional and burr-free metallic components. Conventionally, a metal sheet is machined to a certain dimension, and then cleaned and coated with a thin and light-sensitive material, known as a "photoresist." The workpiece is exposed to UV radiation and then submerged into a particular metal corrosive chemical solution. The exposed area is corroded by the chemical and masking material is removed to obtain the desired pattern on the surface. The unmasked portion develops a thick groove of a few microns; however, this method can be employed to make 3-D objects through proper masking and selective etching [62]. Despite PCM of metals, miniaturization sometimes demands complex patterns for high-tech ceramics, therefore a recent research work is highlighted.

PCM was employed to micromachine an alumina (96% purity) substrate. A thermally stable (up to 400°C) and photosensitive cyclized polybutadiene rubber (Japan Synthetic Rubber Co., JSR CBR-M901) was used as a negative-working photoresist. The thickness of

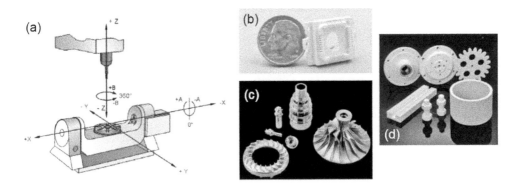

FIGURE 9.14
(a) Schematic view of five-axis CNC [58], CNC machined micro-dimensional accuracy components made of (b) polymer [59], (c) metal [60], and (d) ceramics [61].

the coating after post-baking was 2.0–2.4 μm; details of the photoresist stencil fabrication process is represented by a fluxogram in Figure 9.15a [63]. The etching experiment was conducted in the presence of boiling phosphoric acid at atmospheric pressure and at temperatures ranging from 260°C to 320°C for 15–60 minutes to optimize the process conditions. Figure 9.15b shows the topographical feature and its cross-section etched at 300°C for 30 minutes, within the roughness variation of ±10 μm. In this slot, the inside corner radii were about 0.28 mm and the outside corner about 0.08 mm. This promising technique can be employed to make different patterns on ceramics; however, the type of etchant, the concentration, and the processing time may change from ceramic to ceramic.

9.5.3 Laser Machining

Light amplification by simulated emission of radiation (LASER) is the prime source of LM where systematic material removal takes place by laser beam only. In contrast, laser-assisted manufacturing (LAM) aids the ceramic deformation behavior from brittle to ductile in the presence of a laser and simultaneous material removal takes place with a conventional cutting tool, thus LAM is different from LM. Partial melting, dissociation, decomposition, evaporation, and slow material expulsion in the presence of high density optical energy are the prime issues of such a machining process. However, LM efficiency depends on thermal conductivity, reflectivity, and the specific heat and latent heat of melting and evaporation, and thus it is essential to estimate such properties prior to LM.

A typical schematic representation of the LM of structural ceramics is given in Figure 9.16; however, details of machining steps and processes are discussed elsewhere [64]. Several salient features of LM such as its non-contact process mechanism, thermal behavior of

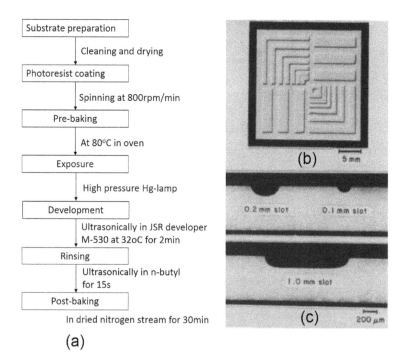

FIGURE 9.15
(a) Fluxogram to make a photoresist stencil for PCM and (b) PCM on alumina [63].

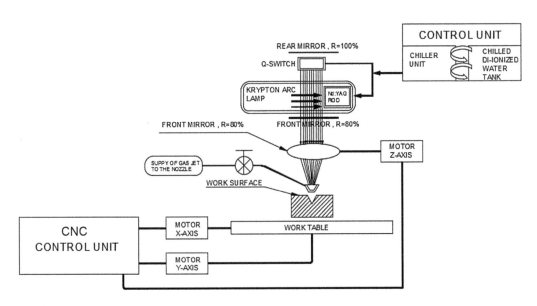

FIGURE 9.16
Schematic of laser machining [64].

materials, and flexibility provide an excellent machining approach over other machining processes. A competitive economical overview of different machining protocols is presented in Table 9.2 [65].

In brief, this non-contact process leads to laser energy transfer through irradiation and eliminates cutting forces, tool wear, and machine vibration. By controlling the input energy and processing speed, material removal can be monitored and this process can avoid maximum tool force, tool chatter, and built-up edge formation. As the LM efficiency depends on a material's thermal conductivity and to some extent its optical properties, it is an excellent tool for ceramic machining. It is capable of drilling, cutting, grooving, welding, and heat treating on the same machine through a multi-axis positioning system or a robot, which encompasses a high degree of reproducibility and output. In principle,

TABLE 9.2

Relative Economic Comparisons of Different Machining Processes [65]

Machining Process	Parameter Influencing Economy				
	Capital Investment	Toolings/ Fixtures	Power Requirements	Removal Efficiency	Tool Wear
Conventional machining	Low	Low	Low	Very low	Low
Ultrasonic machining	Low	Low	Low	High	Medium
Electrochemical machining	Very high	Medium	Medium	Low	Very low
Chemical machining	Medium	Low	High	Medium	Very low
Electrical-discharge machining	Medium	High	Low	High	High
Plasma arc machining	Very low	Low	Very low	Very low	Very low
Laser machining	Medium	Low	Very low	Very high	Very low

two classes of lasers are common in LM: continuous wave (CW) and pulsed mode (PM—nano, pico, and femto second lasers). CW acronym implies continuous pumping of the laser emits incessant ligh; however, pulse laser follows power-off period between two successive pulses. Thus, PM is the preferred LM approach to control precise micromachining compared to CW [66]. Some classic lasers such as CO_2, Nd:YAG (fabrication is discussed in Chapter 3), copper vapor laser (CVL), and excimer (ArF, KrF, XeF, XeCl) lasers are employed in the precious micromachining of structural ceramics. Each class of laser system utilizes different wavelengths of absorption, for example, the wavelength of CO_2, Nd:YAG, CVL, and excimer lasers is 10.6 μm in the far infrared region of the electromagnetic spectrum, 1.06 μm in the near-infrared region of the spectrum, 0.511 μm near visible wavelength, and 0.193–0.351 μm in the ultraviolet to near-ultraviolet spectra, respectively. To understand LM efficacy, the machining and the relevance of the processing conditions of some components are considered and discussed.

High density 50 ± 1 μm diameter hole arrays on 250 μm thick alumina on a 60 μm pitch without any defects was drilled using 20 ns pulse-duration CVL of power 3 W at a 10 kHz pulse frequency (Figure 9.17a). This micromachined ceramic substrate is predominately used in semiconductor test probe-cards. In a single operation, the micron-scale drilling, cutting, and milling of a 0.3 mm thick alumina sheet was performed using CVL, and 45° chamfers along the edge of the bar were created using controlled depth ablation, as shown in Figure 9.17b. The depth variation was adjusted through the total number of pulses used [67].

The high precision, 3-D Nd:YAG LM of slots, grooves, threads, and complex patterns in ceramic workpieces prefers to accommodate two or more laser beams. Typical microscale turning threads in Si_3N_4 and cut gears from an $SiC\omega/Al_2O_3$ composite were machined and are shown in Figure 9.18 [68,69].

9.5.4 Electrochemical Discharge Machining

Electrochemical discharge machining (ECDM) is a non-traditional hybrid technology of ECM and electric discharge machining (EDM) for the machining of electrically non-conducting materials including glass and ceramics [70]. ECM is an advanced machining technique and is the inverse of galvanic coating or the deposition process for electrically conducting workpieces. Almost all kinds of metals can be electrochemically machined, in preference to nickel- and titanium-based alloys and hardened materials. In this precise

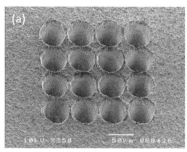 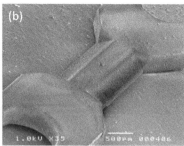

FIGURE 9.17
(a) Array of 50 μm diameter holes drilled in 250 μm thick alumina on a 60 μm pitch using CVL at 3 W with a 10 kHz pulse frequency and a six second drill time; (b) examples of drilling, cutting, and milling in alumina ceramic using CVL [67].

control machining, the metal workpiece is locally dissolved (machining) in the presence of the requisite electricity (electro) and chemistry (chemical) in order to achieve a predefined complex 3-D shape. EDM is an advanced electrothermal machining process where a spark is generated by electrical energy and material is removed thermally instead of chemically.

A typical schematic representation of the ECDM process is given in Figure 9.19a, in which machining tool is connected with a cathode and an auxiliary electrode used as an anode, both are connected to a dc (0-120V) source.The workpiece is placed just below the tool, and the entire assembly including the workpiece, tool, and auxiliary electrode is immersed in an electrolyte (say NaOH) solution in the machining chamber. In this process, the thermal erosive effects of the electrical discharge (ED) action follow an electrochemical (EC) reaction. Electrolysis is initiated in the presence of an applied voltage in the tool and auxiliary electrode. Depending on the electrolyte concentration, H_2 bubbles are generated in the electrolyte chamber beyond a critical applied voltage, and coalesce to form an insulating gas film around the tool. Sparks appear due to EDs in the gas layer that eventually facilitate the ignition of the workpiece through the formation of high thermal energy followed by material removal through the melting/evaporating ratio of the workpiece material. In a recent article, a ceramic workpiece made of alumina and silica was drilled using a WC tool at the applied voltage of 51 V. A cross-section and top view of the machined ceramic are illustrated in Figure 9.19b [71]. To differentiate the ECM, EDM, and ECDM processes, competitive process conditions, outputs, and economic aspects are illustrated in Table 9.3.

FIGURE 9.10
(a) Turning of threads in Si_3N_4 and (b) a gear shape cut in $SiC\omega/Al_2O_3$ composite [68, 69].

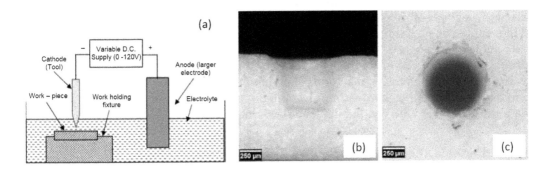

FIGURE 9.19
(a) Schematic representation of the ECDM process. (b, c) ECDM machined cross-section and top view of a ceramic workpiece using a WC tool at 51 V [71].

TABLE 9.3

Competitive Analysis of ECM, EDM, and ECDM

Aspects	ECM	EDM	ECDM
Principle	Anodic dissolution	Spark erosion	Both
Power	10,000 A, 2–3 V, dc	200 A, 50–200 V, dc	3–4 A, 40 V, dc
Workpiece	Conductive	Conductive	Conductive and non-conductive
Material removal rate	Less	Less	Five times faster than EDM
Tool wear	Less than both	Higher than ECDM	Less than EDM
Accuracy	Higher than EDM	Dimensional accurate (±0.03 mm)	Dimensional accurate (±0.01 mm)
Surface finish	Higher than both	Less	Higher than EDM
Cost	Less initial than ECDM	Less initial than ECDM	High

9.5.5 Electron Beam Machining

Electron beam machining is a thermal process that uses a focused beam of high-velocity electrons to perform high-speed drilling and cutting operations. A typical schematic representation of such a unit is illustrated in Figure 9.20a. Using an electron beam gun in pulse mode produces free electrons at the cathode and high-velocity particles move through the minute spot size (≤1micron). The tantalum or tungsten cathode filament causes thermoionic emission near to 150 kV, and the released electrons flow in the direction of the workpiece. This electron movement and machining operation is carried out in a vacuum to avoid collision with air molecules that would result in a loss of energy and a reduction in machining efficiency. When high-velocity electrons strike the workpiece, their kinetic energy is converted into heat energy, which starts flash heating, melting, and vaporizing of the workpiece. This high-speed action can perforate very small circular (up to 0.025 mm) and non-circular holes in metals and ceramics. Moreover, a predetermined cutting path can be controlled by systematically diverting the electron beam or moving the working table.

E-beam machining of a 200 cm^2 green alumina sheet composed of 90% Al_2O_3, 6.7% polyvinyl butyral, and 3.3% dioctyl phthalate was performed using 100 keV electrons at a 1 mA beam current. In this work, the author reported the possibility of forming a line 35 μm wide and 50 μm deep and a 30 μm via hole at a distance of 100 μm. A typical conceptual machining process is shown in Figure 9.20b, where a laminate polymer sheet is used as a guide to machine micro-dimensional lines and holes; eventually, grooves are filled with a conductive metal. The electron beam is focused on the green sheet and the temperature is raised through absorbing the energy above the boiling temperature of the organic binder in the ceramic green sheet, which facilitates the rapid expansion of vacuum and thus material removal, forming a "void" in the green sheet. Repeating this well-controlled e-beam sequence, one can drill vertical holes as well as delineate a groove laterally in such a green ceramic sheet. This precious machining method allows accurate dimensions to be made before further screening, laminating, and sintering. Such a machined pattern composed of vias and lines filled with conducting materials is shown in Figure 9.20c [72].

9.5.6 Ultrasonic Micromachining

USM is a non-thermal process that is of particular interest for cutting non-conductive brittle materials such as engineering workpieces with low ductility and hardness above 40 HRC (230 HV), for example, silicon nitride, inorganic glasses, and nickel/titanium alloys.

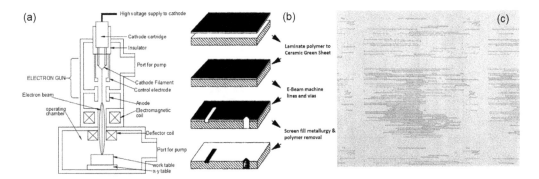

FIGURE 9.20
(a) Schematic representation of an electron beam machining apparatus; (b) e-beam machining of the ceramic green sheet process flow to make a 35 µm wide line and 30 µm diameter vias; and (c) optical photograph of a 200 cm² e-beam machined ceramic green sheet with via holes and line grooves filled with conducting material [72].

Employing such a technique, one can machine a hole as small as 76 µm in diameter with a diameter to depth ratio of 1:3. In USM, a low-frequency electrical signal is applied to a transducer, which converts the electrical energy into a high-frequency (~20 KHz) mechanical vibration along the longitudinal axis with an amplitude of 5–50 µm (Figure 9.21a). Typical power ratings range from 50 to 3,000 W and can reach 4 kW in some machines. A controlled static load is employed on the tool and an abrasive slurry composed of a mixture of abrasive materials such as silicon carbide, diamond, alumina, and boron carbide in water or a suitable chemical solution is pumped around the cutting zone, as described in Figure 9.21b. The vibration of the tool causes the abrasive particles held in the slurry between the tool and the workpiece to impact the workpiece surface causing material removal by microchipping. This machining protocol follows three critical steps: (a) major material removal—mechanical abrasion by direct impact of the abrasive particles on the workpiece; (b) minor material removal—microchipping through continuous rubbing by the free-moving abrasives; and (c) another minor material removal due to cavitation-induced erosion and chemical effects [73]. Despite such conventional ultrasonic machining, rotary ultrasonic machining (RUM) is very popular because it employs a combined rotational and vibrational motion to achieve a fast, high quality, and precious dimension for a variety of glass and ceramics. In RUM, a rotating (up to 8,000 rpm) core drill with metal-bonded diamond abrasives is ultrasonically vibrated in the axial direction while the spindle is fed toward the workpiece at a constant pressure.

At the same time, the coolant that is pumped through the core of the drill washes away the swarf, prevents jamming of the frill, and keeps it cool. However, the material removal rate (MRR) and average surface roughness (R_a) depends on the material properties and process parameters, including type, shape, and size of abrasive grains; amplitude of vibration; materials porosity; hardness; and toughness. Some ultrasonic micromachined specimens consist of square cavities, round through holes, and crossing beams in a 4-inch borosilicate wafer as shown in Figure 9.21c and Figure 9.21d. [74]. An advance pencil tool machined the carbon fiber reinforced composite for mechanical device. Extremely round to within 0.021 mm (21 µm) and 254 mm long (often longer) rods of quartz, glass, sapphire, ruby, etc., can be machined using RUM, as shown in Figure 9.21e [75].

In addition to the foregoing prime machining protocols, there are several machining processes such as mechanical micromachining, focus ion beam (FIB) milling, abrasive cutting, chemical mechanical polishing (CMP), and thermomigration, which have been well adopted, and their details are discussed elsewhere [76].

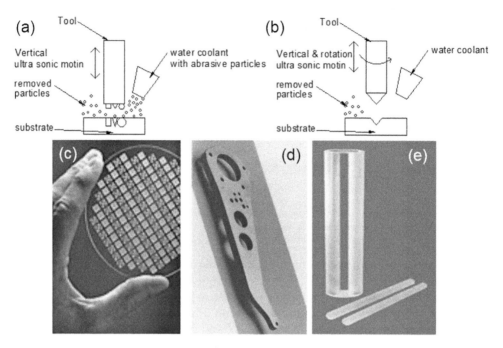

FIGURE 9.21
Schematic representation of (a) USM, (b) RUM, (c) square cavities of borosilicate glass, (d) carbon fiber composite acceleration lever, holes, and outline profile machined by USM, and (e) extremely round 10 inch long sapphire machine by RUM [74, 75].

9.6 Grinding and Polishing

Surface grinding is a very precise machining process in which a stationary, rotating, or abrasive wheel is used to finish the surface. It can be used to finish both metallic and non-metallic workpieces. A rotating abrasive wheel gives a more refined appearance by removing oxides and impurities on the workpiece surface. The high temperatures encountered at the ground surface create residual stresses and a thin martensitic layer may form on the part surface. Examples of surface grinding processes are horizontal spindle surface grinding, vertical spindle surface grinding, and disc grinding. Finishing is one of the most important industrial processes that modifies the surface of a workpiece to achieve a notable and brilliant finish to the workpiece or job. Finishing processes improve appearance, corrosion resistance, chemical resistance, wettability, wear resistance, hardness, etc. [77]. It is also used to modify the thermal as well as the electrical conductivity of the material and controls the surface friction. Finishing is categorized into two types: adding or altering finishing includes a number of finishing processes such as calendering, blanching, cladding, corona treatment, and diffusion processes such as carburizing and nitriding, electroplating, galvanizing, case hardening, etc. Removing or reshaping finishing is done by abrasive blasting, sand blasting, electropolishing, flaw polishing, grinding, and peening such as shot peening, laser peening, buffling, and lapping.

Polishing is also known as metalworking that provides the workpiece with an aesthetic look and an ultra-smooth surface using abrasive grains, a work wheel, or a leather strop.

High-precision polishing of a ceramic sample or workpiece is a challenging task as it possesses high hardness and thus systematic removal of high to low asperities results in a properly polished surface. For example, to make a nanoscale surface finish of a high dense (>98%) sintered alumina disk, consecutive machine polishing of a targeted surface by 1,200 and 600 diamond (sometime SiC) grits that reduce the surface roughness (R_a ~1 µm) up to the micron level, intermediate polishing must be followed. However, extensive sequential polishing by 25, 10, 6, 3, and finally 1 µm diamond paste results in a nanoscale (R_a ~0.1 µm) surface finish. One can see one's eyebrows on this surface! It is essential to avoid forming scratches during the early stage of polishing, because they are troublesome to remove at the final stage of polishing. Employing the requisite load, revolutions per minute, and time are important criteria to obtain an effective polished surface for such specimens.

9.7 Conclusions

A precise ceramic fabrication protocol has opened up new horizons to the possibilities of a wide range of micro-dimensional 3-D objects with coatings of a few nanometers and has proved to be a highly promising substitute for metals and polymers, with a high life expectancy in elevated temperatures and harsh environments.

Recent upcoming technological survey predicts AM market offering a bunch of solutions through different printers, materials, interfacing software and services. In direct 3DP technology, the droplets of build material is deposited on to a substrate, whereas the indirect bind inject 3DP follows a definite viscous liquid binder/s jet printed onto thin layers of powders to build up layer by layer through gluing the particles together in the target to develop of a CAD originated model. When considering SLS, the thermal energy fuses within a confined region of the powder bed of the build material, but SLA forms complex ceramics from a mixture of ceramic particles and UV-cured resin. Thus, from aerospace to medical surgery, the AM protocol is booming in the development of centimeter to a few micrometers components without any machining.

In the semiconductor industry, miniaturization and precise processing are the main motivation for the high-end development of ALD leading to atomic-level control of thin film deposition and conformality on high aspect structures consisting of critical geometry and voids. The continuous and pinhole-free nanoscale coating of different class III-IV semiconductors, oxides, and non-oxides has potential in semiconductors, low electron leakage dielectrics for magnetic read or write heads, and diffusion barrier coatings with low gas permeability.

Microinjection molding is an excellent protocol to achieve commercial-grade, micro-dimensional, complex objects made from polymer, metal, and ceramic. However, their precision die manufacturing is a challenging task for a consistent and high-throughput component. Proper powder selection and mass flow behavior control the shape of a ceramic green compact that eventually develops a definite geometry after systematic drying and sintering. This advanced processing protocol enables the development of several micro-dimensional ceramic components including alumina, zirconia, and ferrites.

Although several high technologies have been developed to fabricate complex micro-dimensional features, load-bearing ceramic components are still under progress through AM, microinjection molding, etc. Direct machining, known as "subtract manufacturing," allows the production of a robust component from sintered ceramics; however, it demands very high precision for different classes of machining including CNC micromachining, PCM, LM, ECDM, electron beam machining, and USM. These methods are commonly used in manufacturing high-tech miniature components made from metal and ceramics that are highly favored in electronics, MEMS, sensors, microgears, cantilevers, and other sophisticated instruments for the aerospace, automotive, energy, and biomedical industries.

References

1. Tay, B. Y., Evans, J. R. G., and Edirisinghe, M. J. 2003. Solid freeform fabrication of ceramics. *International Materials Reviews* 48(6):341–370.
2. Travitzky, N., Bonet, A., Dermeik, B., Fey, T., Filbert-Demut, I., Schlier, L., Schlordt, T., and Greil, P. 2014. Additive manufacturing of ceramic-based materials. *Advanced Engineering Materials* 16(6):729–754.
3. ASTM F2792-12a. 2015. *Standard Terminology for Additive Manufacturing Technologies*. ASTM International: West Conshohocken, PA.
4. Eckel, Z. C., Zhou, C., Martin, J. H., Jacobsen, A. J., Carter, W. B., and Schaedler, T. A. 2016. Additive manufacturing of polymer-derived ceramics. *Science* 351(6268):58–62.
5. Zocca, A., Colombo, P., Gomes, C. M., and Günster, J. 2015. Additive manufacturing of ceramics: Issues, potentialities, and opportunities. *Journal of the American Ceramic Society* 98(7):1983–2001.
6. Chua, C. K., Leong, K. F., and Lim, C. S. 2006. *Rapid Prototyping: Principles and Applications*. World Scientific Publishing Co Pte Ltd.
7. Bourell, D. L., Marcus, H. L., Barlow, J. W., and Beaman, J. J. 1992. Selective laser sintering of metals and ceramics. *International Journal of Powder Metallurgy (Princeton, New Jersey)* 28:369–381.
8. Gibson, I., Rosen, D., and Stucker, B. 2010. *Additive Manufacturing Technologies 3D Printing, Rapid Prototyping, and Direct Digital Manufacturing*. Springer: New York.
9. Ainsley, C., Reis, N., and Derby, B. 2002. Freeform fabrication by controlled droplet deposition of powder filled melts. *Journal of Materials Science* 37(15):3155–3161.
10. Carboprint: ExOne Using Binder Jet 3D Printing to Produce Carbon and Graphite Components with SGL Group, https://3dprint.com/205381/exone-sgl-group-carboprint/.
11. Silva, D. N., Gerhardt de Oliveira, M., Meurer, E., Meurer, M. I., Lopes da Silva, J. V., and Santa-Bárbara, A. 2008. Dimensional error in selective laser sintering and 3D-printing of models for craniomaxillary anatomy reconstruction. *Journal of Cranio-Maxillofacial Surgery* 36(8):443–449.
12. Sallica-Leva, E., Jardini, A. L., and Fogagnolo, J. B. 2013. Microstructure and mechanical behavior of porous Ti–6Al–4V parts obtained by selective laser melting. *Journal of the Mechanical Behavior of Biomedical Materials* 26:98–108.
13. Eshraghi, S. and Das, S. 2010. Mechanical and microstructural properties of polycaprolactone scaffolds with one-dimensional, two-dimensional, and three-dimensional orthogonally oriented porous architectures produced by selective laser sintering. *Acta Biomaterialia* 6(7):2467–2476.
14. Atwood, C. L., Maguire, M. C., Pardo, B. T., and Bryce, E. A. 1996. *Rapid Prototyping Applications for Manufacturing*.
15. Cooper, K., and Faulkner, L. 2001. *Rapid Prototyping Technology*. CRC Press: Boca Raton.
16. An, D., Li, H., Xie, Z., Zhu, T., Luo, X., Shen, Z., and Ma, J.. 2017. Additive manufacturing and characterization of complex Al2O3 parts based on a novel stereolithography method. *International Journal of Applied Ceramic Technology* 14(5):836–844.
17. Wang, X., Jiang, M., Zhou, Z., Gou, J., and Hui, D. 2017. 3D printing of polymer matrix composites: A review and prospective. *Composites Part B: Engineering* 110:442–58.

18. Travitzky, N. 2012. Processing of ceramic–metal composites. *Advances in Applied Ceramics* 111(5–6):286–300.
19. Cawley, J. D. 1999. Solid freeform fabrication of ceramics. *Current Opinion in Solid State and Materials Science* 4(5):483–489.
20. Denry, I. and Kelly, J. R. 2014. Emerging ceramic-based materials for dentistry. *Journal of Dental Research* 93(12):1235–1242.
21. Halloran, J. W. 2016. Ceramic stereolithography: Additive manufacturing for ceramics by photopolymerization. *Annual Review of Materials Research* 46(1):19–40.
22. Ngo, T. D., Kashani, A., Imbalzano, G., Nguyen, K. T. Q., and Hui, D. 2018. Additive manufacturing (3D printing): A review of materials, methods, applications and challenges. *Composites Part B: Engineering* 143:172–196.
23. Tofail, S. A. M., Koumoulos, E. P., Bandyopadhyay, A., Bose, S., O'Donoghue, L., and Charitidis, C. 2018. Additive manufacturing: Scientific and technological challenges, market uptake and opportunities. *Materials Today* 21(1):22–37.
24. Gao, W., Zhang, Y., Ramanujan, D., Ramani, K., Chen, Y., Williams, C. B., Wang, C. C. L., Shin, Y. C., Zhang, S., and Zavattieri, P. D. 2015. The status, challenges, and future of additive manufacturing in engineering. *Computer-Aided Design* 69:65–89.
25. Knez, M., Nielsch, K., and Niinistö, L. 2007. Synthesis and surface engineering of complex nanostructures by atomic layer deposition. *Advanced Materials* 19(21):3425–3438.
26. Puurunen, R. L. 2005. Surface chemistry of atomic layer deposition: A case study for the trimethylaluminum/water process. *Journal of Applied Physics* 97(12).
27. George, S. M. 2010. Atomic layer deposition: An overview. *Chemical Reviews* 110(1):111–131.
28. Ville, M., Markku, L., Mikko, R., and Riikka L. Puurunen 2013. Crystallinity of inorganic films grown by atomic layer deposition: Overview and general trends. *Journal of Applied Physics* 113.
29. Jones, A. C., and O'Brien, P. 1997. *CVD of Compound Semiconductors: Precursor Synthesis, Development and Applications.* Wiley-VCH Verlag GmbH: Weinheim.
30. Ohno, H., Ohtsuka, S., Ishii, H., Matsubara, Y., and Hasegawa, H. 1989. Atomic layer epitaxy of GaAs using triethylgallium and arsine. *Applied Physics Letters* 54(20):2000–2002.
31. Putkonen, M., Sajavaara, T., and Niinistö, L. 2000. Enhanced growth rate in atomic layer epitaxy deposition of magnesium oxide thin films. *Journal of Materials Chemistry* 10(8):1857–1861.
32. Vehkamäki, M., Hatanpää, T., Hänninen, T., Ritala, M., and Leskelä, M. 1999. Growth of SrTiO3 and BaTiO3 thin films by atomic layer deposition. *Electrochemical and Solid-State Letters* 2(10):504–506.
33. Hausmann, D., Becker, J., Wang, S., and Gordon, R. G. 2002. Rapid vapor deposition of highly conformal silica nanolaminates. *Science* 298(5592):402–406.
34. Ritala, M., Leskelä, M., Dekker, J. P., Mutsaers, C., Soininen, P. J., and Skarp, J. 1999. Perfectly conformal TiN and Al_2O_3 films deposited by atomic layer deposition. *Chemical Vapor Deposition* 5(1):7–9.
35. Viirola, H., and Niinistö, L. 1994. Controlled growth of antimony-doped tin dioxide thin films by atomic layer epitaxy. *Thin Solid Films* 251(2):127–135.
36. Hiltunen, L., Leskelä, M., Mäkelä, M., Niinistö, L., Nykänen, E., and Soininen, P. 1988. Nitrides of titanium, niobium, tantalum and molybdenum grown as thin films by the atomic layer epitaxy method. *Thin Solid Films* 166:149–154.
37. Ritala, M., Kalsi, P., Riihelä, D., Kukli, K., Leskelä, M., and Jokinen, J. 1999. Controlled growth of TaN, Ta_3N_5, and TaOxNy thin films by atomic layer deposition. *Chemistry of Materials* 11:1712–1718.
38. Alén, P., Ritala, M., Arstila, K., Keinonen, J., and Leskelä, M. 2005. Atomic layer deposition of molybdenum nitride thin films for Cu metallizations. *Journal of the Electrochemical Society* 152(5):G361–G366.
39. Klaus, J. W., Ott, A. W., Dillon, A. C., and George, S. M. 1998. Atomic layer controlled growth of Si3N4 films using sequential surface reactions. *Surface Science* 418(1):L14–L19.
40. Mårlid, B., Ottosson, M., Pettersson, U., Larsson, K., and Carlsson, J.-O. 2002. Atomic layer deposition of BN thin films. *Thin Solid Films* 402(1–2):167–171.

41. Tanskanen, J. T., Bakke, J. R., Bent, S. F., and Pakkanen, T. A. 2010. ALD growth characteristics of ZnS films deposited from organozinc and hydrogen sulfide precursors. *Langmuir* 26(14):11899–11906.

42. Thimsen, E., Peng, Q., Martinson, A. B. F., Pellin, M. J., and Elam, J. W. 2011. Ion exchange in ultrathin films of Cu_2S and ZnS under atomic layer deposition conditions. *Chemistry of Materials* 23(20):4411–4413.

43. Bakke, J. R., Jung, H. J., Tanskanen, J. T., Sinclair, R., and Bent, S. F. 2010. Atomic layer deposition of CdS films. *Chemistry of Materials* 22(16):4669–4678.

44. McCarthy, R. F., Weimer, M. S., Emery, J. D., Hock, A. S., and Martinson, A. B. F. 2014. Oxygen-free atomic layer deposition of indium sulfide. *ACS Applied Materials and Interfaces* 6(15):12137–12145.

45. Powder Injection Molding Market Forecast (2018–2023), https://industryarc.com/Report/15989/powder-injection-molding-market.html.

46. Nishiyabu, Kazuaki. 2012. Micro metal powder injection molding. *Some Critical Issues for Injection Molding*, p. 105. Kinki University: Japan.

47. German, R. M. 1990. *Powder Injection Molding*. Metal Powder Industries Federation.

48. Löhe, D., and Haußelt, J. Advanced micro & nanosystems volume 3. *Microengineering of Metals and Ceramics*.

49. Suplicz, A., Szabo, F., and Kovacs, J. G. 2013. Injection molding of ceramic filled polypropylene: The effect of thermal conductivity and cooling rate on crystallinity. *Thermochimica Acta* 574:145–150.

50. A review of metal injection molding- process, optimization. *Defects and Microwave Sintering on WC-CO Cemented Carbide*. doi:10.1088/1757-899X/226/1/012162.

51. Edirisinghe, M. J., and Evans, J. R. G. 1986. Review: Fabrication of engineering ceramics by injection moulding. II. Techniques. *International Journal of High Technology Ceramics* 2(4):249–278.

52. Zauner, R. 2006. Micropowder injection molding. *Microelectronic Engineering* 83(4–9):1442–1444.

53. German, R. M., and Bose, A. 1997. *Injection Molding of Metals and Ceramics*. Metal Powder Industries Federation: Princeton, NJ.

54. Piotter, V., Gietzelt, T., and Merz, L. 2003. Micro powder-injection moulding of metals and ceramics. *Sadhana* 28(1–2):299–306.

55. Nishiyabu, K. 2012. *Micro Metal Powder Injection Molding*. Kinki University: Japan.

56. Sutter, M. 2009. *Micro-Injection Molding Puts Ceramics in Top Form, Small Precision Tools*. Carl Hanser Verlag: München, p. 1.

57. McGeough, J. A. 2002. *Micromachining of Engineering Materials*. Dekker: New York.

58. Manufacturing Alliance Associates Inc. Image of 5-axis cnc mill. http://www.mfgaa.com/5axis.table.table.jpg.

59. CCT Precision Machining, Texas, https://www.cctprecision.com/micromachining.

60. Owens Industries, Ultra Precision Machining Services, http://www.owensind.com/ProductImages.

61. Good Fellow Ceramic and Glass Division, http://www.goodfellow-ceramics.com/products/ceramics/shapal-hi-m/.

62. Allen, D. M. 1986. *Principles and Practice of Photochemical Machining and Photoetching*. Adam Hilgar.

63. Making, E., Sato, T., and Yamada, Y. 1987. *Photoresist for Photochemical Machining of Alumina Ceramic, Precision Engineering*. Butterworth & Co (Publishers) Ltd.

64. Kuar, A. S., Doloi, B., and Bhattacharyya, B. 2006. Modelling and analysis of pulsed Nd:YAG laser machining characteristics during micro-drilling of zirconia (ZrO_2). *International Journal of Machine Tools and Manufacture* 46(12–13):1301–1310.

65. Pandey, P. C., and Shan, H. S. 1980. *Modern Machining Processes*. McGraw-Hill: New Delhi.

66. Islam, M. U., and Campbell, G. 1993. Laser machining of ceramics: A review. *Materials and Manufacturing Processes* 8(6):611–630.

67. Knowles, M. R. H., Rutterford, G., Karnakis, D., and Ferguson, A. 2007. Micro-machining of metals, ceramics and polymers using nanosecond lasers. *The International Journal of Advanced Manufacturing Technology* 33(1–2):95–102.

68. Liu, J. S., Li, L. J., and Jin, X. Z. 1999. Accuracy control of three-dimensional Nd:YAG laser shaping by ablation. *Optics and Laser Technology* 31(6):419–423.

69. Islam, M. U. 1996. An overview of research in the fields of laser surface modification and laser machining at the Integrated Manufacturing Technologies Institute, NRC. *Advanced Performance Materials* 3(2):215–238.

70. Wüthrich, R. 2009. *Micromachining Using Electrochemical Discharge Phenomenon*. William Andrew Publishing: Oxford, UK/Burlington, MA.

71. Behroozfar, A., and Razfar, M. R. 2016. Experimental study of the tool wear during the electrochemical discharge machining. *Materials and Manufacturing Processes* 31(5):574–580.

72. Sarfaraz, M. A., and You-Wen, N. S. 1993. Sandhu, Electron beam machining of ceramic green-sheets for multilayer ceramic electronic packaging applications. *Nuclear Instruments and Methods in Physics Research Section B* 82:116–120.

73. Thoe, T. B., Aspinwall, D. K., and Wise, M. L. H. 1998. Review on ultrasonic machining. *International Journal of Machine Tools and Manufacture* 38(4):239–255.

74. Bullen Ultrasonics Inc, Ultrasonic Machining, https://www.Ceramicindustry.Com /Articles/ 88824-Ultrasonic-Machining.

75. Machining Ceramics with Rotary Ultrasonic Machining, https://www.ceramicindustry.com/ articles/83906-machining-ceramics-with-rotary-ultrasonic-machining.

76. Kuljanic, E. Advanced Manufacturing systems and technology. Proceedings of the Seventh International Conference.

77. Eckart Uhlmann Ioan, T. D., and Marinescu, D. 2015. *Handbook of Ceramics Grinding and Polishing*. Elsevior. ISBN: 9781455778591.

Structural and Functional Prototypes

Sangeeta Adhikari and Debasish Sarkar

CONTENTS

10.1 Introduction

Ceramic prototype development requires an understanding of the fascinating world of materials, processing principles, and the technology involved to achieve the targeted properties. The densification during ceramic processing has an intense relationship with its inherent properties. The properties of products relate to their classification as highly dense ceramics (~100%), categorically transparent ceramics, and porous ceramics that has less than 70% relative density; all others are dense ceramics. Thus, it is expected that the performance of the ceramics strongly depends on the processing conditions, and their functionality can be sub-categorized on the mode of utility of different archetypes.

The processing of advanced ceramics adopts basic approaches such as integration or disintegration of the pure precursors. Integration is bottom-up processing, whereas disintegration is top-down processing to form ceramics. The processing for bottom-up approach follows extended frameworks of a single molecule or particle. However, top-down approach follows structural decomposition of bulk to either particles or predefined thin films. The limitation in top-down approaches is the need for high-profile processing to synthesize high crystalline materials with polyatomic or molecular assemblies, in which the orientation and assembly of individual building blocks are responsible for control over the resultant shape and properties. It is worth mentioning that the systematic approach, comprehensive, and in-depth analysis of the product is quite crucial in the application aspect. To accomplish the process' reliably, one has to understand how the combinations of synthesis steps, processing parameters, microstructure, change in property, and finally the functionalization of the resultant works. Thus, conventional manufacturing techniques are being continuously improved and modified to gratify the miniature trend that will eventually facilitate the production of commercially viable ceramic components for versatile technological applications.

With consideration to properties and applications, ceramic prototypes are sub-categorized as structural and functional components. Structural ceramics take account of both the physical and mechanical aspects of engineering ceramics. Ceramics consisting of a robust structure and mechanical responses are classified as structural ceramics. These ceramics have excellent mechanical response with high thermodynamic stability under stressful conditions such as erosion, corrosion, and high-temperature surroundings. The great bond strength in such ceramics allows them to be active in several applications, including automobile pistons, hybrid bearings, tiles for aerospace vehicles, and tribological applications, such as mineral processing units, thermal barrier coatings for gas turbines, abrasives for polishing and grinding, cutting tools for machining, etc. The foremost research in this field involves understanding the microstructure as an outcome of processing parameters and essential correlation with mechanical properties, and eventually testing the performance in a working environment. In contrast with structural ceramics, functional ceramics deal with the relationship between processing of ceramics, their microstructure, and resulting properties. They include different forms of films (bulk, thin, and thick) of inorganic matter, non-metallic solid nanostructures, and nanomaterials fabricated through interesting, sophisticated, and advanced materials processing routes. Functionally active ceramics offer an opportunity to develop special materials to advance technological and targeted applications. In consideration of emerging technological importance and market demand, the fabrication protocol and working principles of commercial-grade structural and functional prototypes are discussed.

10.2 Structural Prototypes

10.2.1 Ceramic Piston

A ceramic piston valve is a displacement pump working on a single motor mechanism to aspirate and dispense combustible fluids in a precise manner through an internal ceramic piston. The major utility of the ceramic in such system is to prevent heat losses through the chamber walls during combustion. The design is simple and the lifeline of the instrument increases due to an internal sealing mechanism provided with a polished ceramic piston that eventually reduces the wear rate. It is designed to fulfill the application of individual pipetting, and is divided into two parts: pump drive module and pump head module. Based on the fluids, each of these modules can be dressed or swapped with a spare pump head as shown in the materials configuration in Figure 10.1a, where "C" stands for ceramic

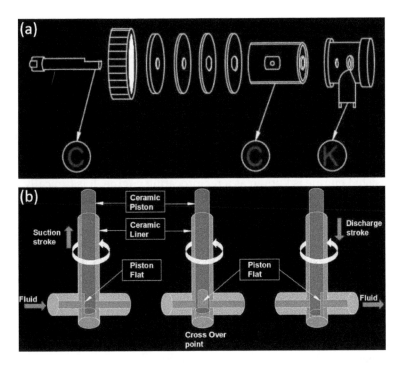

FIGURE 10.1
(a) Pump head materials configuration, and (b) schematic of ceramic piston valve operation.

(e.g., alumina), and 'K' stands for fluorocarbon (polyvinylidene difluoride, PVDF) polymer. Excellent physical strength, abrasion resistance, chemical resistivity, and dimensional stability at operating temperature and pressure are essential criteria of components when these are subject to direct contact with fluids. Commercially available, the components of a pump head and its materials are designated with particular codes based on trade names; details on materials and the different size combinations needed in order to obtain cost-effective ceramic-based pistons are described elsewhere [1].

In brief, the materials used for the configuration include ceramic, stainless steel 316 (SS316), carbon, and fluoropolymers. Most high-value piston and cylinder case liners use ceramics such as silicon nitride and fused crystalline alumina due to their excellent resistivity toward chemicals, mechanical resistance against common abrasives, and high thermal stability. The pistons are usually manufactured following a production process. This process includes (i) raw materials selection; (ii) processing of materials by several methods such as granulating, plasticizing, pulverizing, and spray drying; (iii) casting/molding includes processes such as dry pressing, extrusion, injection molding, and isostatic pressing; (iv) green machining takes account of milling, turning, drilling, and cutting of the machined samples; (v) sintering; (vi) finishing is carried out though practices such as grinding, lapping, polishing, metallizing, glazing, and self-assembly; and (vii) finally, the quality assurance of the finished product is a common practice to achieve the targeted piston geometry and dimension. However, the brittle ceramic piston and liners can undergo fracture under sudden impact and exposure to very dry fluids such as hexane. Thus, polyvinylidene difluoride (PVDF), a fluorocarbon polymer, has high chemical tolerance and is resistant to autoclave atmosphere and sensitive to organic solvents such as esters and ketones, frequently used for tube fittings and cylinder cases. Another fluoropolymer, ethylene tetrafluoroethylene (E-TFE), commercially named Tefzel, is also used in the production of cylinder cases for head modules due to its potential chemical resistivity against acids, bases, and solvents. The chemically and physically strengthened characteristics of SS316 are explored with a view to using it in the production of pistons, tube fittings, and cylinder cases. Conversely, SS316 is not the right choice as piston material when the cylinder liner is made up of ceramic, as later it presents higher hardness than steel. The main disadvantage of SS316 is its surface abrasion in presence of strong acids, bases, and also some halides. Sintered crystalline carbon can be used to make liners in combination with either ceramic or SS316, which are used for pistons. They exhibit sensitivity toward strong oxidants and abrasive materials.

A piston-assisted engine is known as a reciprocating engine. It is typically a heat engine and uses single or multiple reciprocating pistons to convert pressure into rotating motion. The ceramic piston opens flat to the inlet port, creates suction when pulled back, and fills the pump chamber with the fluid. The crossover takes place when the inlet port is sealed and the outlet port is opened. The ports are not opened simultaneously and are also not interconnected. Thus, upon forcing down the piston, this opens flat at the outlet port discharging the fluid present in the pump chamber. After emptying the pump chamber, the outlet port is sealed. This process repeatedly employs a suction stroke and disposes of the fluid to create effective combustion. The mechanistic functioning of the valve is schematically presented in Figure 10.1b [2].

These ceramic pistons improve accuracy by controlling doses with ultra-high precision. They are cheaper to maintain and have an easy handling of slurries without any of the clog/jam issues associated with other pumps. The lightweight and easy installation in desired orientations (vertical and horizontal) eases to reverse the direction of the motor without requiring any pipe change, simplifying the path of the fluid. These piston valves are flexible enough to be equipped with AC, DC, and stepper motors for unified integration. Such systems are mostly used in high-performance diesel engines and internal combustion engines. However, the utility of ceramic pistons is not limited to combustion engines. They can also be used as pumps and dispensers in medical equipment, for accurate metering in paint industries, as dispensers for insulating and encapsulating materials in electric motor industries, and for laboratory instrumentations such as chromatographic systems, quality controllers, automotives, packaging industries for food and dairy, pharmaceutical packaging, cosmetics, and many more.

10.2.2 Cutting Tools

Ceramic exhibits and retains hardness under high temperature and shows lower reactivity to steel, which sanctions its acceptability as a material to produce cutting tools. The life of a tool depends on the speed of cutting implemented without any wear and deformation. The major drawback is that the conventional metallic materials used for tool fabrication are not resistant toward thermal and mechanical shock for long periods. Thus, ceramic oxides, carbides, and nitrides are the three major categories of ceramic tool materials. Recently, nano-polycrystalline diamond and silicon nitride have joined the tool material league. Cutting tools are primarily of two types; single-point and multipoint, based on the cutting requirements. Shaping, planning, and turning are carried by single-point tools, whereas grinding, milling, and drilling are carried out by multipoint tools.

Among other oxides, aluminum oxide (Al_2O_3) is a well-known ceramic used for cutting tools. It holds interesting mechanical and chemical properties, resulting in high-wear resistance, hardness, and low chemical reactivity in comparison to steel. Hot hardness, high abrasion resistance, and a lower tendency to adhere to metal surfaces during machining makes alumina-based cutting tools beyond par. Titanium carbide addition to Al_2O_3 ceramic increases mechanical strength at room-temperature conditions but adversely affects strength, with an increase in temperature near 800°C [3]. Thus, research and development demand high-thermal-resisting cutting tools with excellent mechanical properties from the perspective of consistent life for cutting tools. Single-phase nano-polycrystalline diamond (N-PCD) is another ceramic cutting tool that exhibits ultra-hardness with high mechanical strength and thermal stability. It also offers high cutting precision. The synthesis process follows direct conversion of graphite under high temperature and pressure without any binder. Uniform doping of boron atoms in the diamond lattice can enhance the mechanical properties and also show electrical conductivity in the doped material. It enables machining of the cutting edges by electric discharge. The electrical conductivity property in boron-doped nano-polycrystalline diamond increases the resistance toward abrasive wearing and also tribo-electrical wear occurs due to tribo-microplasma damages [4]. Silicon nitride (Si_3N_4) is a non-oxide ceramic extensively used for cutting tools. However, poor sinterability holds back its potential in comparison to alumina-based ceramics. Sinterability is enhanced by forming composites with other materials of similar competency (For e.g., Al_2O_3, SiC, TiC, and others). Nanocomposite of Si_3N_4/SiC made by in situ carbothermal reduction method is a well-accepted protocol where grains of SiC are formed by adding fine silica during the process of sintering. The lifetime is increased by three times after machining cutting inserts [5]. A study was made on machining tests in order to understand the wear behavior between SiC whisker-reinforced alumina and Ti[C, N]-mixed alumina on martensitic SS-grade 410 and EN 24 steel pieces. Alumina mixed with Ti[C, N] has lower flank wear than the SiC reinforced alumina and performs better in cutting martensitic stainless steel [6].

Very recently, a tool for micro-milling was developed with N-PCD without binder to machine micro-textured surfaces on silicon carbide molds. The three-dimensional cutting on the N-PCD chip was carried out by laser (Figure 10.2a and 10.2b). The wear on the tool was much smaller with N-PCD than with single-crystal diamond, when SiC cutting

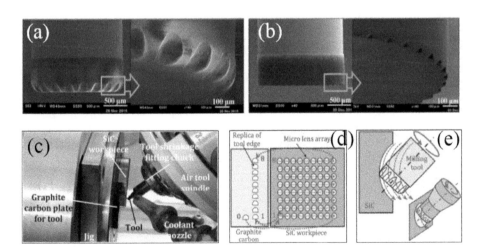

FIGURE 10.2
SEM image of laser-fabricated micro-N-PCD milling tools with (a) round edges of 0.2 mm radius and (b) sharp edges; (c) view of preliminary and micro-texturing experiments; (d) tool wear evaluation method; and (e) mold-machining method [7].

experiments were conducted. An image of a laser-fabricated micro-N-PCD milling tool is presented with a prelim on a micro-texturing experiment in Figure 10.2c, as well as the mold machining methods has been shown in Figure 10.2d and 10.2e. Very fine microstructures were achieved from the SiC mold after precise machining through the N-PCD tool [7].

These cutting tools are fabricated through several ceramic processing methods such as hot pressing, cold isostatic pressing followed by hot isostatic pressing, and reaction bonding techniques followed by sintering to achieve high grain-bonding in a tool work piece [8]. The bonding characteristics to attain cutting tool properties are further enhanced by permutation and combination of ceramic materials and their evaluation in terms of both structural and functional applicability. The development of high-powered machineries to fabricate tools has eased the process, ensuring smooth machining. Alterations in tool fabrication methodologies such as additive manufacturing (AM), with incorporation of modern technology tweaking and practices during machining, can start a revolution in this field [9].

10.2.3 Grinding Media

The ceramic grinding media is mostly used in the process where homogeneous mixing and dispersion play a crucial role. High efficiency in grinding must be in harmony with the lower abrasion rate of the grinding media in order to avoid contamination. For applicability in the medical and pharmaceutical industries, non-ferrous and high wear resistance grinding media is required to avoid undesired impurities from the grinding of the media itself. In majority, these media are used for grinding, crushing, de-agglomeration, deburring, polishing, and shot peening, and also as fillers, proppants, spacers, and refractory bed materials. These grinding media can come in different forms such as balls, cylinders, pellets, and pebbles, depending on the type of crushing/grinding under the influence of dry or wet processing in the mill. Most of the grinding processes use alumina and steatite as the common materials. There are several types of ceramics used as basic materials to fabricate the grinding media, such as zirconium silicate, ceria zirconium oxide, yttria zirconium oxide, glass beads, and alumina and steatite ceramics. Their schematic view is shown in Figure 10.3 [10]. Alumina is considered as a replacement for grinding media where contamination by steel balls is difficult to deal with. It is hard, chemically inert, corrosion resistant, thermal resistant, non-magnetic, non-toxic, and non-porous – these are just a few of its unique properties. It is commonly used in planetary ball mills, and attritions and jar mills. This class of grinding media is useful in making pigments, inks, paints, and epoxy resins. Its durability, non-reactivity, and cost-effectiveness make it a good grinding media. Alumina beads are mostly applicable in the mixing of ceramic masses and in glazing as well.

Another ceramic that can be used as grinding media is zirconium silicate, which has a strong molecular bond, formed due to a reaction between zirconia and silica. It produces

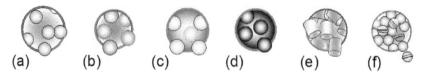

FIGURE 10.3
Grinding media composed of (a) zirconium silicate, (b) ceria zirconium oxide, (c) yttria zirconium oxide, (d) glass beads, (e) alumina, and (f) steatite.

wear-resistant materials with high density. These are fabricated by spray granulation followed by iso-pressing or compact pressing. The green bodies obtained after pressing are sintered in a kiln to obtain a non-porous and tough working surface. This is extremely suitable for milling slurries with low to intermediate density, and it minimizes contamination during milling [11]. Grinding balls of ceria-doped zirconium oxide are mostly used to refine metals such as gold and silver. The major specificity is the content of tetragonal zirconia polycrystalline, which makes it a tighter, and tougher material exhibiting high strength with high wear resistance. These are even stronger than zirconium silicate balls as the difference in hardness is huge. The physical characteristics of these grinding media is presented in Table 10.1.

In comparison to alumina, yttria-stabilized zirconia imparts durable grinding by reducing the grinding time through high impact, as the former has less density than the latter. They are mostly suitable for high-velocity operations and for grinding under wet conditions. Glass beads are another, cleaner grinding medium that preserves the formulation integrity. Thinner suspensions with lower viscosity use glass beads grinding media. These are fabricated through a traditional coating technique where layers of metal alloy or a metal compound film are mixed with the glass surface to change the optical performance of the glass. In another method, a glass cane is heated in a gas furnace until it becomes red hot, and is then inserted in a pressing machine with a definite mold to get the desired shape of molten glass. The pressing of a few beads is only possible in one heating process. The glass is allowed to cool slowly and the excess glass on beads is removed through a two-stage tumbling process, during which the excess pieces are removed and the imperfections on the surface of the glass beads are smoothed. These beads are recommended for use in pharmaceuticals, magnetic coatings, coloring agents, and also in the paper and rubber industries.

Steatite, having the composition $MgO-SiO_2$, is an excellent grinding medium due to its hardness, which is greater than that of glass beads. It is also more resistant than they are to corrosion and wear [13]. It is chemically resistant against all alkalis and acids, which makes it an effective grinding medium to be used in wet conditions for particle size reduction. Moreover, it does not affect the composition and color of the milled product, which is why it is a faster, cleaner, and more reliable grinding medium. However, grinding materials with higher hardness than steatite increase the impurity content. Steatite is usually fabricated through a pelletizing, dry pressing process, and round beads are formed by press-rim. These are preferentially used for milling low-viscosity materials that do not require a high grinding force. Steatite increases the efficiency of milling and it is particularly useful where the generation of heat during the process is prohibitive and detrimental. Other ceramic materials that are joining the list as potential grinding media are SiC, Si_3N_4, and tungsten carbide (WC). Silicon carbide and nitride are expensive grinding media. They are used especially to grind same-composition compounds. WC is the costliest, densest and hardest of all the grinding media, and thus provide excellent impact during milling, but its chemical interference (presence of cobalt binder) in the presence of acids and other solvents acts as a major drawback [14].

10.2.4 Spark Plug

Spark plug is a device placed in the head of the internal combustion engine cylinder. It undergoes ignition by means of electric spark. In other words, the main function of the spark plug is to offer a workplace for an electric spark that is hot enough to ignite the fuel and air mixture inside the chamber of combustion. The spark plug consists of a center electrode, an insulator (e.g., alumina), a metal shell, and a ground electrode or side electrode,

TABLE 10.1

Physical Characteristics of the Ceramic Grinding Media (HV = Vickers Hardness) [12]

Grinding Media	Alumina	Zirconium Silicate	Ceria Zirconium Oxide	Yttria-Stabilized Zirconia	Glass Beads	Steatite Ceramic Beads
Shape	Spherical and Cylinders	Spherical	Spherical	Spherical	Spherical	Spherical and Cylinders
Composition	Aluminum Oxide	Zirconia and Silica	Zirconia and Ceria	Zirconia and Yttria	Glass	Magnesia and Silica
Color	White	White	Brown	White	Clear	Off white
Specific Density (g/cm³)	3.53	3.7	6.2	6.0	2.5	2.6
Hardness	1800 HV	1050 HV	1300 HV	1350 HV	550 HV	625 HV

where a thick metal wire along the length of the plug is the center electrode that conducts electricity between the ignition cable and the other end of the plug. A ceramic insulator surrounds the center electrode as a case exposing the top and bottom portions. The metal shell is a threaded, hexagonal-shaped case that allows the installation of the spark plug in the tapped socket in the head of the engine cylinder. The ground electrode is a thick, nickel-alloy wire connected to a metal shell extending to the center of the electrode. Depending on the engine types, the side and ground electrodes are about a few inches (0.020–0.080 inches) apart to allow the spark to jump across through this gap.

The manufacturing of spark plugs takes place via assembly of the components in six steps, as shown in Figure 10.4. At the beginning, a proper hollow steel shape is obtained by extrusion. These shapes are named blanks. Following this step, blanks are machined and knurled, which is a part of the forming operation. A side electrode with a partial bend is attached to the machined blanks. A ceramic insulator is pressure-molded through the center with a hollow bore. The electrical welding of the center electrode and of the terminal stud then takes place, followed by the insertion of the insulator into the hollow bore. The sealing of this assembly is carried under extreme pressure. Eventually, machine shaping is performed in order to obtain the exact shape for the center electrode, and the final bending of the side electrode is carried out [15].

The ceramic insulator used in spark plugs is a high alumina (Al_2O_3), which experiences high voltage without any electrical breakdown and should fit tightly in the packaging windows of modern engines. This acts as the only barrier between the electrodes; however, recent spark plugs use alumina as an active insulation material. Porcelain, sillimanite, and steatite were also explored as insulating materials. Technological advancement in the production of calcined alumina has made a breakthrough in the development of spark plug insulators. The preparation of the insulator material involves ball milling of the ceramic materials under wet grinding conditions. Controlled grinding followed by spray drying creates a free-flowing ceramics to pour inside the rubber molds in order to obtain high-density green components after pressing. A specialized press applies hydraulic pressure to produce green insulator blanks. During this process, the dimensions of the hollow part are controlled very rigidly to precisely accommodate the center electrodes. Special contouring grinding machines provide the final shape of the pressed insulators, which are then sintered in a tunnel kiln at a temperature above 1,500°C, and glaze is applied thereafter. Usually, the performance of the spark plug is affected by the insulator material, thus it is worth remembering that there should not be any electrical flashover or leakage from the terminal of the plug to the shell.

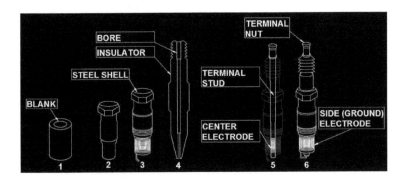

FIGURE 10.4
Part component and steps involved to make the assembly of the spark plug for internal combustion engines.

10.2.5 Prosthesis

The terminology "prosthesis" implies the replacement of an amputated body part with a synthetic or artificially developed device that is body friendly, either inside or outside. The development in ceramic materials has improved the present scenario of prosthetic implants, enhancing their applicability to the body [16]. The most widely applied prosthetics made from ceramics use alumina for hip and knee replacement, zirconia for dental prosthesis, calcium phosphate for bone tissue engineering, bone fillers, bio-glass or glass ceramic for orthopedics, hydroxyapatite and tri-calcium phosphate materials for bone grafting and coating over the implants, and so on. Each of these materials has its specific properties toward being used as an implant and exhibits its limitations in different ways, as discussed in Table 10.2.

There has been a major breakthrough in prosthesis with hip-joint transplantation, thanks to the use of bearings. Such bearings, made of ceramics, were introduced for long-term survivorship, to minimize the wear of the total hip. The main reason to use a ceramic prosthesis is to minimize the wear during articulating motions, increasing the lifetime of

TABLE 10.2

Properties, Applications, and Limitations of a Few Major Ceramics Used as Prosthesis

Ceramics	Properties	Applications and Failure
Alumina (Al_2O_3)	High density, purity, low wear rate, and low coefficient of friction. Bio inertness, chemical inertness, high hardness, resistance to abrasion, biocompatibility.	Artificial total hip, knee, shoulder, elbow, and wrist replacements; vertebrae spacers and extensors; alveolar bone replacements; mandibular reconstruction; endosseous tooth replacement implants, and orthodontic anchors **Limitations:** Implant failure can be due to instability or mechanical and prosthetic design problems.
Zirconia (ZrO_2)	Biocompatibility, high mechanical strength, fracture toughness, good osteointegration, non-cytotoxic, corrosion resistance, excellent wear resistance.	Total hip replacement (ball head) and dental. **Limitations:** Unstable due to shape change, i.e., shape changes from one form to another; and lower strength in comparison to titanium dental implants.
Bio-glass and glass ceramic	Biocompatible, bioactive, non-toxic and osteoinductive properties.	Orthopedic and dental prosthesis. **Limitations:** Sensitive to small defects and flaws that can cause catastrophic failure due to lower fracture toughness.
Calcium phosphate	Stable in body fluid under conditions of dry or moist air up to 1,200°C. Highly bioactive, non-toxic, bioresorbable, osteoconductive and osteotransductive.	Bone tissue engineering, bone fillers and dentistry. **Limitations:** Very poor mechanical properties.
Hydroxyapatite	Good osteointegration, bioactive, high strength, and non-resorbable.	Bone grafts and fillers; coatings for metal implants. **Limitations:** Very poor mechanical strength.
Tricalcium phosphate	Osteoconductive, biocompatible, and resorbable material.	Capping agent; apical barrier; cleft palate; apexification; implant coating, and vertical bone defect. **Limitations:** The potential in osteogenesis is limited and they exhibit brittleness.

the implant and restricting ion leaching as well as osteolysis. Optimized zirconia (5 wt% 3Y-TZP)-toughened alumina (ZTA) is an excellent choice for hip replacement as it enhances wear resistance and minimizes ion leaching in in vivo arthroplasty (Figure 10.5a) [17].

Compared to polymers and metals, ZTA ceramic has higher fracture toughness than alumina, and its inherent structure allows adsorption of water on the surface through hydrogen linking between oxygen atoms in alumina and water molecules.

It imparts higher wettability due to its hydrophilic structure, resulting in better fluid film lubrication compared to other polymer or metal orthopedics. Bio-grade polycrystalline alumina and zirconia are used to fabricate the femoral head and acetabular socket by uniaxial pressing followed by sintering at the high temperature of 1,600°C for six hours in order to produce unassembled parts as shown in Figure 10.5b. Similar compositions were used to prepare different specimens to measure the mechanical and biological responses and justify the processing conditions in the prospect of hip prosthesis [18, 19].

The treatment of intra-oral defects such as missing teeth, parts of teeth, or a missing structure in a jaw and palate can be successfully effected with dental prosthesis. This speciality is called prosthodontics. Zirconia dental ceramics is mostly used in combination with other metals such as hafnium, yttrium, and oxides such as alumina to improve structural properties. The dental blanks of zirconia are an innovation in contrast to metals. Their non-metallic nature removes the danger of corrosion. The zirconia systems used as prosthesis materials include yttrium-doped tetragonal zirconia polycrystals (3Y-TZP),

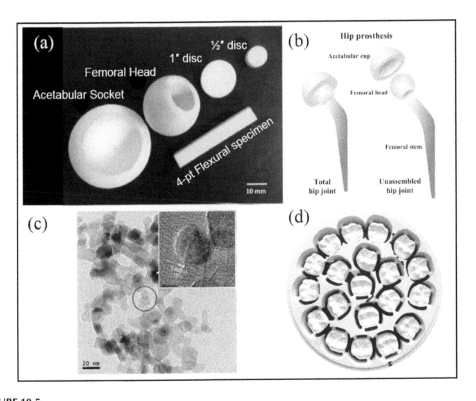

FIGURE 10.5

(a) ZTA-based ceramic components and lab-scale specimens prepared through uniaxial pressing of powders from optimized 95Al$_2$O$_3$-5ZrO$_2$ sintered at 1,600°C for six hours with 800 ppm MgO composition (wt%) [17], (b) assembled and unassembled parts in hip prosthesis, (c) TEM images of zirconia nanoparticles prepared by chemical method [20], (d) commercial zirconia dental blanks and teeth obtained through CAD/CAM machining [22].

magnesium-doped partially stabilized zirconia (Mg-PSZ), and zirconia-toughened alumina (ZTA). In this aspect, the zirconia nanoparticles can be synthesized through a chemical method, and acquire a particle size of 20–50 nm, as shown in Figure 10.5c [20]. The incorporation of zirconia particles into an adhesive layer can increase strength and stimulate mineralization after implantation. Zirconia-based compounds are osseoconductive and exhibit constructive interaction with the soft tissues [21]. They also reduce the formation of plaque on the surface of the implant, which promotes healing and successful implantation. The mechanical properties of dental implants strongly depend on the grain size, which is eventually affected by the processing conditions. It is reported that 3Y-TZP, having grain size above 1 micron, is less stable when subjected to tetragonal or monoclinic transformation, whereas below 0.2 microns, toughening is impossible due to bargained fracture toughness, thus a grain size window of 0.2–1 µm is an excellent choice for sintered zirconia-based dental ceramics. Single-piece oral endosseous implants manufactured by soft machining of pre-sintered blanks, followed by high-temperature sintering and then hard machining to obtain the required structure, are a popular adaptation for dental patients. The pre-sintered blanks are generally obtained by dry pressing and slip casting methods, and are then fired, polished, and shaped according to their application [20]. Advancements in technology have simplified the manufacturing process for fabricating ceramic dental prosthesis and screws, by means of computer-aided designing/manufacturing (Figure 10.5d) [22].

Tissue engineering and stem cell research, through the use of ceramic materials, have found treatments for orofacial fractures, bone augmentation, cartilage regeneration of the temporomandibular joint, pulp repair, periodontal ligament regeneration, and implant osseointegration. Bionic eyes or ceramic microdetectors are another achievement made in prosthesis that can cure blindness. These make use of ultra-thin films, which are stacked in hexagonal arrays mimicking the rod and cone arrangement. Earlier, silicon nanowires were used for this purpose, but atrophy of retina was a major problem. However, the oxide thin film formed using atomic oxygen as natural oxidant exhibits a porous structure that allows the flow of nutrients from the back to the front of the eye, preventing the existing problems with silicon nanowires. Moreover, such integration does not require any wire connections or encapsulation [23].

10.2.6 Stationery Components

Despite the major dynamic components discussed above, other structural components made up of ceramic materials, such as scissors, knives, and mirrors, have extensive market outlets and applications. In the field of stationery cutting tools, zirconium oxide powders are mostly used to fabricate ceramic scissors. Scissors made up of mullite ($3Al_2O_3.2SiO_2$) are also available but limited. These can be used for a wide range cutting that includes soft papers, plastics, art materials, fishing lines, and also clamshell packaging. As their blades have excellent hardness, their lifetime is up to ten times longer than that of steel. Resistance against alkaline and acidic substances has increased the kitchen and industrial applicability of ceramic-based scissors.

The typical ceramic scissor is fabricated following these steps, as depicted [24]. At first, the sub-nanometer-ranged ZrO_2 powder is poured into a scissor-shaped mold. Afterward, uniaxial pressing is performed in the range of ~250 MPa pressure without addition of adhesives, in order to get a base shape for the scissor blade. This green scissor is nearly 20% larger than the targeted dimension. The green specimen is then further inserted into a tied polyvinyl chloride vacuum bag and kept in a closed container. Room-temperature isostatic pressure greater than 200 MPa is applied after the addition of an oil and water mixture for a prolonged time. High density with low porosity and low deformation are the results, after high-tech sintering under controlled and uniform heating temperature and atmosphere. The fourth step is to attach a handle to the fabricated ceramic blades through advanced injection molding to avoid any contamination from glue and to impart strong bonding. The advantage of using injection molding is that it achieves stronger bonding than thermal bonding and is more sanitary than other methods.

Making a perfect blade in both cutting edge and surface finish is important, and takes place via three steps: in the beginning, the blades are ground flat with a diamond wheel on both sides. Immediately after, the edge of the blades is carefully ground. This is called round grinding, and is followed by a preliminary edge produced through open-edge grinding. Then, the edge of the blade and the tip are sharpened to an angle of 11°–13° and 35°–42°, respectively, and the final stage, polishing, is performed to achieve cutting-quality sharpness. The assembly of the blades is carried out with the help of a rivet, and measures are considered for the flexible movement of the blades, to provide effective operation. Finally, the complete scissor set-up has to undergo a quality test to confirm it is a defect-free product, as depicted in Figure 10.6a. A similar fabrication protocol can be followed in order to make a zirconia knife (Figure 10.6b) [25].

Ceramic mirrors are another stationery component manufactured using technical ceramics, such as silicon carbide, silicon, cordierite, and carbon-SiC materials. Mirror fabrication by carbon/silicon carbide (C/SiC) is recognized as a potential candidate because of superior properties such as thermal expansion, high specific stiffness, high thermal conductivity, dimensional stability, electrical conduction, resistance toward corrosion and chemicals, and ultra-lightweight competence with no porosity. Conventional glass mirrors are fragile. To overcome this lacuna, C/SiC can be used for the construction of complex space mirrors, as shown in Figure 10.6c. Shrinkage in SiC material during sintering leads to structural complexity, and thus it can be reduced through addition of an optimum content of carbon-fiber reinforcement. Such mirrors are mostly used in space shuttles and space mirrors. Opto-mechanical space equipment requires high stiffness, and good thermal and electrical conductivity in combination with low thermal expansion coefficient [26].

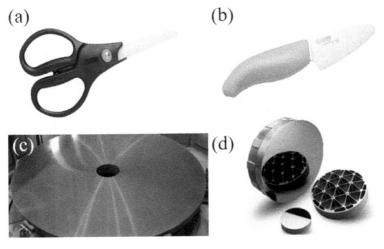

FIGURE 10.6
Stationery ceramic components: (a) zirconia scissors, (b) zirconia knife, (c) carbon fiber-reinforced SiC ceramic mirror, and (d) ultra-SiC sintered silicon carbide.

An image of ultra-SiC sintered silicon carbide is shown in Figure 10.6d. These mirrors are fabricated for the purpose of airborne sensing, such as gimbaled multi-spectral sensors for remote sensing, high-energy laser communications. They are also used as fast steering mirrors and have accessibility to ultraviolet, visible, and infrared wavelengths to match the space telescope applications [27].

C/SiC material processing can be done by several methods, including pressureless sintering, hot pressing, hot isostatic pressing, and chemical vapor deposition. The production of the green body from raw material is carried out by molding the carbon fiber and phenolic resin, followed by carbonization at 1,000°C to form felt, and graphitization at 2,100°C in vacuum [28]. The green body is then milled and mounted on a furnace where molten silicon is infiltrated at high temperature. The molten Si simultaneously reacts with the carbon matrix and carbon fiber surface, resulting in a silicon carbide matrix. This is known as SiC infiltration process. This infiltrated mirror blank is shaped to requirement, and the optical surface is polished through a cladding layer formed by chemical vapor deposition. Surface lapping, optical polishing, and reflective coating perform final mirror-surface touching. High optical performance is achieved by plasma etching of the surface through ion-beam polishing. The sub-components can be joined in the stage of green-body preparation to form the large-shaped structure of C/SiC. This allows us to manufacture monolith mirrors and then assemble them to form larger structures. One of the major applications of C/SiC is in the production of ultra-lightweight scanning mirrors used in meteosat second-generation spacecrafts. These mirrors accomplish thermo-mechanical stiffness with high optical performance and low mass. Thus, they can resist the scorching heat of the sun as well as the intense cold of the space. C/SiC mirrors can also be arranged with C/SiC precision structures to build a thermally optical system, such as all-ceramic telescopes.

10.3 Functional Prototypes

10.3.1 Solid Oxide Fuel Cell (SOFC)

SOFC is an environment-friendly electrochemical energy conversion device that offers the delivery of high electrical efficiency by using a wide range of fuels. Typically, the reaction between fuel and oxidant takes place through a solid electrolyte layer that diffuses the oxide ions/protons to produce electricity through completing the external circuit attached to anode and cathode electrodes. In this system, hydrogen is used as common fuel in combination with oxygen as oxidant. The massive functionality of ceramics has great importance in SOFC, where both the cathode and anode are made up of porous ceramics and the dense ion-conducting ceramics are used as electrolytes. Raw-material preparation, purification, and their proper sizing facilitate the fabrication of dense green compacts of solid electrolytes. To refine and properly mix the materials, milling is performed in the presence/absence of binders or slurry agents. In order to develop commercial-grade green compact, several protocol such as tape casting, calendaring, slip casting, pressuring, extrusion, dip coating, printing, thermal spraying, and vapor deposition have been adopted. The form is then conditioned by drying, bisque firing, and sintering to fulfill the customer demand. In brief, the proper-shaped samples are conditioned by drying at 100–200°C to obtain a green substrate, followed by intermediate bisque firing at 300–500°C and sintering at high temperatures ranging from 500 to 1,200°C depending on the materials used for fabrication. Usually, low-volume production can be performed in batch furnaces, whereas a continuous furnace is considered for a high volume of production.

Usually, each of the fabricated component electrode is assembled through two approaches, namely, planar and tubular. In the planar approach, the cells are stacked in layers, whereas in the tubular approach the electrode components are stacked, forming tubes of either type of SOFC. Recently, an efficient flat-tube-type SOFC has been introduced to get the combined advantages of the planar and tubular types. The tubular type has the advantage of resistance against thermal stress, while the planar type has the advantage of higher power density compared to tubular cell when the cells are stacked. Some virtual models of flat, tubular, and flat-tube types are illustrated in Figure 10.7a [29, 30]. These assemblies are then carefully checked to avoid probable leaks and conditioned again for better bonding. The oxide ions motion in the electrolyte are crucial because oxygen activity difference in the electrodes is the prime working principle of SOFC. The dissociated oxide ions from oxygen under electron consumption migrate through the electrolyte to the layer of anode, where they react with fuel hydrogen forming water. The electron released creates a current flow through the external circuit [16].

The choice of materials for electrodes depends on the functionality of each individual electrode. The requirement for cathode material is to have a triple-phase boundary with active exchange of surface oxygen, electronic conductivity, and porosity. It should be able to transport gas and pick up the current. Conventionally, the most used ceramics are perovskites, $La_{1-x}Sr_xMnO_3$ (LSM), $La_{1-x}Sr_xFeO_3$ (LSF), and $La_{1-x}Sr_xCo_{1-y}Fe_yO_3$ (LSCF). Despite this, the anode should be stable towards redox reactions, electrocatalytically active, and sulfur tolerant. Cermets and $Ni-Y_{0.15}Zr_{0.85}O_{1.93}$ (Ni-YSZ) are the most explored anode materials. The electrolytes used in SOFC are solid state, and should be capable of transporting oxygen ions. They should have high oxygen ion conductivity and electrical insulation. They

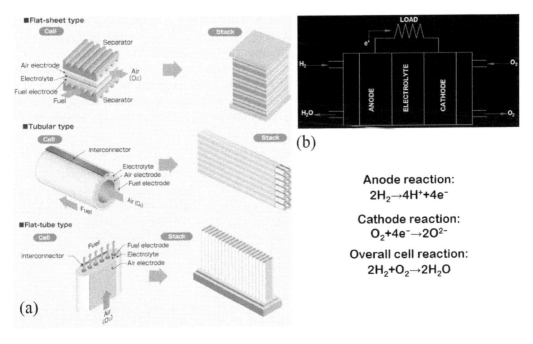

FIGURE 10.7
Assembly of SOFC in (a) planar stack, tubular stack, and flat-tube stack, and (b) single SOFC with anode, cathode, and overall cell reactions.

should also be gas tight and impart mechanical stability. A typical SOFC consists of anode and cathode; their reaction mechanism is illustrated in Figure 10.7b. Yttrium-stabilized zirconia ($Y_{0.15}Zr_{0.85}O_{1.93}$, 8-YSZ) is the most common solid-state electrolyte, although a different class of electrolytes is in the forefront to develop SOFC [29]. The thickness of each electrode and electrolyte varies depending on the fabrication technique. The thickness of the anode support varies from 300 to 2,000 μm, and is prepared through methods such as tape casting or die pressing. Alternatively, ceria anodes are also developed in comparison to common existing anodes such as $La_{0.8}Sr_{0.2}Cr_{0.5}Mn_{0.5}O_3$ (LSCM) and $Ce_{0.6}Gd_{0.4}O_{1.8}$. During the fabrication of cells, electrolytes and anodes are preferentially co-fired in stack. The porosity of the materials can be controlled by adding pore former and also by controlling the shrinkage. The gas transport through anode demands a coarse and openly porous structure. On the other hand, the material for solid-state oxide electrolyte should be such that there should not be much loss while transporting the oxygen ions with separation of two gaseous atmospheres. Electrolytes experience chemical stability to resist oxygen gradient pressure while simultaneously being exposed to both reducing and oxidizing conditions. The operating temperature of SOFC can be reduced by decreasing the electrolyte thickness by a few microns to nanometeres, and thus in this aspect the thin film made from nanopowders is a promising technique [16, 31, 32]. More complex cell concepts may be developed in the future through minimization of the operating temperature of SOFC. SOFC fabrication costs can be further reduced by using spray pyrolysis and thermal spray techniques.

10.3.2 Dye-Sensitized Solar Cells (DSSC)

DSSC is a solar harvesting renewable energy resource that makes use of a semiconductor material as light absorber and the charge transport layer imprinted by Grätzel [33, 34]. Typically, in a DSSC, separation of the charge generated at dye-semiconductor interface occurs followed by transport of the charge through the semiconductor-electrolyte interface. The operating principle of DSSC (Figure 10.8a) is fairly simple, and follows the photonic absorption by the sensitizer S (Equation 10.1), leading to the sensitizer excitation (S $\rightarrow$ S*), where an electron gets injected into the conduction band of a semiconductor material, leaving the sensitizer at an oxidized state of S$^+$ as presented in Equation 10.2. The electron injected then flows through the network of semiconductor material, arriving at spinal contact, then moves through the external circuit reaching the counter electrode,

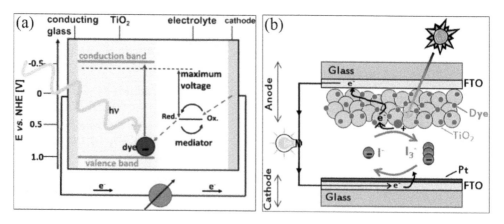

FIGURE 10.8
(a) Operating principle, and (b) schematic representation with components of DSSC.

where it reduces the redox mediator to regenerate the sensitizer, completing a complete circuit (Equations 10.3 and 10.4) [35].

$$S_{(\text{photon absorbed})} + h\upsilon \rightarrow S^*_{(\text{photon absorbed})} \qquad (10.1)$$

$$S^*_{(\text{photon adsorbed})} \rightarrow S^+_{(\text{photon absorbed})} + e^-_{(\text{injected})} \qquad (10.2)$$

$$I_3^- + 2\,e^-_{(\text{cathode})} \rightarrow 3\,I^-_{(\text{cathode})} \qquad (10.3)$$

$$S^+_{(\text{photon absorbed})} + \frac{3}{2}I^- \rightarrow S_{(\text{photon absorbed})} + \frac{1}{2}I_3^- \qquad (10.4)$$

Other reactions that result in loss of cell efficiency are the probable recombination of the injected electrons with the oxidized sensitizer or with the oxidized redox couple at the surface of TiO_2, as shown in equations 10.5 and 10.6. These reactions contribute to reducing the cell efficiency.

$$S^+_{(\text{photon absorbed})} + e^-_{(\text{TiO}_2)} \rightarrow S_{(\text{photon absorbed})} \qquad (10.5)$$

$$I_3^- + e^-_{(\text{TiO}_2)} \rightarrow 3I^-_{(\text{anode})} \qquad (10.6)$$

In reality, DSSC is constituted of five basic components:

1. Transparent conducting oxide (TCO) substrate that acts as a mechanical support.
2. Semiconducting material (for the case above, TiO_2).
3. A sensitizer that can absorb visible energy is adsorbed on the semiconductor surface (ruthenium complexes, Ru(4,4'-dicarboxy-2,2'-bipyridine ligand)₃);
4. A redox mediator which is the electrolyte (triiodide/iodine complex) in DSSC.
5. A counter electrode (typically platine/platinum) to complete the circuit by regenerating the redox mediator as marked in Figure 10.8b. The photoelectrode is the dye-sensitized semiconductor material over TCO.

Typically, the process for DSSC fabrication starts with photoanode and photocathode preparation. Photoanode preparation includes the preparation of TiO_2 paste from nanoscale particles, drop-cast or spin-coated on the confined region of the coating area on TCO, followed by annealing. Then, the annealed electrode is allowed to soak in the dye solution (ruthenium complex) for 10 minutes, followed by washing with water and ethanol to remove the excess solution. The photocathode can be prepared by taking a glass substrate and coating one of the surfaces with platinum through sputtering; otherwise, a carbon film can be scratched using a graphite pencil; this is then annealed at 450°C for few minutes and washed with ethanol. Both the photoelectrodes are then assembled when the carbon/platinum-coated cathode is placed over the dye-sensitized photoanode. Both the electrodes can be clipped using alligator clips to complete the circuit assembly, and the electrolyte (triiodide/iodine) is poured neatly between the two electrodes. To have homogeneous distribution of electrolytes, each of the clips is opened and closed, alternatively. Now, the DSSC device is tested under a light source, and conductivity is checked using a multimeter. The materials used for photoanodes are mostly nanocrystalline wide-band gap n-type semiconductors such as TiO_2, SnO_2, ZnO, and others [36]. As already discussed, the injected electron can follow two pathways: either it can travel to the back contact, or it can move to the dye oxidized for further recombination. These recombinations can be overcome when the oxidized dye is neutralized more quickly by the mediated electrolyte.

Ceramic material plays a role when the surface of the photoanode is passivated by compact high-band ceramic materials such as Al_2O_3, HfO_2, SiO_2, MgO, and ZrO_2 [37]. An interesting analysis clarified the influence of HfO_2 either on the coating of TCO or on the coating of the particles, and these details are depicted in Figure 10.9. The bottom panel represents the band diagram for each photoanode and their possible electron recombination processes, such as TiO_2/electrolyte, TiO_2/dye, and ITO/electrolyte, respectively, in presence of HfO_2 [38]. These compact layers act as blocking layers to control the recombination process. To achieve greater cell efficiency, HfO_2 was used as the blocking layer in one of the DSSC studies, where an atomic layer deposition technique was employed to deposit HfO_2 on the TiO_2 surface, and also on the conducting substrate indium-doped tin oxide (ITO). The various types of recombination process have been schematically represented with the design of fabricated anodes and their corresponding band diagram for each photoanode, with incorporation of HfO_2 blocking layer. The mechanism reveals that layering TiO_2 nanoparticles with HfO_2 reduces the recombination sharply [39].

Transition metal nitrides and carbides are known to exhibit Pt-like behavior and are alternatively used in hydrogenation, dehydrogenation, and methanol oxidation processes [40]. WC, MoC, TiC, SiC, NbC, Cr_3C_2, and Ta_4C_3 are some of the best-known carbides used as counter electrodes [41]. Beside the carbides, the nitrides and oxides of the above-mentioned transition elements are also well explored. However, both carbides and nitrides facilitate properties such as high hardness and melting point with high thermal and electrical conductivity as well. They are the low-cost counter electrode catalysts that have replaced the noble metal platinum (Pt) in DSSCs and other versatile fields of applications as well.

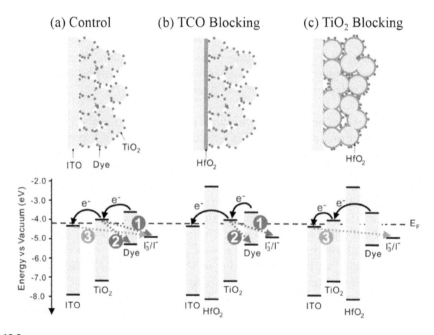

FIGURE 10.9
(a) Control consisting of TiO_2 nanoparticles on ITO, (b) a TCO blocking layer consisting of HfO_2 deposited on ITO, and (c) a TiO_2 blocking layer consisting of HfO_2 deposited on TiO_2 nanoparticles.

10.3.3 Energy Storage Devices

Considering modernization, batteries and supercapacitors have evolved to be highly efficient energy storage devices. Batteries have a very crucial role in today's society, especially where work involves portable devices. These devices are single and compact systems that fulfill high voltage requirements, high energy density, and also a prolonged life cycle of the devices [42]. Portable electronic devices such as cellular phones, laptops, and other electronic gadgets make extensive use of such properties. In this context, systems with benefits of volumetric and gravimetric energy density have evolved. They have variable operating range and safety when compared to traditional liquid electrolyte systems. There still remain fundamental issues with the understanding of the phenomena at the electrochemical interface. Looking into these solid-state battery systems (SSB), the structure consists of cathode, anode, electrolyte, and current collector as shown in Figure 10.10a [43].

The electrolyte in these systems acts as both separator and ionic conductor. Since solid-state systems are much simpler than liquid-based systems, they reduce the cost of sophisticated packaging, thereby reducing the fabrication cost as well. In consideration to the Li-battery systems, the working principle is such that, during charging process, Li ion transfers from the cathode's crystal lattice to the anode through solid-state ionic conducting electrolyte.

However, during discharging, the intercalated Li ions de-intercalate from the anode and get transferred to the cathode via the electrolyte, where electrons passing through the external circuit keep the device in working. There are several reaction steps involved

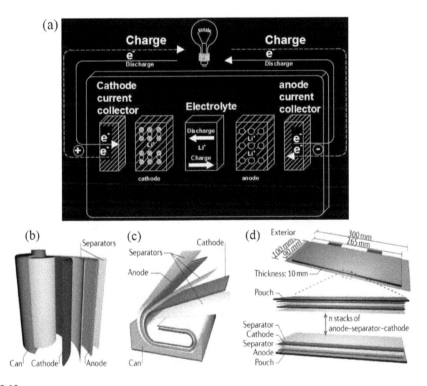

FIGURE 10.10
(a) Schematic representation of solid-state battery (SSB). Commercial configurations of SSB, (b) cylindrical-type cell, (c) prismatic-type cell, (d) pouch-type cell. The pouch dimensions are denoted, along with the internal configuration for n anode–separator–cathode stacks [44].

in working mechanisms at the electrolyte/electrode interface. Some of the representative commercial cell structures are presented in Figure 10.10b [44].

In SSB systems, the solid electrolyte has a significant role as it is in contact with both the anode and the cathode. The solid electrolytes used are either films of polymer or electrolytes composed of inorganic-based glassy or ceramic materials. In the present world of SSBs, the individual components in SSB systems are made up of ceramic-based materials. The solid electrolytes explored in this context are Li_3N, lithium phosphorous oxy-nitride (LiPON), Li_2S-based glass, sodium superionic conductor, NaSICON-type oxides such as $Na_3Zr_2Si_2PO_{12}$, $Li_7La_3Zr_2O_{12}$, and $Li_{0.05-3x}La_{0.5+x}TiO_3$, etc. [45]. Inorganic ceramic electrolytes have come to attention due to their easy production, mechanical stability, and high ionic conduction. Each of these electrolytes possesses some advantages and disadvantages: for instance, glass-based electrolytes impart isotropic properties, which facilitates fast ion migration, but their sensitivity toward humidity limits their application [46]. Recently, Suzuki et al. have developed a honeycomb-like structure of $Li_{0.55}La_{0.35}TiO_3$ (LLT) with micro-sized holes as a three-dimensional electrolyte sandwiched between sol-gel-derived $LiCoO_2$ and $Li_4Mn_5O_{12}$ as cathode and anode material, respectively. The active materials impregnated in the precursor sol provide a good contact between active material and LLT, which was operated successfully at 1.1V with a discharge capacity of 7.3Ah/cm², attributed to the minimization of internal cell resistance [47].

An ideal confined geometry is supposed to increase the volume-per-unit area of the active material while keeping minimum thickness for fast Li diffusion. Thus, the three-dimensional SSBs were developed comprising thermally coated SiO_2 on Si (001) substrates with Ti/Pt as cathode-current collector. A coating of $LiCoO_2$ was carried out through sputtering to achieve a nominal thickness of ~300 nm, followed by annealing at 700°C for two hours under ambient atmosphere. Further, the samples were sputter coated with LiPON electrolyte with thickness ~500 nm at 200°C. Finally, a deposition of Si with ~100 nm thickness took place and, on top of that, Cu deposition of ~400 nm was carried out as current collector. A hole of 0.5mm diameter of stainless steel shadow was masked such that each of these individual set of components acts as a microbattery. Details are illustrated in Figure 10.11 [48]. Although the power performance was not high, the development of such systems, with improvement in uniform coating, can lead to high-performance of fabricated SSBs.

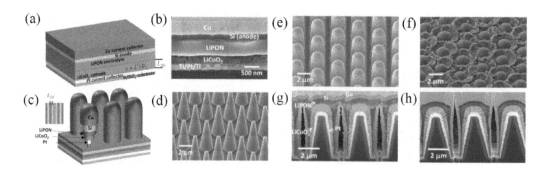

FIGURE 10.11

(a) Schematic of a planar thin-film SSB, (b) focused ion beam (FIB) cross-section of a thin-film SSB, (c) schematic of a 3D SSB, (d) Si conical microcolumns used as scaffolding for 3D SSB, (e) Ti/Pt and $LiCoO_2$ deposition on structure (c), (f) LiPON, Si, and Cu current collector deposition, and (g, h) FIB cross sections of 3D SSBs with nominally 500 and 250 nm thick LiPON on structure (d), respectively [48].

Mostly, garnet-structured solid electrolytes are chosen due to their high ionic conduction with stable electrochemical properties such as $Li_7La_3Zr_2O_{12}$ (LLZO). However, consideration should be given to their moisture-resistance properties, as well as to their sintering temperature. The most common NaSICON electrolytes are $LiGe_2(PO_4)_3$, (LGP), $LiTi_2(PO_4)_3$, and (LTP). LTP, upon Al substitution ($Li^{1+x}Al_xTi_{2-x}(PO_4)_3$ (LATP) with Ti shows high ionic conductivity with reduced temperature of sintering and a stabilized high crystalline structure. In order to avoid adverse reactions at the interface of electrode and electrolyte, a proper electrode must be chosen to guarantee that the anode and cathode have low and high potential, respectively, without causing decomposition of the electrolyte. In relation to LATP as electrolyte, $Li_4Ti_5O_{12}$ (LTO) is the best anode material with a potential of 1.55V. The compounds with olivine structures, such as $LiMPO_4$ (M = Fe, Mn, Ni, Co, etc.), fall under the category of promising cathode materials due to their low loss capacity and high structure stability [49]. Multifunctional integrated supercapacitors are the new emerging type of electrochemical energy storage devices that functionalize as photodetectors with activity monitoring and chemical sensing. The integration of photodetectors can save both space and energy. This can meet the demand of lightweight, portable electronics as well [50].

The incorporation of a photodetector in a supercapacitor can result in a durable integrated device characterized by fast charging through photo collector. Photosensitive metallic oxide semiconductors are mostly used to support this incidence. However, in recent times, an integrated system was developed using Si nanowire array as electrode. This fabrication protocol is depicted in Figure 10.12. The set-up was fabricated using a thin deposition layer of silver on silicon vapor employing electron beam evaporation followed by annealing at 800°C to produce nanoparticles of Ag. Sequentially, a bi-layer of Ti/Au was deposited on the Si wafer. Ultrasonication was performed to remove the deposited Ag on the target to achieve a mesh of Ti/Au bi-layer. The wafer was finally etched to obtain a vertically aligned Si nanowire array on the silicon substrate. To build positive and negative electrodes, interconnected carbon walls were formed by drop casting a sugar solution and annealing at 700°C for two hours in a protected (Ar/H_2) environment. The mesh-like structure allows the UV light to pass through acting as UV detector. The micro-supercapacitor was assembled combining the two fragments of carbon wall interconnected Si nanowires facing each other, which were separated by a solid-gel electrolyte. The assembly was naturally air-dried to give high performance. The device, upon illumination with UV light, generates photogenerated charge carriers and diffuses into the micro-supercapacitor, eventually increasing the current in the circuit (Figure 10.12f). This is attributed to the lower power consumption by the device in presence of a UV detector. Si nanowire arrays favor fast UV detection and high-energy storage capacity [51].

10.3.4 Superconductors

Superconductors are those materials that exhibit high conductivity devoid of any resistance and vitality with respect to temperature. Considering their versatile utility, they have been technologically exploited in the medical field for magnetic resonance imaging, in industries for magnetic shielding, sensors, and transducers, in electrical connections for generators, motors, and flywheel energy storage, in transportation for magnetically levitated trains, for magneto hydrodynamic in marine propulsion, for superconducting quantum interference devices (SQUIDs), and for filters in electronics in order to create high speed or quantum computing.

Commercially, magnetic resonance imaging (MRI) is one of the largest community-required high superconductive magnets. This magnet is also the most expensive

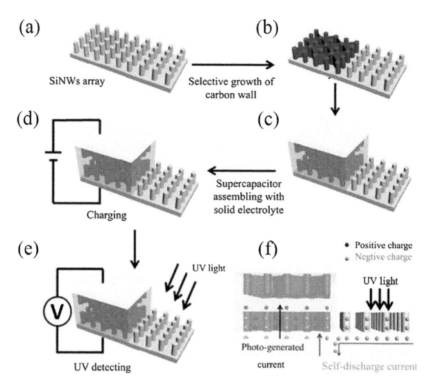

FIGURE 10.12
Process of fabrication and structure of the UV detector [51].

component in an MRI system. Potential and commercial-grade superconductors are MgB_2, $Bi_2Sr_2CaCu_2O_8$ (BSCCO), and rare-earth barium copper oxides (ReBCO) such as yttrium–barium copper oxide ($YBa_2Cu_4O_8$, YBCO). The optimized functional property is achieved through the fulfillment of competing requirements, with cost being the key barrier. Practically, NbTi is the most used superconducting material in MRI systems due to its low-cost and long development, but its employment is limited by low critical temperature and critically low field, which opens the path for MgB_2. For commercial MRI systems, MgB_2 is the closest material that meets the technical requirements. Although MgB_2 is well on the way to competing with NbTi, numerous modifications are needed in order to beat magnet technology. NbTi is a low-temperature superconductor with critical temperature around 10K whereas MgB_2 has not yet been commercialized, but its critical operating temperature is 20K. Conventionally, NbTi magnets are typical wires with a diameter of 0.8mm with 28 of such filaments in each magnet.

Usually, such magnets are produced as mono-components or multi-components. The monofilament processing wires are quite fast and efficient, as shown in Figure 10.13a. In the beginning, NbTi billet is inserted into a Cu-can, which is then heat-treated, followed by hot extrusion and drawing to get a monofilament wire (Phase-I). The production of the multifilament follows the same procedure, but instead of a mono wire insertion, multiple monofilament wires are inserted into the billet as depicted in Figure 10.13b (Phase-II). Figure 10.13c shows the different modules of manufactured superconductor magnets and their design [52]. On the other hand, the production process of MgB_2 is not similar to NbTi. It follows a powder-compaction technology in a tube mold, where stoichiometric Mg and B are filled, followed by heat treatment during in situ formation of MgB_2 wires, but the

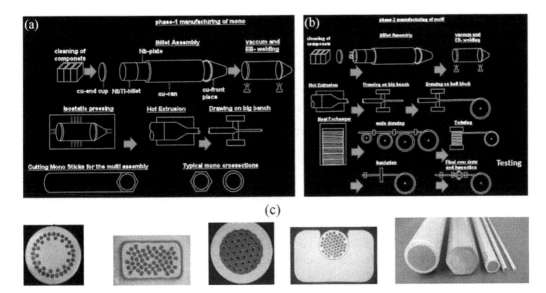

FIGURE 10.13
Manufacturing of (a) monofilament; (b) multifilament superconducting wires, and (c) design-type modules of different superconducting magnets [52].

ex-situ method adopts the compaction of MgB_2 powder in a metallic tube to manufacture superconducting wires. The ex-situ process exhibits fewer superconducting properties compared to the counterpart in situ process [53]. The metallic tube used for filling the powders of MgB_2 in the tube method needs to be selected judiciously to avoid a brittle superconducting core during processing.

NbTi wires are also applicable in particle accelerator systems in high-energy physics, but their high cost still limits their use, stopping it from going from demonstration to commercialization, considering the requirements of substantial magnetic field level and also operating temperature above liquid helium. The operated electronic circuitry for liquid helium is presently toying with an Nb thin-film technology in blend with $Nb/Al/AlO_x/Nb$ Josephson junctions to be applicable in commercial SQUID magnetic field sensors [54]. In BSCCO-based superconductors, different compositions of copper in the system are coded differently, such as $Bi_2Sr_2CaCu_2O_{8-x}$ (Bi-2212) and $Bi_2Sr_2Ca_2Cu_3O_{8-x}$ (Bi-2223). Ag-alloy clad Bi-2212 is the only high-temperature superconductor that can have a twisted configuration which reduces the charging loss hysteretically. Moreover, being electro-magnetically isotropic and multifilamentary, it is not significantly affected by the screening currents induced due to a large radial field in wounded solenoid with anisotropic tape conductors. In this context, this material has competed with other candidates in the high-energy and high-homogeneity magnetic fields, and is majorly applicable in nuclear magnetic resonance and particle accelerator magnets. A persistent current mode is highly desirable for NMR where the superconducting joints in Ag-alloy-clad Bi-2212 are the key component. A study demonstrated proper joint matrix with addition of different amounts of Ag in Bi-2212 and investigated their differential thermal analysis. The study also stated that the process is effective in practical applications, enabling the achievement of Bi-2212 superconducting joints with good superconductive properties [55]. The processing of such superconducting cuprate materials is not sophisticated and does not require complex systems for their synthesis. These can be prepared through a modified co-precipitation process

$(Bi_{2.1}Sr_{1.96}CaCu_xO_{8+\delta}$ (Bi-2212)) [56]. Materials such as Y-Ba-Cu-O, Bi-Ca-Sr-Cu-O, and Tl-Ca-Ba-Cu-O are prepared using a modified citrate precursor process [57]. In thallium-based systems, thallium oxide is added during the final sintering process. The powders obtained from these processes exhibit high reproducibility.

10.3.5 Microelectromechanical Systems (MEMS)

MEMS is a tiny technology consisting of integrated devices/systems that combine both electrical and mechanical properties in a single system. The fabrication of MEMS requires expertise from diverse fields, such as integrated circuit process technology, materials science, electrical and mechanical engineering, chemistry, and chemical engineering, as well as knowledge of fluids, optics, instrumentation, and packaging. MEMS is constituted of four components, namely, microstructures, microsensors, microactuators, and microelectronics [58]. These components range in size from 1 to 100 μm, whereas the complete device size ranges between 20 μm and 1 mm. MEMS are applicable in diverse objects in order to fulfill market demand. Some of the best established and commercialized MEMS are automotive airbag sensors, medical pressure sensors, inkjet printer heads, and digital micromirror devices [59]. Recently, MEMS have also been explored extensively in biomedical applications referred as bioMEMS/biochip, and also for wireless communications. The most common fabrication methods are bulk micromachining, surface micromachining, and high aspect ratio micromachining (HARM). HARM includes a complex technology LIGA, which has been derived from German acronyms where LI-lithographie stands for lithography, G-galvanoformung which is electroforming, and A-abformung refers to molding [60]. The most commonly used substrates are silicon wafers, and other semiconductors such as germanium and gallium arsenide are also used. The film addition to these substrates can be silicon-based compounds (Si_xN_y, SiO_2, SiC), metal and metallic compounds (Au, Cu, Al, GaAs, CdS, IrO_x), organics, and ceramics (Al_2O_3 and complex ceramics).

One of the most successful and commercial MEMS is the miniature disposable pressure sensor that is extensively used in hospitals to monitor blood pressure while in connection with an intravenous (IV) line. This disposable sensor comprises of a silicon substrate that is etched to form a membrane and bonded with the substrate. The working principle is depicted in Figure 10.14a [61]. The schematic demonstrates that the top piezoelectric layer is applied on the surface of the membrane near the edges to convert the mechanical stress to electrical voltage. The pressure applied encompasses the membrane deflection. To fit

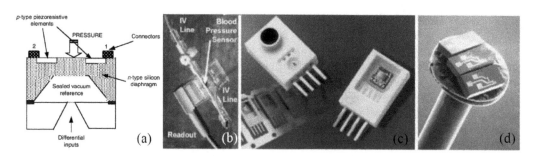

FIGURE 10.14
(a) Schematic of piezoelectric pressure sensors [61], (b) disposable blood pressure sensor connected to an IV line, (c) disposable sensors without connection, and (d) intracardial catheter-tip sensor for monitoring blood pressure during cardiac catheterization (head of the pin) [62].

the design, the sensing element is mounted on the plastic/ceramic base with a plastic cap covering. Figure 10.14b shows the image of the disposable blood pressure sensor which is connected to the IV line and individual disposable sensors are shown in Figure 10.14c. The sensing element and saline solution in the IV line are separated by a gel when in line with connection for monitoring the pressure. A catheter-tip sensor is placed on the head of a pin (Figure 10.14d), which is used as sensor for monitoring blood pressure during a cardiac catheterization process [62].

DNA and protein sensing have become another crucial field in MEMS technology. One of the studies reported thin film microelectromechanical molecular sensors that are fabricated on glass substrates at temperatures lower than 110°C. The MEM structure is comprised of a bi-layer bridge formed by surface machining that is composed of phosphorous-doped amorphous hydrogenated silicon and aluminum. Through the silanization process, the hydroxylated bridge surface specifically binds to DNA. These microbridges are actuated electrostatically, and are measured for resonance frequency in vacuum before and after surface functionalization and also after DNA immobilization. The functionalized molecular layer could be detected with a shift in the resonance frequency [63].

The coupling of MEMS technology with microfluidics can develop lab-on-a-chip devices for various investigation of biological analytes. Such technology has activated the development of point of care (POC) devices. Capturing single cells for their type determination and analysis has been possible with POC devices made using MEMS technology. These devices make use of seeding protocols involving deposition of large cell numbers on chip surface. The challenge in such a system is to reduce the cell loss and increase the seeding efficiency, which depends on the seeding flow, the surface affinity of the chip, and the well geometry [64]. The assessment of diseases could be made through routine detection of cells. These devices can act as clinical markers for detecting DNA/RNA [65] and protein molecules for disease assessment [66], as well as for specific cell detection [67]. MEMS technology opened up a new horizon for the future-generation diagnostic tool.

10.3.6 Ultrasonic Transducer

High-frequency ultrasound imaging is a promising tool for the detailed allocation of superficial anatomical structures. The piezoelectric transducer is used for imaging purposes. Typically, the process is as follows: the transducer is placed on the part requiring imaging, prior to which a gel is applied on the part that helps in transmitting the waves. The waves transmitted bounce back, producing an image. As the process is continuous, it produces time-dependent images and it is less harmful due to non-usage of ionizing radiations. Technological advancement has led to the production of low-cost and portable transducers. However, one has to keep in mind that there are certain drawbacks, such as surface tissue heating and generation of hot pockets. This imaging technology is also used to unveil the flaws that cannot be detected by other means. The sound waves are produced from a piezoelectric transducer with frequencies ranging from 2 to 15MHz [68]. A beam of radio wave is created using the phased array method, which can point the arrays in different directions without moving the antenna. Through this method, a change of depth makes it possible to see and analyse the desired part. The sound waves that are reflected from the parts return back to the transducer, producing real-time images of the parts. The image formation depends on the time it takes the reflection to travel back after the wave transmission, and on the strength of the reflected waves. Once the transducer determines the anatomical parts, it produces the required images. Within a probe there are multiple acoustic transducers to send pulses into the material. When the sound waves come across a material or a medium having a different density, a portion of the sound wave is reflected back and the depth of the part is measured. As the waves pass through the material encountering parts and mediums of different densities, most of the acoustic energy gets lost due to absorption. The change in the speed of sound due to travelling in mediums of different densities is ignored during calculations and assumed to be constant. Doppler ultrasonic imaging is used to observe blood flow and muscle movements in the human body.

Piezoelectric ceramics, mostly ferroelectric materials such as barium titanate ($BaTiO_3$) and lead zirconate titanate (PZT), are used for this purpose. The single crystals of such materials are more expensive than the randomly oriented polycrystals separated by grain boundaries. However, polycrystalline materials also exhibit strong piezoelectric properties along the axes of polarization. The fabrication of such devices is carried out using photolithography on patterned substrates [69].

This innovative protocol facilitates the task of making prototypes capable of very fine control and easy monitoring. A better and advanced photolithography process that employs a versatile dicing technique can produce transducers in nanoscale dimensions in order to promote the device's efficiency [16]. Ozaki et al. have epitaxially grown functionally active $Pb(Zr_{0.52}Ti_{0.48})O_3$ (PZT) thin film on $Si(1\ 1\ 1)/\gamma$-$Al_2O_3(1\ 1\ 1)/SrRuO_3(1\ 1\ 1)$ substrate through micromachining in order to fabricate an array of piezoelectric micromachined ultrasonic transducers (pMUT) for biomedical applications [70]. The control of crystal orientation of PZT films was carried out by incorporating the epitaxial γ-Al_2O_3 film on Si substrate. The fabrication of the pMUT array uses large-scale integration facilities where, initially, 10 nm Al_2O_3 thin film was grown epitaxially on Si substrate (111) by molecular beam epitaxy. Then, on top of the film, a further 100 nm thin Pt film was deposited by RF sputtered technique. An interface of 10 nm $SrRuO_3$ (SRO) was sputtered on Pt film at 700°C. A PZT film was deposited using a chemical solution deposition technique with PZT precursor (Pb:Zr:Ti = 1.15:0.52:0.48). The spin-coated samples were dried at 150°C for five minutes following pyrolysis at 250°C for 5 minutes. To achieve a PZT film with a perovskite phase, rapid thermal annealing was employed to be sintered at 650°C for 90 seconds in an oxygen environment. The PZT thin-film process was repeated for five times to achieve a thickness of 500 nm, where 100 nm per layer was produced in each process. Further, the SRO was RF sputtered to have a 100 nm film, before performing lithography to obtain the desired patterning (Figure 10.15a).

The tops of SRO and PZT films were lithographed to obtain a patterned structure, which is then etched using an inductive coupled plasma reactive ion etching (ICP-RIE) process by means of CHF_3 gas. The bottom part, consisting of $SrRuO_3$/Pt and Al_2O_3, was patterned in presence of Ar gas using the same ICP-RIE process (Figure 10.15b). An insulating silicon nitride (SiN) film of 1 micron was deposited using plasma enhanced chemical vapor deposition at 300°C. Again SiN deposition was ICP-RIE patterned in SiF_6 gas (Figure 10.15c). To make an interconnection, 1.2 μm Al thin film was deposited using an electron beam evaporator, followed by patterning through an RIE technique using BCl_3, Cl_2, and N_2 gases (Figure 10.15d). To obtain matching impedance, a photoresisting (PR) layer of 1 μm was spin-coated, baked, and patterned (Figure 10.15e). To form a diaphragm structure, the Si underneath the pMUT active area was etched away by XeF_2 gas through a topside hole. At last, the wafer was diced to a size of 1 cm² to be used as a pMUT device (Figure 10.15f). The processing steps for fabricating a pMUT device have been shown in Figure 10.15 with the formed pMUT structure (Figure 10.15g) and packaged pMUT chip on a printed circuit board, as shown in Figure 10.15h [70].

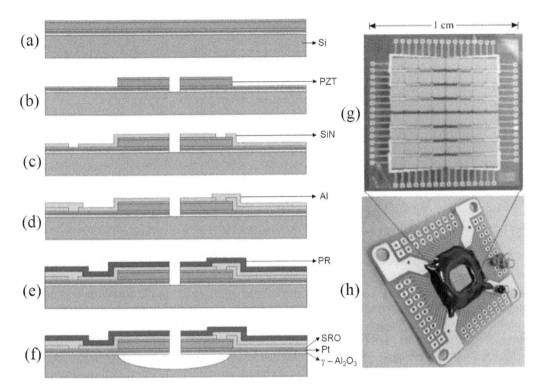

FIGURE 10.15
Processing steps for the fabrication of pMUTs array, (a) fabrication of SRO/PZT(1 1 1)/SRO(1 1 1)/Pt(1 1 1)/ γ-Al$_2$O$_3$/Si(1 1 1) structure, (b) patterning of top "SRO" and PZT layers, bottom "SRO," "Pt," and γ-Al$_2$O$_3$ layers, (c) deposition and patterning of SiN layer, (d) deposition and patterning of "Al" layer, (e) deposition and patterning of PR layer, and (f) etching of "Si" underneath active area. Fabricated ultrasonic transducer: (g) pMUT array and (h) packaged pMUT chip on a printed circuit board (PCB).

10.3.7 Nuclear Fuel

The application of materials in the nuclear industry is exceptionally important throughout the entire fuel cycle, starting from the fabrication of fuel to the stabilization of fuel wastes. The materials become more crucial as the global community commences to look at the advanced nuclear reactor systems that can minimize the waste while increasing the proliferation resistance. The advanced reactor concepts use highly durable and heat-resistant materials to operate under a high temperature. The electrical power generation in the nuclear fission process uses advanced ceramic fuels such as uranium oxide (UO$_2$), plutonium oxide (PuO$_2$), or a mix of both oxides. These materials impart high melting temperature and improved corrosion resistance; moreover, they are chemically compatible with the cladding materials and also exhibit dimensional stability in a blend with fission products during irradiation. All of these properties can increase the marginal safety over metallic fuels, easing the fabrication process. In most power reactors, fuels are used as pellets. They are fabricated through a conventional cold pressing process followed by sintering at high temperature. These pellets are then machined to get the appropriate size, reviewed, and encapsulated in stainless steel or Zircaloy-clad to form fuel rods. These fuel rods are bundled to be loaded into the nuclear reactor. The control over the chemistry of the fuel and microstructure governs the reactor's operation. The dimensional stability during irradiation of fuel can be made precise through strict control over of the residual porosity. The process uses compacted milled powder, which is granulated and repressed prior to the sintering process. The waste generated is made into suitable ceramic waste forms such as 'Synroc', which is a multiphase form, and pyrochlores such as $Gd_2Zr_2O_7$ sand zirconolite as single-phase forms [71]. All of these processes are carried out in a radiation-controlled environment, in closed chambers and with protective clothing.

10.3.8 Other Functional Components

Other commercial prototypes include diesel particulate filters and filters in wastewater treatments, diesel engines, catalytic converters, oxygen sensors, and more. The most common ceramic material used as catalytic converter core is cordierite, which increases the filtration efficiency with a cost lower than that of the precipitator [72]. Ceramics are traditionally used to disperse the compressed air in the form of fine bubbles into sewage for the activated sludge biological purification process. They are also used to inject ozone into the drinking water in the wastewater treatment [73]. Automotive catalytic converter cores are honeycomb ceramic monolith made up of ceria (CeO_2) or zirconia (ZrO_2) to be used as oxygen storage promoters in the catalytic converters. These honeycomb structures are in candle forms, which consist of a ceramic film, a ceramic fiber matrix, and the catalyst particles. Such structures are much more efficient than the conventionally used air filter in bag form. The main application of such filters is to control the vehicle emission. The honeycomb structure is prepared through "polymeric sponge method," as shown in Figure 10.16a (details in Chapter 4). In this method, ceramic suspension is soaked all over the polymeric sponge made up of polyurethane until the internal pores are filled completely [74, 75]. The sponge impregnated with suspension is passed through rollers to remove excessive suspension and forms a thin ceramic coating. The slurry at this stage should be sufficiently fluidic, but the wet coating should be viscous enough to avoid the dripping of the ceramic suspension [76].

Thus, the suspension characterized by shear-thinning behavior can efficiently coat the polymeric template [77]. The sponge is then air-dried, and this is followed by oven drying or microwave heating. The oven drying is carried out at 100–700°C, which is completed in a time ranging between 15 minutes and about 6 hours. After the drying process, the organics are removed by burning the sponge in inert atmosphere from 350 to 800°C. The heating rate is slow to avoid the blowing up of the ceramic structure. Finally, the structure is heated at 1,000–1,700°C to densify under controlled heating rate, and the cycle of firing depends on the desired ceramic composition. For example, an alumina-made structure requires heating at 1,350°C for about five hours to achieve a complete, dense monolith structure. Ceramic-based oxygen sensors are commonly used to control air or fuel ratio for automobile exhausts where zirconia is the active material [78]. On the other hand, the plastic forming process (Figure 10.16a), otherwise known as extrusion process, is also followed to obtain monoliths. This process is mostly used to obtain highly symmetrical structures such as rods, channels, tubes, and also honeycomb structures. The typical microstructures of macroporous SiC ceramics prepared by replica technique have been presented in Figure 10.16b–e, which feature different pore structures due to a difference in processing protocol [79–82]. In order to reduce the emissions in light duty systems, cordierite-made gasoline particulate filters are used, as shown in Figure 10.16f [83]. Such systems enable a high wash-coated loading with an excellent drop in pressure to outperform the catalytic performance. Another commercially available ceramic honeycomb filter, shown in Figure 10.16g, is popular to filter impurities during liquid metal casting. It eventually improves the quality of casting, and also reduces the porosity during casting by improving the flow of molten metal [84].

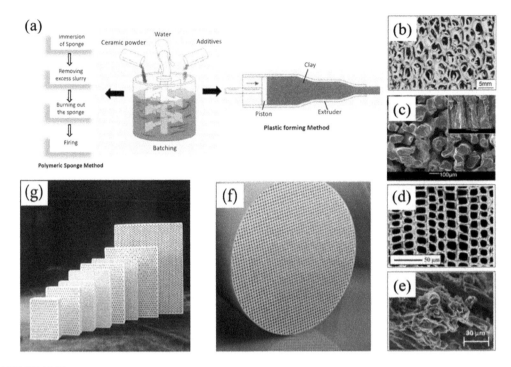

FIGURE 10.16
(a) Processing of ceramic monolith filters using polymeric sponge/replica method [74]. Typical microstructures of macroporous SiC ceramics produced via the replica technique: (b) SiC foams produced by impregnation of polyurethane sponges with SiC slurry [79], (c) SiC foams produced by CVD process [80], (d) SiC foams produced by infiltration of wood-derived carbon foams with Si vapor [81], (e) SiC foams produced by infiltration of wood-derived carbon foams with Si vapor [82], (f) cordierite-made gasoline particulate filters for reduced emissions in light duty systems [83], (g) cordierite-made ceramic honeycombs with different shapes [84].

10.4 Conclusions

The performance of the commercialized product depends on the customer's application duration, on the operating conditions, and on the manner by which it meets expectations. A significant improvement in the processing, reducing time and cost consumption, can develop more miniature and micropatterned components for cost-effective commercialization. Although there are many optimized ceramic materials and their processed end-product is applicable in versatile fields, consistence in performance is also essential to encounter and monitor for any particular zone of interest. Many of the new processing techniques with advanced systems are discussed to avoid loss in contents. The utility of structural and functional ceramic prototypes in microsystems has always been of interest because of their mechanical and tribological properties, resistance toward heat and chemicals, and other significant properties that satisfy the requirements which are not fulfilled by other metals and polymers. The present guidelines for designing might support the development, but there is still a lacuna. Each of these designs is an experience taken from macro-range systems. These designs and technology are slowly being transferred to micro-components through materials' anisotropy, and are progressively influencing the systems in multi-dimensional concepts.

References

1. Fluid Metering, Inc., http://fluidmetering.com/materials-construction.html.
2. Fluid Metering, Inc., http://fluidmetering.com/meteringPump-Animation.html.
3. Cheng, M., Liu, H., Zhao, B., Huang, C., Yao, P., and Wang, B. 2017. Mechanical properties of two types of Al_2O_3/TiC ceramic cutting tool material at room and elevated temperatures. *Ceramics International* 43(16):13869–13874.
4. Sumiya, H., Ikeda, K., Arimoto, K., and Harano, K. 2016. High wear-resistance characteristic of boron-doped nano-polycrystalline diamond on optical glass. *Diamond and Related Materials* 70:7–11.
5. Sajgalik, P., Hnatko, M., Lences, Z., Dusza, J., and Kasiarova, M., 2009. Chapter 15 In situ preparation of Si_3N_4/SiC nanocomposites for cutting tools application. *Progress in Nanotechnology*, 107–112. John Wiley & Sons, Inc: New York.
6. Senthil Kumar, A., Raja Durai, A. R., and Sornakumar, T. 2006. Wear behaviour of alumina based ceramic cutting tools on machining steels. *Tribology International* 39(3):191–197.
7. Suzuki, H., Okada, M., Asai, W., Sumiya, H., Harano, K., Yamagata, Y., and Miura, K. 2017. Micro milling tool made of nano-polycrystalline diamond for precision cutting of SiC, CIRP. *Annals* 66:93–96.
8. Vleugels, J. 2006. Fabrication, wear and performance of ceramic cutting tools. *Advances in Science and Technology* 45: 1776–1785.
9. Ezugwu, E. O. 1994. Manufacturing methods of ceramic cutting tools. *Key Engineering Materials* 96: 19–32.
10. Fox Industries, http://foxindustries.com/products/grinding-media/.
11. Weber, U. 2010. The effect of grinding media performance on milling a water-based color pigment. *Chemical Engineering and Technology* 33(9):1456–1463.
12. Fox Industries, http://foxindustries.com/products/grinding-media/zirconium-silicate-grinding-beads.
13. Vela, E., Peiteado, M., García, F., Caballero, A. C., and Fernández, J. F. 2007. Sintering behaviour of steatite materials with barium carbonate flux. *Ceramics International* 33(7):1325–1329.
14. Kwon, Y. S., Kim, H.-T., Choi, D. W., and Kim, J. S. 2005. Mechanical properties of binderless WC produced by SPS process. *Novel Materials Processing by Advanced Electromagnetic Energy Sources*, pp. 275–279.
15. How Products are made, Spark Plug Forum, http://www.madehow.com/Volume-1/Spark-Plug.html.
16. Sarkar, D. 2019. Chapter 9 Nanostructured ceramics for health. *Nanostructured Ceramics: Characterization and Analysis*. CRC Press: Boca Raton, FL.
17. Sarkar, D., Reddy, B. S., and Basu, B. 2018. Implementing statistical modeling approach towards development of ultrafine grained bioceramics: Case of ZrO_2-toughened Al_2O_3. *Journal of the American Ceramic Society* 101(3):1333–1343.
18. Sarkar, D., Sambi Reddy, B. S., Mandal, S., RaviSankar, M., and Basu, B. 2016. Uniaxial compaction-based manufacturing strategy and 3D microstructural evaluation of near-net-shaped ZrO_2-Toughened Al_2O_3 acetabular socket. *Advanced Engineering Materials* 18(9):1634–1644.
19. Sarkar, D., Mandal, S., Reddy, B. S., Bhaskar, N., Sundaresh, D. C., and Basu, B. 2017. ZrO_2–toughend Al2O3-based near-net shaped fermoral head: Unique fabrication approach, 3D microstruture, burst strength and muscle cell response. *Materials Science and Engineering: C* 77:1216–1227.
20. Sarkar, D., Swain, S. K., Adhikari, S., Reddy, B. S., and Maiti, H. S. 2013. Synthesis, mechanical properties and bioactivity of nanostructrued zirconia. *Materials Science and Engineering: C* 33:3413–3417.
21. Takahashi, T., Gonda, T., Mizuno, Y., Fujinami, Y., and Maeda, Y. 2017. Reinforcement in removable prosthodontics: A literature review. *Journal of Oral Rehabilitation* 44(2):133–143.

22. 3D CAD/CAM, Dental Milling, http://www.mtabdental.com/medical/dental/zirconica/milling-machine/sintering oven/index.html.

23. Ignatiev, A. 2002. Bionic eyes-ceramic micro detectors may cure blindness. *Materials World* 10(7):31–32.

24. The Complete Process of How to Make a Ceramic Knife, https://kikusumiknife.com/ceramic-knife.

25. Kyocera Advanced Ceramics, https://kyoceraadvancedceramics.com.

26. Zhang, G., Zhao, W., Zhao, R., Bao, J., Dong, B., Cui, C., Wang, X., and Cao, Q. 2016. Fabricating large-scale mirrors using reaction-bonded silicon carbide, *The International Society for Optics and Photonics*. http://spie.org/newsroom/6582-fabricating-large-scale-mirrors-using-reaction-bonded-silicon-carbide?SSO=1.

27. Coorstek, https://www.coorstek.com/english/industries/transportation/aerospace/optics-imaging.

28. Zhou, H., Zhang, C.-R., Cao, Y.-B., and Zhou, X.-G. 2006. Lightweight C/SiC mirrors for space application. *2nd International Symposium on Advanced Optical Manufacturing and Testing Technologies, SPIE*, p. 6.

29. Mahmud, L. S., Muchtar, A., and Somalu, M. R. 2017. Challenges in fabricating planar solid oxide fuel cells: A review. *Renewable and Sustainable Energy Reviews* 72:105–116.

30. Osaka Gas, Residential SOFC system, http://www.osakagas.co.jp/en/rd/fuelcell/sofc/sofc/system.html.

31. Park, J.-S., Kim, D.-J., Chung, W.-H., Lim, Y., Kim, H.-S., and Kim, Y.-B. 2017. Rapid, cool sintering of wet processed yttria-stabilized zirconia ceramic electrolyte thin films. *Scientific Reports* 7(1):12458.

32. Kim, K. J., Park, B. H., Kim, S. J., Lee, Y., Bae, H., and Choi, G. M. 2016. Micro solid oxide fuel cell fabricated on porous stainless steel: A new strategy for enhanced thermal cycling ability. *Scientific Reports* 6:22443.

33. Nazeeruddin, M. K., Baranoff, E., and Grätzel, M. 2011. Dye-sensitized solar cells: A brief overview. *Solar Energy* 85(6):1172–1178.

34. Grätzel, M. 2003. Dye-sensitized solar cells. *Journal of Photochemistry and Photobiology C: Photochemistry Reviews* 4(2):145–153.

35. Bella, F., Gerbaldi, C., Barolo, C., and Grätzel, M. 2015. Aqueous dye-sensitized solar cells. *Chemical Society Reviews* 44(11):3431–3473.

36. Gong, J., Liang, J., and Sumathy, K. 2012. Review on dye-sensitized solar cells (DSSCs): Fundamental concepts and novel materials. *Renewable and Sustainable Energy Reviews* 16(8):5848–5860.

37. Shaikh, J. S., Shaikh, N. S., Mali, S. S., Patil, J. V., Pawar, K. K., Kanjanaboos, P., Hong, C. K., Kim, J. H., and Patil, P. S. 2018. Nanoarchitectures in dye-sensitized solar cells: Metal oxides, oxide perovskites and carbon-based materials. *Nanoscale* 10(11):4987–5034.

38. Li, L., Xu, C., Zhao, Y., Chen, S., and Ziegler, K. J. 2015. Improving performance via blocking layers in dye-sensitized solar cells based on nanowire photoanodes. *ACS Applied Materials and Interfaces* 7(23):12824–12831.

39. Li, L., Chen, S., Xu, C., Zhao, Y., Rudawski, N. G., and Ziegler, K. J. 2014. Comparing electron recombination via interfacial modifications in dye-sensitized solar cells. *ACS Applied Materials and Interfaces* 6(23):20978–20984.

40. Wu, M., Lin, X., Wang, Y., Wang, L., Guo, W., Qi, D., Peng, X., Hagfeldt, A., Grätzel, M., and Ma, T. 2012. Economical Pt-free catalysts for counter electrodes of dye-sensitized solar cells. *Journal of the American Chemical Society* 134(7):3419–3428.

41. Wu, M., and Ma, T. 2014. Recent progress of counter electrode catalysts in dye-sensitized solar cells. *The Journal of Physical Chemistry C* 118(30): 16727–16742.

42. Raja, M., Sanjeev, G., Prem Kumar, T., and Manuel Stephan, A. M. 2015. Lithium aluminate-based ceramic membranes as separators for lithium-ion batteries. *Ceramics International* 41(2):3045–3050.

43. Sousa, R., Sousa, J. A., Ribeiro, J. F., Goncalves, L. M., and Correia, J. H. 2013. All-solid-state batteries: An overview for bio applications. *2013 EEE. 3rd Portuguese Meeting in Bioengineering (ENBENG)*, pp. 1–4.

44. Choi, J. W., and Aurbach, D. 2016. Promise and reality of post-lithium-ion batteries with high energy densities. *Nature Reviews Materials* 1(4):16013.

45. Sun, C., Liu, J., Gong, Y., Wilkinson, D. P., and Zhang, J. 2017. Recent advances in all-solid-state rechargeable lithium batteries. *Nano Energy* 33:363–386.

46. Mizuno, F., Hayashi, A., Tadanaga, K., and Tatsumisago, M. 2005. New highly ion-conductive crystals precipitated from Li2S–P2S5 glasses. *Advanced Materials* 17:918–921.

47. Suzuki, Y., Munakata, H., Kajihara, K., Kanamura, K., Sato, Y., Yamamoto, K., and Yoshida, T. 2009. Fabrication of three-dimensional battery using ceramic electrolyte with honeycomb structure by sol-gel process. *ECS Transactions* 16:37–43.

48. Talin, A. A., Ruzmetov, D., Kolmakov, A., McKelvey, K., Ware, N., El Gabaly, F., Dunn, B., and White, H. S. 2016. Fabrication, testing, and simulation of all-solid-state three-dimensional Li-Ion batteries. *ACS Applied Materials and Interfaces* 8(47):32385–32391.

49. Wei, X., Rechtin, J., and Olevsky, E. 2017. The fabrication of all-solid-state lithium-ion batteries via spark plasma sintering. *Metals* 7(9):372.

50. Chen, M., Yang, Y., Chen, D., and Wang, H. 2018. Recent progress of unconventional and multifunctional integrated supercapacitors. *Chinese Chemical Letters* 29(4):564–570.

51. Sun, L., Wang, X., Zhang, K., Zou, J., Yan, Z., Hu, X., and Zhang, Q. 2015. Bi-functional electrode for UV detector and supercapacitor. *Nano Energy* 15:445–452.

52. Holm, M., Oy, L.P. 2011. Industrial Manufacturing of Low Temperature Superconducting (LTS) wires European Summer School on Superconductivity 2011, Harjattula Mansion: Turku, Finland. http://www.prizz.fi/sites/default/files/tiedostot/linkki2ID686.pdf.

53. Vinod, K., Kumar, R. G. A., and Syamaprasad, U. 2007. Prospects for MgB 2 superconductors for magnet application. *Superconductor Science and Technology* 20(1):R1–R13.

54. Tolpygo, S. K., and Amparo, D. 2010. Fabrication-process-induced variations of Nb/Al/AlO x /Nb Josephson junctions in superconductor integrated circuits. *Superconductor Science and Technology* 23(3):034024.

55. Chen, P., Trociewitz, U. P., Davis, D. S., Bosque, E. S., Hilton, D. K., Kim, Y., Abraimov, D. V., Starch, W. L., Jiang, J., Hellstrom, E. E., and Larbalestier, D. C. 2017. Development of a persistent superconducting joint between Bi-2212/Ag-alloy multifilamentary round wires. *Superconductor Science and Technology* 30(2):025020.

56. Shengnan, Z., Chengshan, L., Qingbin, H., Jianqing, F., and Pingxiang, Z. 2017. Influence of Cu content in precursor powders on the phase evolution and superconducting properties of Bi-2212 superconductors. *Rare Metal Materials and Engineering* 46(3):585–590.

57. Liu, R. S., Wang, W. N., Chang, C. T., and Wu, P. T. 1989. Synthesis and characterization of high- T c superconducting oxides by the modified citrate gel process. *Japanese Journal of Applied Physics* 28(2, No. 12):L2155–L2157.

58. Spearing, S. M. 2000. Materials issues in microelectromechanical systems (MEMS). *Acta Materialia* 48(1):179–196.

59. Fedder, G. K., Howe, R. T., Tsu-Jae King Liu, T. J. K., and Quevy, E. P. 2008. Technologies for Cofabricating MEMS and electronics. *Proceedings of the IEEE* 96(2):306–322.

60. Zhan, D., Han, L., Zhang, J., Shi, K., Zhou, J. Z., Tian, Z. W., and Tian, Z. Q. 2016. Confined chemical etching for electrochemical machining with nanoscale accuracy. *Accounts of Chemical Research* 49(11):2596–2604.

61. Eren, H. 2004. Chapter 158 Sensors. *Electronic Portable Instruments-Design and Applications*, pp. 1–31. CRC Press: Boca Raton, FL.

62. An Introduction to MEMS, Prime Faraday Technology Watch. 2002. Loughborough University: Loughborough, UK. http://www.lboro.ac.uk/microsites/mechman/research/ipm-ktn/pdf/Technology_review/an-introduction-to-mems.pdf.

63. Adrega, T., Prazeres, D. M. F., Chu, V., and Conde, J. P. 2006. Thin-film silicon MEMS DNA sensors. *Journal of Non-Crystalline Solids* 352(9–20):1999–2003.

64. Nikkhah, M., Strobl, J. S., Schmelz, E. M., Roberts, P. C., Zhou, H., and Agah, M. 2011. MCF10A and MDA-MB-231 human breast basal epithelial cell co-culture in silicon micro-arrays. *Biomaterials* 32(30):7625–7632.

65. Eicher, D., and Merten, C. A. 2011. Microfluidic devices for diagnostic applications. *Expert Review of Molecular Diagnostics* 11(5):505–519.

66. Hartman, M. R., Ruiz, R. C. H., Hamada, S., Xu, C., Yancey, K. G., Yu, Y., Han, W., and Luo, D. 2013. Point-of-care nucleic acid detection using nanotechnology. *Nanoscale* 5(21):10141–10154.

67. Huang, S.-B., Wu, M.-H., Lin, Y.-H., Hsieh, C.-H., Yang, C.-L., Lin, H.-C., Tseng, C.-P., and Lee, G.-B. 2013. High-purity and label-free isolation of circulating tumor cells (CTCs) in a microfluidic platform by using optically-induced-dielectrophoretic (ODEP) force. *Lab on a Chip* 13(7):1371–1383.

68. Shung, K. K. 2009. High frequency ultrasonic imaging. *Journal of Medical Ultrasound* 17(1):25–30.

69. Smith, R., Arca, A., Chen, X., Marques, L., Clark, M., Aylott, J., and Somekh, M. 2011. Design and fabrication of ultrasonic transducers with nanoscale dimensions. *Journal of Physics: Conference Series* 278:012035.

70. Ozaki, K., Matin, A., Numata, Y., Akai, D., Sawada, K., and Ishida, M. 2014. Fabrication and characterization of a smart epitaxial piezoelectric micromachined ultrasonic transducer. *Materials Science and Engineering: B* 190:41–46.

71. Marra, J. 2011. Advanced ceramic materials for next-generation nuclear applications. *IOP Conference Series: Materials Science and Engineering* 18(16):162001.

72. Adler, J. 2005. Ceramic diesel particulate filters. *International Journal of Applied Ceramic Technology* 2(6):429–439.

73. Wu, S., Qi, Y., Yue, Q., Gao, B., Gao, Y., Fan, C., and He, S. 2015. Preparation of ceramic filler from reusing sewage sludge and application in biological aerated filter for soy protein secondary wastewater treatment. *Journal of Hazardous Materials* 283:608–616.

74. How Products Are Made, Ceramic Filter, http://www.madehow.com/Volume-5/Ceramic-Filter.html.

75. Eom, J.-H., Kim, Y.-W., and Raju, S. 2013. Processing and properties of macroporous silicon carbide ceramics: A review. *Journal of Asian Ceramic Societies* 1(3):220–242.

76. Ishizaki, K., Sheppard, L., Okada, S., Hamasaki, T., and Huybrechts, B. 2018. Ceramic transactions. *Porous Materials* 31.

77. Studart, A. R., Gonzenbach, U. T., Tervoort, E., and Gauckler, L. J. 2006. Processing routes to macroporous ceramics: A review. *Journal of the American Ceramic Society* 89(6):1771–1789.

78. Moos, R., and Brief, A. 2005. A brief overview on automotive exhaust gas sensors based on electroceramics. *International Journal of Applied Ceramic Technology* 2(5):401–413.

79. Yao, X., Tan, S., Huang, Z., and Jiang, D. 2006. Effect of recoating slurry viscosity on the properties of reticulated porous silicon carbide ceramics. *Ceramics International* 32(2):137–142.

80. Liu, G., Dai, P., Wang, Y., Yang, J., and Zhang, Y., 2011. Fabrication of wood-like porous silicon carbide ceramics without templates, Journal of the European Ceramic Society, 31(5):847–85

81. Vogli, E., Sieber, H., and Greil, P. 2002. Biomorphic SiC-ceramic prepared by Si-vapor phasein-filtration of wood. *Journal of the European Ceramic Society* 22(14–15):2663–2668.

82. Almeida Streitwieseer, D., Popovska, N., and Gerhard, H. 2006. Optimization of the ceramization process for the production of three-dimensional biomorphic porous SiC ceramics by chemical vapor infiltration (CVI). *Journal of the European Ceramic Society* 26(12):2381–2387.

83. Environmental Technologies, Gasoline Particulate Filters for Reduced Emissions in Light-Duty Systems, https://www.corning.com/emea/de/products/environmental-technologies/products/ceramic-particulate-filters/corning-dura-trap-gc-filters.html.

84. Ceramic Honeycomb - Molten Filtration, Metal Induceramic, http://www.induceramic.com/industrial-ceramic-product/ceramic-honeycomb-molten-metal-filtration.

Summary

Radical changes brought on by fast progressing disruptive technologies and the rapid developments of the newer dimensions of ceramics has given birth to the need of writing this book. The manufacturing sectors, including ceramics, are at a crossroad due to the adoption of high-level automation and the implementation of niche and unconventional technologies. The relentless quest of the human mind for improvement never allows science and technology to rest in peace. So, the paradigm shifts. In an effort to keep knowledge seekers abreast of the latest trends, developments, and additions in the field of ceramics there has been an attempt to capture some of these trends in carefully selected chapters. This book starts with introductory information on *manufacturing excellence in ceramic industry* followed by *particle management*, which is one of the most important steps for ceramic processing. In brief, the processing protocols have been taken into account for the manufacturing of a wide range of bulk ceramics starting from extremely high dense (*transparent ceramics*) to low dense ceramics (*porous ceramics*) followed by *ceramic coatings*. Considering this information, afterward, traditional and advanced ceramics including *refractories and failures, whitewares and glazes, glasses and properties, miniaturization of complex ceramics*, and *structural and functional prototypes* are discussed. The topics were chosen and completed through brain-storming discussions with distinguished industry personalities, academia, and researchers around the globe.

Key Features

1. Provides a brief overview on modern manufacturing excellence; design of experiment and particle management in the perspective of both traditional and advanced ceramics.

2. Deals with basic understanding on particle morphology, processing protocol, and composition to synchronize in order to obtain extremely high dense to porous ceramics.

3. Excellent combination of examples and manufacturing outline of ceramic coatings, to scaling up the performance and durability of bulk ceramics for a wide horizon of service environments.

4. Processing and shortcomings of refractories, whitewares, glasses, additive manufacturing, micromachining, and prototypes are highlighted on the basis of market demand.

5. Materials scientists/engineers working in the field of ceramics can accomplish a strong interface for mutual communication between product development and properties.

6. Concise text help both student and professionals to grasp fast knowledge of individual subject before technical interview and discussion.

I hope this book will help students to gain deeper insights on the topics covered here. This reflects on many complex aspects, which may not be available in other commonly available contemporary books. This could be an extension to their basic knowledge of ceramics. It would be a great pleasure and satisfaction, if it helps the students in their quest for knowledge.

Index